KU-534-488

SCHAUM'S
OUTLINE OF

s Cross
Centre
Inn Road
1X

We

FLUID
MECHANICS

SCHAUM'S OUTLINE OF

FLUID MECHANICS

MERLE C. POTTER, Ph.D.

Professor Emeritus of Mechanical Engineering
Michigan State University

DAVID C. WIGGERT, Ph.D.

Professor Emeritus of Civil Engineering
Michigan State University

Schaum's Outline Series

McGRAW-HILL

New York   Chicago   San Francisco   Lisbon   London
Madrid   Mexico City   Milan   New Delhi   San Juan
Seoul   Singapore   Sydney   Toronto

MERLE C. POTTER received a BS in Mechanical Engineering and an MS in Engineering Mechanics from Michigan Technological University; he obtained an MS in Aerospace Engineering and a Ph.D. in Engineering Mechanics from The University of Michigan. He coauthored *Fluid Mechanics*, *The Mechanics of Fluids*, *Thermodynamics for Engineers*, *Thermal Sciences*, *Differential Equations*, and *Advanced Engineering Mathematics*, in addition to numerous exam review books. His research involved fluid flow stability and energy-related projects. He is professor emeritus of Mechanical Engineering at the Michigan State University.

DAVID C. WIGGERT received a BS, MS, and Ph.D. from The University of Michigan. He coauthored *The Mechanics of Fluids*. His research involved fluid transients, fluid structure interaction, and groundwater flow and mass transport. Dr Wiggert is professor emeritus of Civil and Environmental Engineering at the Michigan State University.

Schaum's Outline of
FLUID MECHANICS

1 2 3 4 5 6 7 8 9 0 CUS CUS 0 1 3 2 1 0 9 8 7

ISBN 978-0-07-148781-8
MHID 0-07-148781-6

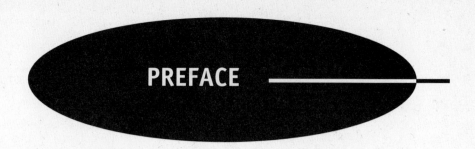

PREFACE

This book is intended to accompany a text used in that first course in fluid mechanics which is required in all mechanical engineering and civil engineering departments, as well as several other departments. It provides a succinct presentation of the material so that the students more easily understand those difficult parts. If an expanded presentation is not a necessity, this book can be used as the primary text. We have included all derivations and numerous applications, so it can be used with no supplemental material. A solutions manual is available from the authors at MerleCP@sbcglobal.net.

We have included a derivation of the Navier–Stokes equations with several solved flows. It is not necessary, however, to include them if the elemental approach is selected. Either method can be used to study laminar flow in pipes, channels, between rotating cylinders, and in laminar boundary layer flow.

The basic principles upon which a study of fluid mechanics is based are illustrated with numerous examples, solved problems, and supplemental problems which allow students to develop their problem-solving skills. The answers to all supplemental problems are included at the end of each chapter. All examples and problems are presented using SI metric units. English units are indicated throughout and are included in the Appendix.

The mathematics required is that of other engineering courses except that required if the study of the Navier–Stokes equations is selected where partial differential equations are encountered. Some vector relations are used, but not at a level beyond most engineering curricula.

If you have comments, suggestions, or corrections or simply want to opine, please e-mail me at: merlecp@sbcglobal.net. It is impossible to write an error-free book, but if we are made aware of any errors, we can have them corrected in future printings. Therefore, send an email when you find one.

MERLE C. POTTER
DAVID C. WIGGERT

CONTENTS

Chapter 1

Basic Information

1.1 INTRODUCTION

Fluid mechanics is encountered in almost every area of our physical lives. Blood flows through our veins and arteries, a ship moves through water and water flows through rivers, airplanes fly in the air and air flows around wind machines, air is compressed in a compressor and steam expands around turbine blades, a dam holds back water, air is heated and cooled in our homes, and computers require air to cool components. All engineering disciplines require some expertise in the area of fluid mechanics.

In this book we will present those elements of fluid mechanics that allow us to solve problems involving relatively simple geometries such as flow through a pipe and a channel and flow around spheres and cylinders. But first, we will begin by making calculations in fluids at rest, the subject of fluid statics. The math requirement is primarily calculus but some differential equation theory will be used. The more complicated flows that usually are the result of more complicated geometries will not be presented in this book.

In this first chapter, the basic information needed in our study will be presented. Much of it has been included in previous courses so it will be a review. But, some of it should be new to you. So, let us get started.

1.2 DIMENSIONS, UNITS, AND PHYSICAL QUANTITIES

Fluid mechanics, as all other engineering areas, is involved with physical quantities. Such quantities have dimensions and units. The nine basic dimensions are mass, length, time, temperature, amount of a substance, electric current, luminous intensity, plane angle, and solid angle. All other quantities can be expressed in terms of these basic dimensions, e.g., force can be expressed using Newton's second law as

$$F = ma \qquad (1.1)$$

In terms of dimensions we can write (note that F is used both as a variable and as a dimension)

$$F = M\frac{L}{T^2} \qquad (1.2)$$

where F, M, L, and T are the dimensions of force, mass, length, and time. We see that force can be written in terms of mass, length, and time. We could, of course, write

$$M = F\frac{T^2}{L} \qquad (1.3)$$

Units are introduced into the above relationships if we observe that it takes 1 N to accelerate 1 kg at 1 m/s² (using English units it takes 1 lb to accelerate 1 slug at 1 ft/sec²), i.e.,

$$N = kg \cdot m/s^2 \qquad lb = slug\text{-}ft/sec^2 \qquad\qquad (1.4)$$

These relationships will be used often in our study of fluids. Note that we do not use "lbf" since the unit "lb" will always refer to a pound of force; the slug will be the unit of mass in the English system. In the SI system the mass will always be kilograms and force will always be newtons. Since weight is a force, it is measured in newtons, never kilograms. The relationship

$$W = mg \qquad\qquad (1.5)$$

is used to calculate the weight in newtons given the mass in kilograms, where $g = 9.81$ m/s² (using English units $g = 32.2$ ft/sec²). Gravity is essentially constant on the earth's surface varying from 9.77 to 9.83 m/s².

Five of the nine basic dimensions and their units are included in Table 1.1 and derived units of interest in our study of fluid mechanics in Table 1.2. Prefixes are common in the SI system so they are presented in Table 1.3. Note that the SI system is a special metric system; we will use the units presented

Table 1.1 Basic Dimensions and Their Units

Quantity	Dimension	SI	Units	English	Units
Length l	L	meter	m	foot	ft
Mass m	M	kilogram	kg	slug	slug
Time t	T	second	s	second	sec
Temperature T	Θ	kelvin	K	Rankine	°R
Plane angle		radian	rad	radian	rad

Table 1.2 Derived Dimensions and Their Units

Quantity	Dimension	SI units	English units
Area A	L^2	m²	ft²
Volume V	L^3	m³ or L (liter)	ft³
Velocity V	L/T	m/s	ft/sec
Acceleration a	L/T^2	m/s²	ft/sec²
Angular velocity Ω	T^{-1}	s⁻¹	sec⁻¹
Force F	ML/T^2	kg·m/s² or N (newton)	slug-ft/sec² or lb
Density ρ	M/L^3	kg/m³	slug/ft³
Specific weight γ	M/L^2T^2	N/m³	lb/ft³
Frequency f	T^{-1}	s⁻¹	sec⁻¹
Pressure p	M/LT^2	N/m² or Pa (pascal)	lb/ft²
Stress τ	M/LT^2	N/m² or Pa (pascal)	lb/ft²
Surface tension σ	M/T^2	N/m	lb/ft
Work W	ML^2/T^2	N·m or J (joule)	ft-lb
Energy E	ML^2/T^2	N·m or J (joule)	ft-lb
Heat rate $\dot{Q}$	ML^2/T^3	J/s	Btu/sec

Table 1.2   Continued

Quantity	Dimension	SI units	English units
Torque T	ML^2/T^2	N·m	ft-lb
Power $\dot{W}$	ML^2/T^3	J/s or W (watt)	ft-lb/sec
Mass flux $\dot{m}$	M/T	kg/s	slug/sec
Flow rate Q	L^3/T	m^3/s	ft^3/sec
Specific heat c	$L^2/T^2\Theta$	J/kg·K	Btu/slug-°R
Viscosity μ	M/LT	N·s/m^2	lb-sec/ft^2
Kinematic viscosity v	L^2/T	m^2/s	ft^2/sec

Table 1.3   SI Prefixes

Multiplication factor	Prefix	Symbol
10^{12}	tera	T
10^9	giga	G
10^6	mega	M
10^3	kilo	k
10^{-2}	centi	c
10^{-3}	milli	m
10^{-6}	micro	μ
10^{-9}	nano	n
10^{-12}	pico	p

in these tables. We often use scientific notation, such as 3×10^5 N rather than 300 kN; either form is acceptable.

We finish this section with comments on significant figures. In every calculation, well, almost every one, a material property is involved. Material properties are seldom known to four significant figures and often only to three. So, it is not appropriate to express answers to five or six significant figures. Our calculations are only as accurate as the least accurate number in our equations. For example, we use gravity as 9.81 m/s^2, only three significant figures. It is usually acceptable to express answers using four significant figures, but not five or six. The use of calculators may even provide eight. The engineer does not, in general, work with five or six significant figures. Note that if the leading numeral in an answer is 1, it does not count as a significant figure, e.g., 1248 has three significant figures.

EXAMPLE 1.1   Calculate the force needed to provide an initial upward acceleration of 40 m/s^2 to a 0.4-kg rocket.

Solution: Forces are summed in the vertical y-direction:

$$\sum F_y = ma_y$$
$$F - mg = ma$$
$$F - 0.4 \times 9.81 = 0.4 \times 40$$
$$\therefore F = 19.92\,\text{N}$$

Note that a calculator would provide 19.924 N, which contains four significant figures (the leading 1 does not count). Since gravity contained three significant figures, the 4 was dropped.

1.3   GASES AND LIQUIDS

The substance of interest in our study of fluid mechanics is a gas or a liquid. We restrict ourselves to those liquids that move under the action of a shear stress, no matter how small that shearing stress may be. All gases move under the action of a shearing stress but there are certain substances, like ketchup, that do not move until the shear becomes sufficiently large; such substances are included in the subject of rheology and are not presented in this book.

A force acting on an area is displayed in Fig. 1.1. A *stress vector* is the force vector divided by the area upon which it acts. The *normal stress* acts normal to the area and the *shear stress* acts tangent to the area. It is this shear stress that results in fluid motions. Our experience of a small force parallel to the water on a rather large boat confirms that any small shear causes motion. This shear stress is calculated with

$$\tau = \lim_{\Delta A \to 0} \frac{\Delta F_t}{\Delta A} \tag{1.6}$$

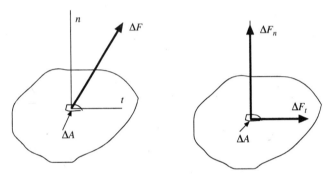

Figure 1.1   Normal and tangential components of a force.

Each fluid considered in our study is continuously distributed throughout a region of interest, that is, each fluid is a *continuum*. A liquid is obviously a continuum but each gas we consider is also assumed to be a continuum; the molecules are sufficiently close to one another so as to constitute a continuum. To determine whether the molecules are sufficiently close, we use the *mean free path*, the average distance a molecule travels before it collides with a neighboring molecule. If the mean free path is small compared to a characteristic dimension of a device (e.g., the diameter of a rocket), the continuum assumption is reasonable. In atmospheric air at sea level, the mean free path is approximately 6×10^{-6} cm and at an elevation of 100 km, it is about 10 cm. So, at high elevations, the continuum assumption is not reasonable and the theory of rarified gas dynamics is needed.

If a fluid is a continuum, the *density* can be defined as

$$\rho = \lim_{\Delta V \to 0} \frac{\Delta m}{\Delta V} \tag{1.7}$$

where Δm is the infinitesimal mass contained in the infinitesimal volume ΔV. Actually, the infinitesimal volume cannot be allowed to shrink to zero since near zero there would be few molecules in the small volume; a small volume ϵ would be needed as the limit in Eq. (1.7) for the definition to be acceptable. This is not a problem for most engineering applications since there are 2.7×10^{16} molecules in a cubic millimeter of air at standard conditions.

So, with the continuum assumption, the quantities of interest are assumed to be defined at all points in a specified region. For example, the density is a continuous function of x, y, z, and t, i.e., $\rho = \rho(x,y,z,t)$.

1.4 PRESSURE AND TEMPERATURE

In our study of fluid mechanics, we often encounter pressure. It results from compressive forces acting on an area. In Fig. 1.2 the infinitesimal force ΔF_n acting on the infinitesimal area ΔA gives rise to the *pressure*, defined by

$$p = \lim_{\Delta A \to 0} \frac{\Delta F_n}{\Delta A} \qquad (1.8)$$

The units on pressure result from force divided by area, that is, N/m^2, the pascal, Pa. A pressure of 1 Pa is a very small pressure, so pressure is typically expressed as kilopascals or kPa. Using English units, pressure is expressed as lb/ft^2 (psf) or lb/in^2 (psi). Atmospheric pressure at sea level is 101.3 kPa, or most often simply 100 kPa ($14.7 \ lb/in^2$). It should be noted that pressure is sometimes expressed as millimeters of mercury, as is common with meteorologists, or meters of water; we can use $p = \rho g h$ to convert the units, where ρ is the density of the fluid with height h.

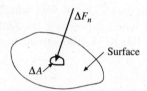

Figure 1.2   The normal force that results in pressure.

Pressure measured relative to atmospheric pressure is called *gage pressure*; it is what a gage measures if the gage reads zero before being used to measure the pressure. *Absolute pressure* is zero in a volume that is void of molecules, an ideal vacuum. Absolute pressure is related to gage pressure by the equation

$$p_{\text{absolute}} = p_{\text{gage}} + p_{\text{atmosphere}} \qquad (1.9)$$

where $p_{\text{atmosphere}}$ is the atmospheric pressure at the location where the pressure measurement is made; this atmospheric pressure varies considerably with elevation and is given in Table C.3 in App. C. For example, at the top of Pikes Peak in Colorado, it is about 60 kPa. If neither the atmospheric pressure nor elevation are given, we will assume standard conditions and use $p_{\text{atmosphere}} = 100$ kPa. Figure 1.3 presents a graphic description of the relationship between absolute and gage pressure. Several common representations of the *standard atmosphere* (at 40° latitude at sea level) are included in that figure.

We often refer to a negative pressure, as at B in Fig. 1.3, as a vacuum; it is either a negative pressure or a *vacuum*. A pressure is always assumed to be a gage pressure unless otherwise stated (in thermodynamics the pressure is assumed to be absolute). A pressure of −30 kPa could be stated as 70 kPa absolute or a vacuum of 30 kPa, assuming atmospheric pressure to be 100 kPa (note that the difference between 101.3 and 100 kPa is only 1.3 kPa, a 1.3% error, within engineering acceptability).

We do not define temperature (it requires molecular theory for a definition) but simply state that we use two scales: the Celsius scale and the Fahrenheit scale. The absolute scale when using temperature in degrees Celsius is the kelvin (K) scale and the absolute scale when using temperature in degrees Fahrenheit is the Rankine scale. We use the following conversions:

$$\begin{aligned} K &= {}^\circ C + 273.15 \\ {}^\circ R &= {}^\circ F + 459.67 \end{aligned} \qquad (1.10)$$

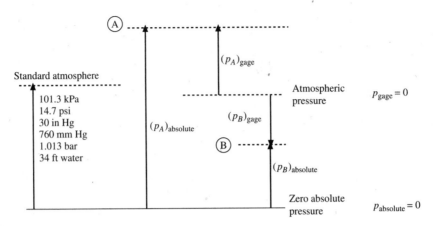

Figure 1.3   Absolute and gage pressure.

In engineering problems we use the numbers 273 and 460, which allows for acceptable accuracy. Note that we do not use the degree symbol when expressing the temperature in degrees kelvin nor do we capitalize the word "kelvin." We read "100 K" as 100 kelvins in the SI system (remember, the SI system is a special metric system).

EXAMPLE 1.2   A pressure is measured to be a vacuum of 23 kPa at a location in Wyoming where the elevation is 3000 m. What is the absolute pressure?
    Solution: Use Appendix C to find the atmospheric pressure at 3000 m. We use a linear interpolation to find $p_{\text{atmosphere}} = 70.6$ kPa. Then,

$$p_{\text{abs}} = p_{\text{atm}} + p = 70.6 - 23 = 47.6\,\text{kPa}$$

The vacuum of 23 kPa was expressed as −23 kPa in the equation.

1.5   PROPERTIES OF FLUIDS

A number of fluid properties must be used in our study of fluid mechanics. Mass per unit volume, density, was introduced in Eq. (1.7). We often use weight per unit volume, the *specific weight γ*, related to density by

$$\gamma = \rho g \tag{1.11}$$

where g is the local gravity. For water, γ is taken as 9810 N/m^3 (62.4 lb/ft^3) unless otherwise stated. Specific weight for gases is seldom used.

    Specific gravity S is the ratio of the density of a substance to the density of water and is often specified for a liquid. It may be used to determine either the density or the specific weight:

$$\rho = S\rho_{\text{water}} \qquad \gamma = S\gamma_{\text{water}} \tag{1.12}$$

As an example, the specific gravity of mercury is 13.6, which means that it is 13.6 times heavier than water. So, $\rho_{\text{mercury}} = 13.6 \times 1000 = 13\,600\,\text{kg/m}^3$, where we used the density of water to be 1000 kg/m^3, the value used for water if not specified.

    Viscosity can be considered to be the internal stickiness of a fluid. It results in shear stresses in a flow and accounts for losses in a pipe or the drag on a rocket. It can be related in a one-dimensional flow to the velocity through a shear stress τ by

$$\tau = \mu \frac{du}{dr} \tag{1.13}$$

where we call du/dr a *velocity gradient*, where r is measured normal to a surface and u is tangential to that surface, as in Fig. 1.4. Consider the units on the quantities in Eq. (1.13): the stress (force divided by an area) has units of N/m^2 (lb/ft^2) so that the viscosity has the units $N \cdot s/m^2$ ($lb\text{-}sec/ft^2$).

To measure the viscosity, consider a long cylinder rotating inside a second cylinder, as shown in Fig. 1.4. In order to rotate the inner cylinder with the rotational speed Ω, a torque T must be applied. The velocity of the inner cylinder is $R\Omega$ and the velocity of the outer cylinder is zero. The velocity distribution in the gap h between the cylinders is essentially a linear distribution as shown, so that

$$\tau = \mu \frac{du}{dr} = \mu \frac{R\Omega}{h} \qquad (1.14)$$

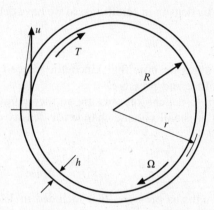

Figure 1.4 Fluid being sheared between two long cylinders.

We can relate the shear to the applied torque as follows:

$$T = \text{stress} \times \text{area} \times \text{moment arm}$$
$$= \tau \times 2\pi RL \times R$$
$$= \mu \frac{R\Omega}{h} \times 2\pi RL \times R = 2\pi \frac{R^3 \Omega L \mu}{h} \qquad (1.15)$$

where the shear acting on the ends of the long cylinder has been neglected. A device used to measure the viscosity is a *viscometer*.

In this introductory book, we focus our attention on *Newtonian fluids*, those that exhibit a linear relationship between the shear stress and the velocity gradient, as in Eqs. (1.13) and (1.14), as displayed in Fig. 1.5. Many common fluids, such as air, water, and oil are Newtonian fluids. *Non-Newtonian fluids* are classified as *dilatants*, *pseudoplastics*, and *ideal plastics* and are also displayed.

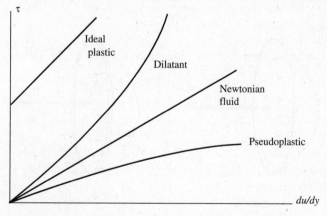

Figure 1.5   Newtonian and Non-Newtonian fluids.

A very important effect of viscosity is to cause the fluid to stick to a surface, the *no-slip condition*. If a surface is moving extremely fast, as a satellite entering the atmosphere, this no-slip condition results in very large shear stresses on the surface; this results in extreme heat which can burn up entering satellites. The no-slip condition also gives rise to wall shear in pipes resulting in pressure drops that require pumps spaced appropriately over the length of a pipe line transporting oil or gas.

Viscosity is very dependent on temperature. Note that in Fig. C.1 in App. C, the viscosity of a liquid decreases with increased temperature but the viscosity of a gas increases with increased temperature. In a liquid the viscosity is due to cohesive forces but in a gas it is due to collisions of molecules; both of these phenomena are insensitive to pressure so we note that viscosity depends on temperature only in both a liquid and a gas, i.e., $\mu = \mu(T)$.

The viscosity is often divided by density in equations, so we have defined the *kinematic viscosity* to be

$$v = \frac{\mu}{\rho} \tag{1.16}$$

It has units of m^2/s (ft^2/sec). In a gas we note that kinematic viscosity does depend on pressure since density depends on both temperature and pressure.

The volume of a gas is known to depend on pressure and temperature. In a liquid, the volume also depends slightly on pressure. If that small volume change (or density change) is important, we use the *bulk modulus B*:

$$\mathrm{B} = \mathcal{V} \left.\frac{\Delta p}{\Delta \mathcal{V}}\right|_T = \rho \left.\frac{\Delta p}{\Delta \rho}\right|_T \tag{1.17}$$

The bulk modulus has the same units as pressure. It is included in Table C.1 in App. C. For water at 20°C, it is about 2100 MPa. To cause a 1% change in the volume of water, a pressure of 21 000 kPa is needed. So, it is obvious why we consider water to be incompressible. The bulk modulus is also used to determine the speed of sound in water. It is given by

$$c = \sqrt{B/\rho} \tag{1.18}$$

This yields about $c = 1450$ m/s for water at 20°C.

Another property of occasional interest in our study is *surface tension σ*; it results from the attractive forces between molecules, and is included in Table C.1. It allows steel to float, droplets to form, and small droplets and bubbles to be spherical. Consider the free-body diagram of a spherical droplet and a bubble, as shown in Fig. 1.6. The pressure force inside the droplet balances the force due to surface tension around the circumference:

$$p\pi r^2 = 2\pi r \sigma$$

$$\therefore p = \frac{2\sigma}{r} \tag{1.19}$$

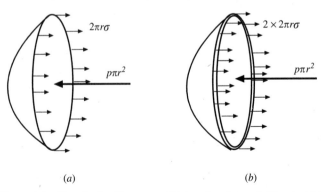

(a)   (b)

Figure 1.6 Free-body diagrams of (a) a droplet and (b) a bubble.

Note that in a bubble there are two surfaces so that the force balance provides

$$p = \frac{4\sigma}{r} \qquad (1.20)$$

So, if the internal pressure is desired, it is important to know if it is a droplet or a bubble.

A second application where surface tension causes an interesting result is in the rise of a liquid in a capillary tube. The free-body diagram of the water in the tube is shown in Fig. 1.7. Summing forces on the column of liquid gives

$$\sigma \pi D \cos \beta = \rho g \frac{\pi D^2}{4} h \qquad (1.21)$$

where the right-hand side is the weight W. This provides the height the liquid will climb in the tube:

$$h = \frac{4\sigma \cos \beta}{\gamma D} \qquad (1.22)$$

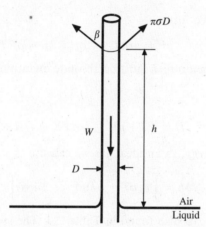

Figure 1.7 The rise of a liquid in a small tube.

The final property to be introduced in this section is vapor pressure. Molecules escape and reenter a liquid that is in contact with a gas, such as water in contact with air. The *vapor pressure* is that pressure at which there is equilibrium between the escaping and reentering molecules. If the pressure is below the vapor pressure, the molecules will escape the liquid; it is called *boiling* when water is heated to the temperature at which the vapor pressure equals the atmospheric pressure. If the local pressure is decreased to the vapor pressure, vaporization also occurs. This can happen when liquid flows through valves, elbows, or turbine blades, should the pressure become sufficiently low; it is then called *cavitation*. The vapor pressure is found in Table C.1 in App. C.

EXAMPLE 1.3 A 0.5 m × 2 m flat plate is towed at 5 m/s on a 2-mm-thick layer of SAE-30 oil at 38°C that separates it from a flat surface. The velocity distribution between the plate and the surface is assumed to be linear. What force is required if the plate and surface are horizontal?

Solution: The velocity gradient is calculated to be

$$\frac{du}{dy} = \frac{\Delta u}{\Delta y} = \frac{5 - 0}{0.002} = 2500 \text{ m/(s·m)}$$

The force is the stress multiplied by the area:

$$F = \tau \times A = \mu \frac{du}{dy} \times A = 0.1 \times 2500 \times 0.5 \times 2 = 250 \text{ N}$$

Check the units to make sure the units of the force are newtons. The viscosity of the oil was found in Fig. C.1.

EXAMPLE 1.4 A machine creates small 0.5-mm-diameter bubbles of 20°C water. Estimate the pressure that exists inside the bubbles.

Solution: Bubbles have two surfaces leading to the following estimate of the pressure:

$$p = \frac{4\sigma}{r} = \frac{4 \times 0.0736}{0.0005} = 589\,\text{Pa}$$

where the surface tension was taken from Table C.1.

1.6 THERMODYNAMIC PROPERTIES AND RELATIONSHIPS

A course in thermodynamics and/or physics usually precedes a fluid mechanics course. Those properties and relationships that are presented in those courses that are used in our study of fluids are included in this section. They are of particular use when compressible flows are studied, but they also find application to liquid flows.

The *ideal gas law* takes the two forms

$$p V = mRT \qquad \text{or} \qquad p = \rho RT \tag{1.23}$$

where the pressure p and the temperature T must be absolute quantities. The gas constant R is found in Table C.4 in App. C.

Enthalpy is defined as

$$H = m\tilde{u} + p V \qquad \text{or} \qquad h = \tilde{u} + pv \tag{1.24}$$

where $\tilde{u}$ is the *specific internal energy*. In an ideal gas we can use

$$\Delta h = \int c_p\, dT \qquad \text{and} \qquad \Delta \tilde{u} = \int c_v\, dT \tag{1.25}$$

where c_p and c_v are the specific heats also found in Table C.4. The specific heats are related to the gas constant by

$$c_p = c_v + R \tag{1.26}$$

The ratio of specific heats is

$$k = \frac{c_p}{c_v} \tag{1.27}$$

For liquids and solids, and for most gases over relatively small temperature differences, the specific heats are essentially constant and we can use

$$\Delta h = c_p \Delta T \qquad \text{and} \qquad \Delta \tilde{u} = c_v \Delta T \tag{1.28}$$

For *adiabatic* (no heat transfer) *quasi-equilibrium* (properties are constant throughout the volume at an instant) *processes*, the following relationships can be used for an ideal gas assuming constant specific heats:

$$\frac{T_2}{T_1} = \left(\frac{p_2}{p_1}\right)^{(k-1)/k} \qquad \frac{p_2}{p_1} = \left(\frac{\rho_2}{\rho_1}\right)^{k} \tag{1.29}$$

The adiabatic, quasi-equilibrium process is also called an *isentropic process*.

A small pressure wave with a relatively low frequency travels through a gas with a wave speed of

$$c = \sqrt{kRT} \tag{1.30}$$

Finally, the *first law of thermodynamics* will be of use in our study; it states that when a *system*, a fixed set of fluid particles, undergoes a change of state from state 1 to state 2, its energy changes from

E_1 to E_2 as it exchanges energy with the surroundings in the form of work W_{1-2} and heat transfer Q_{1-2}. This is expressed as

$$Q_{1-2} - W_{1-2} = E_2 - E_1 \qquad (1.31)$$

To calculate the heat transfer from given temperatures and areas, a course on heat transfer is required, so it is typically a given quantity in thermodynamics and fluid mechanics. The work, however, is a quantity that can often be calculated; it is a force times a distance and is often due to the pressure resulting in

$$
\begin{aligned}
W_{1-2} &= \int_{l_1}^{l_2} F\, dl \\
&= \int_{l_1}^{l_2} pA\, dl = \int_{\Psi_1}^{\Psi_2} p\, d\Psi
\end{aligned}
\qquad (1.32)
$$

The energy E considered in a fluids course consists of kinetic energy, potential energy, and internal energy:

$$E = m\left(\frac{V^2}{2} + gz + \tilde{u}\right) \qquad (1.33)$$

where the quantity in the parentheses is the specific energy e. (We use $\tilde{u}$ to represent specific internal energy since u is used for a velocity component.) If the properties are constant at an exit and an entrance to a flow, and there is no heat transferred and no losses, the above equation can be put in the form

$$\frac{V_2^2}{2g} + \frac{p_2}{\gamma_2} + z_2 = \frac{V_1^2}{2g} + \frac{p_1}{\gamma_1} + z_1 \qquad (1.34)$$

This equation does not follow directly from Eq. (1.31); it takes some effort to derive Eq. (1.34). An appropriate text could be consulted, but we will derive it later in this book. It is presented here as part of our review of thermodynamics.

Solved Problems

1.1 Show that the units on viscosity given in Table 1.1 are correct using (a) SI units and (b) English units.

Viscosity is related to stress by

$$\mu = \tau \frac{dy}{du}$$

In terms of units this is

$$[\mu] = \frac{\text{N}}{\text{m}^2}\frac{\text{m}}{\text{m/s}} = \frac{\text{N}\cdot\text{s}}{\text{m}^2} \qquad [\mu] = \frac{\text{lb}}{\text{ft}^2}\frac{\text{ft}}{\text{ft/sec}} = \frac{\text{lb-sec}}{\text{ft}^2}$$

1.2 If force, length, and time are selected as the three fundamental dimensions, what are the dimensions on mass?

We use Newton's second law, which states that

$$F = ma$$

In terms of dimensions this is written as

$$F = M\frac{L}{T^2} \qquad \therefore\ M = \frac{FT^2}{L}$$

1.3   The mean free path of a gas is $\lambda = 0.225m/(\rho d^2)$, where d is the molecule's diameter, m is its mass, and ρ the density of the gas. Calculate the mean free path of air at 10 000 m elevation, the elevation where many commercial airplanes fly. For an air molecule $d = 3.7 \times 10^{-10}$ m and $m = 4.8 \times 10^{-26}$ kg.

Using the formula given, the mean free path at 10 000 m is

$$\lambda = 0.225 \times \frac{4.8 \times 10^{-26}}{0.4136(3.7 \times 10^{-10})^2} = 8.48 \times 10^{-7} \text{ m or } 0.848 \, \mu\text{m}$$

where the density was found in Table C.3.

1.4   A vacuum of 25 kPa is measured at a location where the elevation is 3000 m. What is the absolute pressure in millimeters of mercury?

The atmospheric pressure at the given elevation is found in Table C.3. It is interpolated to be

$$p_{\text{atm}} = 79.84 - \frac{1}{2}(79.84 - 61.64) = 70.7 \text{ kPa}$$

The absolute pressure is then

$$p = p_{\text{gage}} + p_{\text{atm}} = -25 + 70.7 = 45.7 \text{ kPa}$$

In millimeters of mercury this is

$$h = \frac{p}{\rho_{\text{Hg}} g} = \frac{45\,700}{(13.6 \times 1000)9.81} = 0.343 \text{ m or } 343 \text{ mm}$$

1.5   A flat 30-cm-diameter disk is rotated at 800 rpm at a distance of 2 mm from a flat, stationary surface. If SAE-30 oil at 20°C fills the gap between the disk and the surface, estimate the torque needed to rotate the disk.

Since the gap is small, a linear velocity distribution will be assumed. The shear stress acting on the disk will be

$$\tau = \mu \frac{\Delta u}{\Delta y} = \mu \frac{r\omega}{h} = 0.38 \times \frac{r(800 \times 2\pi/60)}{0.002} = 15\,900r$$

where the viscosity is found from Fig. C.1 in App. C. The shear stress is integrated to provide the torque:

$$T = \int_A r\, dF = \int_A r\tau 2\pi r\, dr = 2\pi \int_0^{0.15} 15\,900 r^3\, dr = 10^5 \times \frac{0.15^4}{4} = 12.7 \text{ N}\cdot\text{m}$$

Note: The answer is not given to more significant digits since the viscosity is known to only two significant digits. More digits in the answer would be misleading.

1.6   Water is usually assumed to be incompressible. Determine the percentage volume change in 10 m³ of water at 15°C if it is subjected to a pressure of 12 MPa from atmospheric pressure.

The volume change of a liquid is found using the bulk modulus of elasticity (see Eq. (1.17)):

$$\Delta V = -V\frac{\Delta p}{B} = -10 \times \frac{12\,000\,000}{214 \times 10^7} = -0.0561 \text{ m}^3$$

The percentage change is

$$\% \text{ change} = \frac{V_2 - V_1}{V_1} \times 100 = \frac{-0.0561}{10} \times 100 = -0.561\%$$

This small percentage change can usually be ignored with no significant influence on results, so water is essentially incompressible.

1.7   Water at 30°C is able to climb up a clean glass of 0.2-mm-diameter tube due to surface tension. The water-glass angle is 0° with the vertical ($\beta = 0$ in Fig. 1.7). How far up the tube does the water climb?

The height that the water climbs is given by Eq. (1.22). It provides

$$h = \frac{4\sigma \cos \beta}{\gamma D} = \frac{4 \times 0.0718 \times 1.0}{(996 \times 9.81)0.0002} = 0.147 \, \text{m or } 14.7 \, \text{cm}$$

where the properties of water come from Table C.1 in App. C.

1.8   Explain why it takes longer to cook potatoes by boiling them in an open pan on the stove in a cabin in the mountains where the elevation is 3200 m.

Water boils when the temperature reaches the vapor pressure of the water; it vaporizes. The temperature remains constant until all the water is boiled away. The pressure at the given elevation is interpolated in Table C.3 to be 69 kPa. Table C.1 provides the temperature of slightly less than 90°C for a vapor pressure of 69 kPa, i.e., the temperature at which the water boils. Since it is less than the 100°C at sea level, the cooking process is slower. A pressure cooker could be used since it allows a higher temperature by providing a higher pressure inside the cooker.

1.9   A car tire is pressurized in Ohio to 250 kPa when the temperature is $-15°$C. The car is driven to Arizona where the temperature of the tire on the asphalt reaches 65°C. Estimate the pressure in the tire in Arizona assuming no air has leaked out and that the volume remains constant.

Assuming the volume does not change, the ideal gas law requires

$$\frac{p_2}{p_1} = \frac{mR \cancel{V}_1 T_2}{mR \cancel{V}_2 T_1} = \frac{T_2}{T_1}$$

$$\therefore p_2 = p_1 \frac{T_2}{T_1} = (250 + 100) \times \frac{423}{258} = 574 \, \text{kPa abs or } 474 \, \text{kPa gage}$$

since the mass also remains constant. (This corresponds to 37 lb/in^2 in Ohio and 70 lb/in^2 in Arizona.)

1.10   A farmer applies nitrogen to a crop from a tank pressurized to 1000 kPa absolute at a temperature of 25°C. What minimum temperature can be expected in the nitrogen if it is released to the atmosphere?

The minimum exiting temperature occurs for an isentropic process (see Eq. (1.29)), which is

$$T_2 = T_1 \left(\frac{p_2}{p_1} \right)^{(k-1)/k} = 298 \times \left(\frac{100}{1000} \right)^{0.4/1.4} = 154 \, \text{K or } -119°\text{C}$$

Such a low temperature can cause serious injury should a line break and nitrogen impact the farmer.

Supplementary Problems

1.11   There are three basic laws in our study of fluid mechanics: the conservation of mass, Newton's second law, and the first law of thermodynamics. (a) State an integral quantity for each of the laws and (b) state a quantity defined at a point for each of the laws.

Dimensions, Units, and Physical Quantities

1.12   Verify the SI units presented in Table 1.2 for the following:

    (*a*) Force   (*b*) Specific weight   (*c*) Surface tension
    (*d*) Torque   (*e*) Viscosity   (*f*) Work

1.13   Verify the dimensions presented in Table 1.2 for the following:

    (*a*) Force   (*b*) Specific weight   (*c*) Surface tension
    (*d*) Torque   (*e*) Viscosity   (*f*) Work

1.14   Select the F–L–T system of dimensions and state the dimensions on the following:

    (*a*) Force   (*b*) Specific weight   (*c*) Surface tension
    (*d*) Torque   (*e*) Viscosity   (*f*) Work

1.15   An equation that provides the flow rate in an open channel is given by

$$Q = kAR^{\frac{2}{3}}S^{\frac{1}{2}}$$

where k is a constant, A is the area of the channel, R is a radius, and S is a slope. Determine both the dimensions and the SI units on k.

1.16   Express the following using powers rather than prefixes:

    (*a*) 200 cm^2 (*b*) 500 mm^3 (*c*) 10 μm
    (*d*) 32 MPa   (*e*) 400 kN   (*f*) 5 nN

1.17   Express the following using prefixes rather than powers:

    (*a*) 2×10^{-8} m   (*b*) 5×10^{8} m   (*c*) 2×10^{-5} Pa
    (*d*) 32×10^{8} Pa   (*e*) 4×10^{-6} N   (*f*) 8×10^{11} N

1.18   Quantities are often given in units that are unacceptable when using the SI system of units. Convert each of the following to acceptable SI units:

    (*a*) 60 mi/h   (*b*) 35 lb/in^2 (*c*) 2 g/cm^3
    (*d*) 22 slug/h   (*e*) 20 ft^3/min   (*f*) 50 kW·h

1.19   What force is needed to accelerate a 1500-kg car at 3 m/s^2:

    (*a*) on the horizontal?   (*b*) on a 20° incline?

1.20   An astronaut weighs 850 N on earth. Calculate the weight of the astronaut on the moon, where $g = 5.4$ ft/sec^2.

1.21   Estimate the mean free path of air molecules, using information from Solved Problem 1.3, at an elevation of

    (*a*) 750 m   (*b*) 40 000 m   (*c*) 80 000 m

Pressure and Temperature

1.22   A pressure of 28 kPa is measured at an elevation of 2000 m. What is the absolute pressure in

    (*a*) kPa   (*b*) lb/in^2 (*c*) mm of Hg   (*d*) ft of water

1.23   A gage reads a vacuum of 24 kPa. What is the absolute pressure at

     (*a*) sea level   (*b*) 4000 m   (*c*) 8000 m

1.24   The equation $p(z) = p_0 e^{-gz/RT_0}$ is a good approximation to the pressure in the atmosphere. Estimate the pressure at $z = 6000$ m using this equation and calculate the percent error using the more accurate value found in Table C.3. Assume $p_0 = 100$ kPa and $T_0 = 15°$C.

1.25   A pressure of 20 kPa and a shear stress of 80 Pa act on a 0.8-m²-flat surface. Calculate the normal force F_n, the tangential shear force F_t, and the total force F acting on the surface. Also, calculate the angle the total force makes with respect to a normal coordinate.

1.26   A temperature of 20°C is measured at a certain location. What is the temperature in

     (*a*) kelvins   (*b*) degrees Fahrenheit   (*c*) degrees Rankine

Properties of Fluids

1.27   A fluid mass occupies 2 m³. Calculate the density, specific weight, and specific gravity if the fluid mass is

     (*a*) 4 kg   (*b*) 8 kg   (*c*) 15 kg

1.28   A formula that provides a good estimate of the density in kg/m³ of water is

$$\rho_{water} = 1000 - \frac{(T-4)^2}{180}$$

where the temperature T is in degrees Celsius. Use this formula and find the density of water at 80°C. What is the error?

1.29   The specific weight of a fluid is 11 200 N/m³. Calculate the mass contained in 2 m³

     (*a*) Using the standard gravity.
     (*b*) Using the maximum gravity on the earth's surface.
     (*c*) Using the minimum gravity on the earth's surface.

1.30   The specific gravity of mercury is given by the formula

$$S_{Hg} = 13.6 - 0.0024T$$

where the temperature is in degrees Celsius. What is the specific weight of mercury at 45°C? Calculate the error if $S_{Hg} = 13.6$ were used at 45°C.

1.31   A viscometer, used to measure the viscosity of a liquid, is composed of two 12-cm-long concentric cylinders with radii 4 and 3.8 cm. The outer cylinder is stationary and the inner one rotates. If a torque of 0.046 N·m is measured at a rotational speed of 120 rpm, estimate the viscosity of the liquid. Neglect the contribution to the torque from the cylinder ends and assume a linear velocity profile.

1.32   Water at 20°C flows in a 0.8-cm-diameter pipe with a velocity distribution of $u(r) = 5(1 - r^2/16 \times 10^{-6})$ m/s. Calculate the shear stress on (*a*) the pipe wall, (*b*) at a radius where $r = 0.2$ cm, and (*c*) at the centerline of the pipe.

1.33   SAE-30 oil at 30°C fills the gap between a 40-cm-diameter flat disk rotating 0.16 cm above a flat surface. Estimate the torque needed to rotate the disk at

   (*a*) 200 rpm   (*b*) 600 rpm   (*c*) 1200 rpm

1.34   A 2-m-long, 4-cm-diameter shaft rotates inside an equally long 4.02-cm-diameter cylinder. If SAE-10W oil at 25°C fills the gap between the concentric cylinders, determine the torque and horsepower needed to rotate the shaft at 1200 rpm.

1.35   A 0.1-m^3 volume of water is observed to be 0.0982 m^3 after a pressure is applied. What is that pressure?

1.36   How long would it take a small wave to travel under 22°C water a distance of 800 m?

1.37   The *coefficient of thermal expansion* α_T allows the expansion of a liquid to be determined using the equation $\Delta V = \alpha_T V \Delta T$. Calculate the decrease in 2 m^3 of 40°C water if the temperature is lowered by 10°C. What pressure would be needed to cause the same decrease in volume?

1.38   Estimate the pressure inside a droplet of 20°C water and a bubble of 20°C water if their diameters are

   (*a*) 40 μm   (*b*) 20 μm   (*c*) 4 μm

1.39   How high would 20°C water climb in a 24-μm-diameter vertical capillary tube if it makes an angle of 20° with the wall of the tube?

1.40   Mercury makes an angle of 130° with respect to the vertical when in contact with clean glass. How far will mercury depress in a clean, 10-μm-diameter glass tube if $\sigma_{Hg} = 0.467$ N/m.

1.41   A steel needle of length L and radius r will float in water if carefully placed. Write an equation that relates the various variables for a floating needle assuming a vertical surface tension force.

1.42   Using the equation developed in Supplementary Problem 1.41, determine if a 10-cm-long, 1-mm-diameter steel needle will float in 20°C water. $\rho_{steel} = 7850$ kg/m^3.

1.43   Derive an equation that relates the vertical force T needed to just lift a thin wire loop from a liquid assuming a vertical surface tension force. The wire radius is r and the loop diameter is D. Assume $D \gg r$.

Thermodynamic Properties and Relationships

1.44   Two kilograms of 40°C air is contained in a 4-m^3 volume. Calculate the pressure, density, specific volume, and specific weight.

1.45   The temperature outside a house is −20°C and inside it is 20°C. What is the ratio of the density of the outside air to the density of the inside air? Would infiltration, which results from cracks around the windows, doors, and siding, etc., occur even with no wind causing a pressure difference?

1.46   A car with tires pressurized to 240 kPa (35 lb/in^2) leaves Phoenix with the tire temperature at 50°C. Estimate the tire pressure (in kPa and lb/in^2) when the car arrives in Alaska with a tire temperature of −30°C.

1.47   Estimate the mass and weight of the air contained in a classroom where Thermodynamics is taught. Assume the dimensions to be $3.2 \text{ m} \times 8 \text{ m} \times 20 \text{ m}$.

1.48   Calculate the weight of the column of air contained above a 1-m^2 area of atmospheric air from sea level to the top of the atmosphere.

1.49   A 100 kg body falls from rest from a height of 100 m above the ground. Calculate its maximum velocity when it hits the ground. (a) Use the maximum value for gravity, (b) use the minimum value for gravity, and (c) use the standard value for gravity. (The minimum value is at the top of Mt Everest and the maximum value is at bottom of the lowest trench in the ocean.)

1.50   Air expands from a tank maintained at 18°C and 250 kPa to the atmosphere. Estimate its minimum temperature as it exits.

1.51   Air at 22°C is received from the atmosphere into a 200 cm^3 cylinder. Estimate the pressure and temperature if it is compressed isentropically to 10 cm^3.

1.52   Two cars, each with a mass of 6000 kg, hit head on each traveling at 80 km/h. Estimate the increase in internal energy absorbed by the materials in each car.

1.53   A 6500-kg car is traveling at 90 km/h and suddenly brakes to a stop. If the four brake disks absorb all the energy, estimate the maximum increase in temperature of those disks, assuming the disks absorb the energy equally. The 0.7-cm-thick, 30-cm-diameter disks are made of steel. Use $\rho_{steel} = 7850 \text{ kg/m}^3$ and $(c_p)_{steel} = 0.5 \text{ kJ/kg·°C}$.

1.54   Calculate the speed of sound in: (a) air at 0°C, (b) nitrogen at 20°C, (c) hydrogen at 10°C, (d) air at 100°C, and (e) oxygen at 50°C.

1.55   Lightning is observed and thunder is heard 1.5 s later. About how far away did the lightning occur?

Answers to Supplementary Problems

1.11   (a) The mass flux into a jet engine; the force of air on a window; the heat transfer through a wall. (b) The velocity V; the pressure p; the temperature T.

1.12   (a) $F = ma$. $N = \text{kg·m/s}^2$, etc.

1.13   (a) $F = ma$. $F = ML/T^2$, etc.

1.14   (b) $\gamma = \text{weight/volume} = F/L^3$, etc.

1.15 $L^{1/3}/T$, $\text{m}^{1/3}/\text{s}$

1.16   (a) $2 \times 10^{-2} \text{ m}^2$ (b) $5 \times 10^{-7} \text{ m}^3$ (c) 10^{-5} m (d) $32 \times 10^6 \text{ Pa}$ (e) $4 \times 10^5 \text{ N}$
(f) $5 \times 10^{-9} \text{ N}$

1.17   (a)   20 nm   (b)   500 Mm   (c)   20 μm   (d)   320 MPa   (e)   4 μN   (f)   800 GN

1.18   (a)   96.56 m/s   (b)   241 kPa   (c)   2000 kg/m^3 (d)   0.0892 kg/s
(e) $1.573 \times 10^{-4} \text{ m}^3/\text{s}$ (f)   80 MJ

1.19   (a)   4500 N   (b)   9533 N

1.20   468 N

1.21   (*a*)   0.000308 mm   (*b*)   0.0877 mm   (*c*)   17.5 mm

1.22   (*a*)   107.5   (*b*)   15.6   (*c*)   806   (*d*)   36

1.23   (*a*)   77.3 kPa   (*b*) 37.6 kPa   (*c*)   11.65 kPa

1.24   49.1 kPa, 4.03%

1.25   16 kN, 64 N, 0.229°

1.26   293 K, 68°F, 528°R

1.27   (*a*)   2 kg/m^3, 19.62 N/m^3, 0.002   (*b*)   4 kg/m^3, 39.24 N/m^3, 0.004
           (*c*)   7.5 kg/m^3, 73.6 N/m^3, 0.0075

1.28   968 kg/m^3, -0.4%

1.29   (*a*)   2283 kg   (*b*)   2279 kg   (*c*)   2293 kg

1.30   13.49, -0.8%

1.31   0.1628 N·s/m^2

1.32   (*a*)   2.5 N/m^2 (*b*)   1.25 N/m^2 (*c*)   0 N/m^2

1.33   (*a*)   7.2 N·m, 0.2 hp   (*b*)   21 N·m, 1.81 hp   (*c*)   43 N·m, 7.2 hp

1.34   0.88 N·m, 0.15 hp

1.35   37.8 MPa

1.36   0.539 s

1.37 -0.0076 m^3, 7.98 MPa

1.38   (*a*)   3680 Pa, 7360 Pa   (*b*)   36.8 Pa, 73.6 Pa   (*c*)   7.36 Pa, 14.72 Pa

1.39   1.175 m

1.40 -0.900 m

1.41 $2\sigma > \rho\pi r^2$

1.42   Yes

1.43 $\pi D(2\sigma + \gamma_{\text{wire}}\pi r^2)$

1.44   45 kPa, 0.5 kg/m^3, 2 m^3/kg, 4.905 N/m^3

1.45   1.158, yes

1.46   156 kPa, 22.7 lb/in^2

1.47   609 kg, 5970 N

1.48   100 kN

1.49   44.34 m/s, 44.20 m/s, 44.29 m/s

1.50 $-69.6°$C

1.51   6630 kPa, 705°C

1.52   1.48 MJ

1.53   261°C

1.54   (*a*)   331 m/s   (*b*)   349 m/s   (*c*)   1278 m/s   (*d*)   387 m/s   (*e*) 342 m/s

1.55   515 m

Chapter 2

Fluid Statics

2.1 INTRODUCTION

In *fluid statics*, there is no relative motion between fluid particles, so there are no shear stresses present (a shear results from a velocity gradient). This does not mean that the fluid particles are not moving, but only that they are not moving relative to one another; if they are moving, as in a can of water rotating about its axis, they move as a solid body. The only stress involved in fluid statics is the normal stress, the pressure. It is the pressure acting over an area that gives rise to the forces in problems involving fluid statics. The three types of problems that are presented in this chapter are: (1) fluids at rest, as in the design of a dam; (2) fluids undergoing linear acceleration, as in a rocket; and (3) fluids that are rotating about an axis.

2.2 PRESSURE VARIATION

Pressure is a quantity that acts at a point. But, does it have the same magnitude in all directions at the point? To answer this question, consider Fig. 2.1. A pressure p is assumed to act on the hypotenuse and different pressures p_x and p_y on the other two sides of the infinitesimal element that has a uniform depth dz into the paper. The fluid particle occupying the fluid element could be accelerating, so we use Newton's second law in both the x- and y-directions:

$$\sum F_x = ma_x: \quad p_x\, dy\, dz - p\, ds\, dz \sin\beta = \rho\, \frac{dx\, dy\, dz}{2}\, a_x$$

$$\sum F_y = ma_y: \quad p_y\, dx\, dz - p\, ds\, dz \cos\beta - \rho g\, \frac{dx\, dy\, dz}{2} = \rho\, \frac{dx\, dy\, dz}{2}\, a_y \tag{2.1}$$

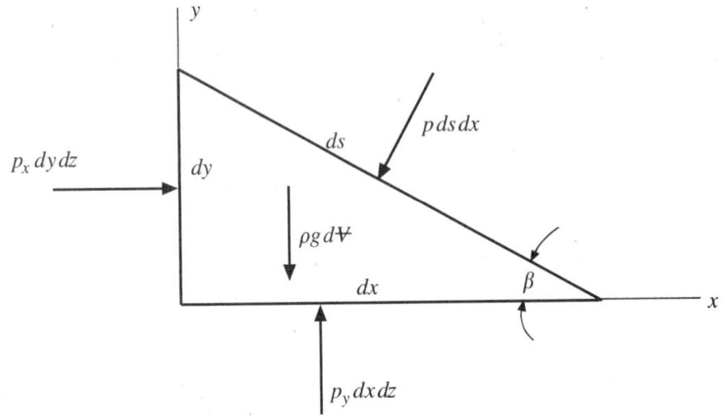

Figure 2.1 Pressure acting on an infinitesimal element.

20

recognizing that $d\mathcal{V} = \dfrac{dx\,dy\,dz}{2}$. From Fig. 2.1, we have

$$dy = ds\sin\beta \qquad dx = ds\cos\beta \tag{2.2}$$

Substituting these into Eq. (2.1), we obtain

$$p_x - p = \rho\frac{dx}{2}a_x$$
$$p_y - p = \rho\frac{dy}{2}(a_y + g) \tag{2.3}$$

Here we see that the quantities on the right-hand sides are infinitesimal, i.e., extremely small, and can be neglected[*] so that

$$p_x = p_y = p \tag{2.4}$$

Since the angle β is arbitrary, this holds for all angles. We could have selected dimensions dx and dz and arrived at $p_x = p_z = p$. So, the pressure is a scalar function that acts equally in all directions at a point in our applications to fluid statics.

In the preceding discussion, pressure only at a point was considered. The pressure variation from point to point will now be investigated. The fluid element of depth dy in Fig. 2.2 can be accelerating as in a rotating container. Newton's second law provides

$$p\,dy\,dz - \left(p + \frac{\partial p}{\partial x}dx\right)dy\,dz = \rho g\,dx\,dy\,dz\,a_x$$
$$p\,dx\,dy - \left(p + \frac{\partial p}{\partial z}dz\right)dx\,dy = -\rho g\,dx\,dy\,dz + \rho g\,dx\,dy\,dz\,a_z \tag{2.5}$$

If the element was shown in the y-direction also, the y-component equation would be

$$p\,dx\,dz - \left(p + \frac{\partial p}{\partial y}dy\right)dx\,dz = \rho g\,dx\,dy\,dz\,a_y \tag{2.6}$$

Equations (2.5) and (2.6) reduce to

$$\frac{\partial p}{\partial x} = -\rho a_x \qquad \frac{\partial p}{\partial y} = -\rho a_y \qquad \frac{\partial p}{\partial z} = -\rho(a_z + g) \tag{2.7}$$

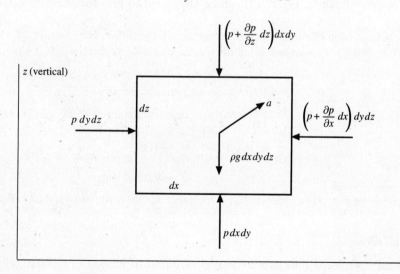

Figure 2.2   Forces acting on an element of fluid.

---

[*] Mathematically, we could use an element with sides Δx and Δy and let $\Delta x \to 0$ and $\Delta y \to 0$.

Finally, the pressure differential can be written as

$$dp = \frac{\partial p}{\partial x}dx + \frac{\partial p}{\partial y}dy + \frac{\partial p}{\partial z}dz$$

$$= -\rho a_x\, dx - \rho a_y\, dy - \rho(a_z + g)dz \qquad (2.8)$$

This can be integrated to give the desired difference in pressure between specified points in a fluid.

In a fluid at rest, there is no acceleration so that the pressure variation from Eq. (2.8) is

$$dp = -\rho g\, dz \qquad \text{or} \qquad dp = -\gamma\, dz \qquad (2.9)$$

This implies that as the elevation z increases, the pressure decreases, a fact that we are aware of in nature; the pressure increases with depth in the ocean and decreases with height in the atmosphere.

Consider the pressure variation in a liquid in which γ is constant. Equation (2.9) allows us to write

$$\Delta p = -\gamma\, \Delta z \qquad (2.10)$$

where Δp is the pressure change over the elevation change Δz. If we desire an expression for the pressure at a distance h below a free surface where the pressure is zero, it would be

$$p = \gamma h \qquad (2.11)$$

where $h = -\Delta z$. Equation (2.11) is used to convert pressure to an equivalent height of a liquid; atmospheric pressure is often expressed as millimeters of mercury (the pressure at the bottom of a 30-in column of mercury is the same as the pressure at the earth's surface due to the entire atmosphere).

If the pressure variation in the atmosphere is desired, then Eq. (2.9) would be used with the ideal gas law $p = \rho RT$ to give

$$dp = -\frac{p}{RT}g\, dz \qquad \text{or} \qquad \int_{p_0}^{p}\frac{dp}{p} = -\frac{g}{R}\int_0^z\frac{dz}{T} \qquad (2.12)$$

where p_0 is the pressure at $z = 0$. If the temperature could be assumed constant over the elevation change, then the above equation could be integrated to obtain

$$p = p_0 e^{-gz/RT} \qquad (2.13)$$

In the troposphere (between the earth's surface and to a height of about 10 km) where the temperature (in kelvins) is $T = 288 - 0.0065z$, Eq. (2.12) can be integrated to give the pressure variation.

EXAMPLE 2.1   Convert 230 kPa to millimeters of mercury, inches of mercury, and feet of water.
  Solution: Equation (2.11) is applied using the specific weight of mercury, which is $13.6\gamma_{\text{water}}$,

$$p = \gamma h \qquad\qquad 230\,000 = (13.6 \times 9800)h$$

$$\therefore\ h = 1.726\,\text{m} \quad \text{or} \quad 1726\,\text{mm of mercury}$$

This is equivalent to $1.726\,\text{m} \times 3.281\frac{\text{ft}}{\text{m}} \times 12\frac{\text{in}}{\text{ft}} = 68.0\,\text{in of mercury}$. Returning to Eq. (2.11) first convert kPa to lb/ft^2:

$$230\,\text{kPa} \times 20.89\frac{\text{lb/ft}^2}{\text{kPa}} = 4805\,\text{psf} \qquad 4805 = 62.4h$$

$$\therefore\ h = 77.0\,\text{ft of water}$$

We could have converted meters of mercury to feet of mercury and then multiplied by 13.6 to obtain feet of water.

2.3   MANOMETERS

A *manometer* is an instrument that uses a column of liquid to measure pressure, rather than using a pressure gage. Let us analyze a typical U-tube manometer attached to a pipe, as shown in Fig. 2.3, to illustrate how to interpret a manometer; this one uses water and mercury. There are several ways to analyze a manometer; this is one way. Identify two points that have the same pressure, i.e., that are at the same elevation in the same liquid, such as points 2 and 3. Then we can write

$$p_2 = p_3$$
$$p_1 + \gamma_{\text{water}}h = p_4 + \gamma_{\text{Hg}}H \qquad\qquad (2.14)$$

Since point 4 is shown to be open to the atmosphere, the pressure there is zero gage pressure: $p_4 = 0$. Thus, the manometer would measure the pressure p_1 in the pipe to be

$$p_1 = \gamma_{\text{Hg}}H - \gamma_{\text{water}}h \qquad\qquad (2.15)$$

Note that a point is positioned at all interfaces. Some manometers will have several fluids with several interfaces. Each interface should be located with a point when analyzing the manometer.

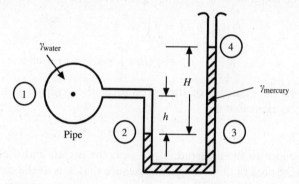

Figure 2.3   A U-tube manometer using water and mercury.

EXAMPLE 2.2   A manometer connects an oil pipeline and a water pipeline as shown in Fig. 2.4. Determine the difference in pressure between the two pipelines using the readings on the manometer. Use $S_{\text{oil}} = 0.86$ and $S_{\text{Hg}} = 13.6$.

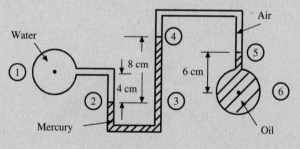

Figure 2.4

Solution: The points of interest have been positioned on the manometer in Fig. 2.4. The pressure at point 2 is equal to the pressure at point 3:

$$p_2 = p_3$$
$$p_{\text{water}} + \gamma_{\text{water}} \times 0.04 = p_4 + \gamma_{\text{Hg}} \times 0.08$$

Note that the heights must be in meters. The pressure at point 4 is essentially the same as that at point 5, since the specific weight of air is negligible compared with that of the oil. So,

$$p_4 = p_5$$
$$= p_{\text{oil}} - \gamma_{\text{oil}} \times 0.06$$

Finally,

$$p_{\text{water}} - p_{\text{oil}} = -\gamma_{\text{water}} \times 0.04 + \gamma_{\text{Hg}} \times 0.08 - \gamma_{\text{oil}} \times 0.06$$
$$= -9800 \times 0.04 + (13.6 \times 9800)0.08 - (0.86 \times 9800)0.06 = 10\,780\,\text{Pa}$$

2.4   FORCES ON PLANE AND CURVED SURFACES

In engineering designs where a liquid is contained by surfaces, such as a dam, the side of a ship, a water tank, or a levee, it is necessary to calculate the forces and their locations due to the liquid on the various surfaces. The liquid is most often water, but it could also be oil or some other liquid. We will develop equations for forces on plane surfaces, but forces on curved surfaces can be determined using the same equations. Examples will illustrate.

Consider the general surface shown in Fig. 2.5. The liquid acts on the plane area shown as a section of the wall; a top view gives additional detail of the geometry. The force on the plane surface is due to the pressure $p = \gamma h$ acting over the area, i.e.,

$$F = \int_A p \, dA = \gamma \int_A h \, dA$$
$$= \gamma \sin \alpha \int_A y \, dA = \gamma \bar{y} A \sin \alpha \qquad (2.16)$$

where $\bar{y}$ is the distance[*] to the centroid of the plane area; the centroid is identified as the point C. Equation (2.16) can also be expressed as

$$F = \gamma \bar{h} A \qquad (2.17)$$

where $\bar{h}$ is the vertical distance to the centroid. Since $\gamma \bar{h}$ is the pressure at the centroid, we see that the magnitude of the force is the area multiplied by the pressure that acts at the centroid of the area. It does not depend on the angle α of inclination. But, the force does not, in general, act at the centroid.

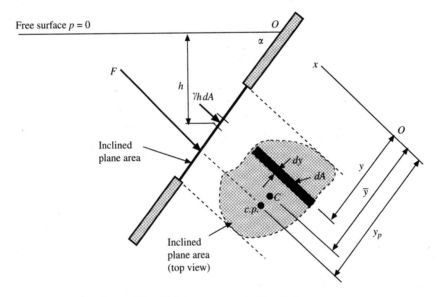

Figure 2.5   The force on an inclined plane area.

Let us assume that the force acts at some point called the *center of pressure*, located by the point (x_p, y_p). To determine where the force acts, we must recognize that the sum of the moments of all the infinitesimal forces must equal the moment of the resultant force, i.e.,

$$y_p F = \gamma \int_A yh \, dA$$
$$= \gamma \sin \alpha \int_A y^2 \, dA = \gamma I_x \sin \alpha \qquad (2.18)$$

---

[*] Recall that $\bar{y} A = \int_A y \, dA$.

where I_x is the second moment[*] of the area about the x-axis. The parallel-axis transfer theorem states that

$$I_x = \bar{I} + A\bar{y}^2 \qquad (2.19)$$

where $\bar{I}$ is the moment of the area about its centroidal axis. So, substitution of Eq. (2.19) into Eq. (2.18) and using the expression for F from Eq. (2.16) results in

$$y_p = \bar{y} + \frac{\bar{I}}{A\bar{y}} \qquad (2.20)$$

This helps us to locate where the force acts. For a horizontal surface, the pressure is uniform over the area so that the pressure force acts at the centroid of the area. In general, y_p is greater than $\bar{y}$. The centroids and second moments of various areas are presented in books on Statics or Strength of Materials. They will be given in the problems in this book.

If the top of the inclined area in Fig. 2.5 was at the free surface, the pressure distribution on that area would be triangular and the force F due to that pressure would act through the centroid of that triangular distribution, i.e., two-thirds the distance from the top of the inclined area.

To locate the x-coordinate x_p of the center of pressure, we use

$$x_p F = \gamma \sin \alpha \int_A xy\, dA$$

$$= \gamma I_{xy} \sin \alpha \qquad (2.21)$$

where I_{xy} is the product of inertia of the area. Using the transfer theorem for the product of inertia, the x-location of the center of pressure is

$$x_p = \bar{x} + \frac{\bar{I}_{xy}}{A\bar{y}} \qquad (2.22)$$

The above equations also allow us to calculate the forces acting on curved surfaces. Consider the curved gate shown in Fig. 2.6(a). The objective of this problem would be to find the force P of the gate on the vertical wall and the forces on the hinge. From the free-body diagrams in Fig. 2.6(b) and 2.6(c), the desired forces can be calculated provided the force F_W, which acts through the center of gravity of the area, can be found. The forces F_1 and F_2 can be found using Eq. (2.17). The forces F_H and F_V are the horizontal and vertical components of the force of the water acting on the gate. If a free-body diagram of only the water above the gate was identified, then we would see that

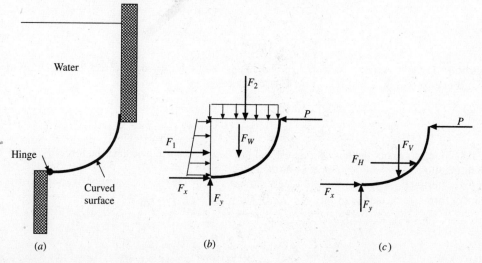

(a) (b) (c)

Figure 2.6   Forces on a curved surface: (a) the gate, (b) the water and the gate, and (c) the gate only.

---

[*] Recall the second moment of a rectangle about its centroidal axis is $bh^3/12$.

$$F_H = F_1 \quad \text{and} \quad F_V = F_2 + F_W \tag{2.23}$$

Often, the gate is composed of a quarter circle. In that case, the problem can be greatly simplified by recognizing that the forces F_H and F_V, when added together as a vector, must act through the center of the quarter circle, since all the infinitesimal forces due to the water pressure on the gate that makes up F_H and F_V act through the center. So, for a gate that has the form of a part of a circle, the force components F_H and F_V can be located at the center of the circular arc. An example will illustrate.

A final application of forces on surfaces involves *buoyancy*, i.e., forces on floating bodies. *Archimedes' principle* states that there is a buoyancy force on a floating object equal to the weight of the displaced liquid, written as

$$F_B = \gamma V_{\text{displaced liquid}} \tag{2.24}$$

Since there are only two forces acting on a floating body, they must be equal and opposite and act through the center of gravity of the body (the body could have density variations) and the centroid of the liquid volume. The body would position itself so that the center of gravity and centroid would be on a vertical line. Questions of stability arise (does the body tend to tip?), but are not considered here.

EXAMPLE 2.3   A 60-cm square gate has its top edge 12 m below the water surface. It is on a 45° angle and its bottom edge is hinged as shown in Fig. 2.7(a). What force P is needed to just open the gate?

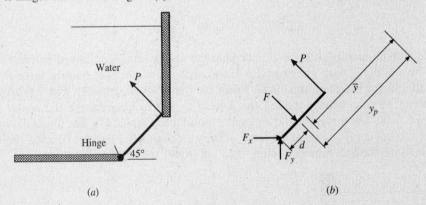

(a)   (b)

Figure 2.7

Solution: The first step is to sketch a free-body diagram of the gate so the forces and distances are clearly identified. It is done in Fig. 2.7(b). The force F is calculated to be

$$F = \gamma \bar{h} A$$
$$= 9810 \times (12 + 0.3\sin 45°)(0.6 \times 0.6) = 43\,130\,\text{N}$$

We will take moments about the hinge so that it will not be necessary to calculate the forces F_x and F_y. Let us find the distance d where the force F acts from the hinge:

$$\bar{y} = \frac{\bar{h}}{\sin 45°} = \frac{12 + 0.3\sin 45°}{\sin 45°} = 17.27\,\text{m}$$

$$y_p = \bar{y} + \frac{\bar{I}}{A\bar{y}} = 17.27 + \frac{0.6 \times 0.6^3/12}{(0.6 \times 0.6)17.27} = 17.272\,\text{m}$$

$$\therefore d = \bar{y} + 0.3 - y_p \cong 0.3\,\text{m}$$

Note: The distance $y_p - \bar{y}$ is very small and can be neglected because of the relatively large 12 m height compared with the 0.6 m dimension. So, the force P can be calculated:

$$P = \frac{0.3F}{0.6} = 21\,940\,\text{N}$$

Note again that all dimensions are converted to meters.

EXAMPLE 2.4   Consider the gate in Fig. 2.8 to be a quarter circle of radius 80 cm with the hinge 8 m below the water surface. If the gate is 1 m wide, what force P is needed to hold the gate in the position shown?

Solution: Let us move the forces F_H and F_V of Fig. 2.6(c) to the center of the circular arc, as shown in Fig. 2.8. This is allowed since all the force components that make up the resultant vector force $\mathbf{F}_H + \mathbf{F}_V$ pass through the center of the arc. The free-body diagram of the gate would appear as in Fig. 2.8. If moments are taken about the hinge, F_x, F_y, and F_V produce no moments. So,

$$P = F_H$$

a rather simple result compared with the situation if we used Fig. 2.6(c). The force P is

$$P = \gamma \bar{h} A = 9810 \times (8 - 0.4)(0.8 \times 1)$$
$$= 93\,200\,\text{N}$$

where $F_H = F_1$ and F_1 is the force on the vertical area shown in Fig. 2.6(b).

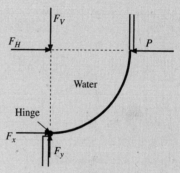

Figure 2.8

2.5 ACCELERATING CONTAINERS

The pressure in a container accelerating with components a_x and a_z is found by integrating Eq. (2.8) between selected points 1 and 2 to obtain

$$p_2 - p_1 = -\rho a_x(x_2 - x_1) - \rho(a_z + g)(z_2 - z_1) \tag{2.25}$$

If points 1 and 2 lie on a constant-pressure line (e.g., a free surface) such that $p_2 = p_1$, as in Fig. 2.9, and $a_z = 0$, Eq. (2.25) allows an expression for the angle α:

$$0 = -\rho a_x(x_2 - x_1) - \rho g(z_2 - z_1)$$
$$\tan \alpha = \frac{z_1 - z_2}{x_2 - x_1} = \frac{a_x}{g} \tag{2.26}$$

If a_z is not zero, then it is simply included. The above equations allow us to make calculations involving linearly accelerating containers. The liquid is assumed to be not sloshing; it is moving as a rigid body. An example will illustrate.

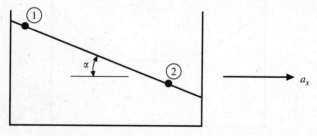

Figure 2.9   A linearly accelerating container.

To determine the pressure in a rotating container, Eq. (2.8) cannot be used, and so it is necessary to derive the expression for the differential pressure. Refer to the infinitesimal element of Fig. 2.10. A top view of the element is shown. Newton's second law applied in the radial r-direction provides, remembering that $a_r = r\Omega^2$,

$$pr\,d\theta\,dz - \left(p + \frac{\partial p}{\partial r}dr\right)(r + dr)d\theta\,dz + p\,dr\,dz\sin\frac{d\theta}{2} + p\,dr\,dz\sin\frac{d\theta}{2} = \rho r\,d\theta\,dr\,dz\,r\Omega^2 \qquad (2.27)$$

Expand the second term carefully, use $\sin d\theta/2 = d\theta/2$, neglect higher-order terms, and simplify Eq. (2.27) to

$$\frac{\partial p}{\partial r} = \rho r\Omega^2 \qquad (2.28)$$

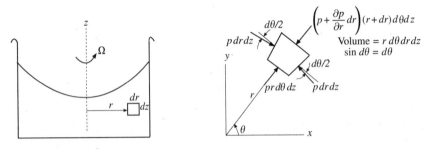

Figure 2.10   The rotating container and the top view of the infinitesimal element.

This provides the pressure variation in the radial direction and our usual $dp = -\rho g\,dz$ provides the pressure variation in the z-direction. Holding z fixed, the pressure difference from r_1 to r_2 is found by integrating Eq. (2.28):

$$p_2 - p_1 = \frac{\rho\Omega^2}{2}(r_2^2 - r_1^2) \qquad (2.29)$$

If point 1 is at the center of rotation so that $r_1 = 0$, then $p_2 = \rho\Omega^2 r_2^2/2$. If the distance from point 2 to the free surface is h as shown in Fig. 2.11, so that $p_2 = \rho gh$, we see that

$$h = \frac{\Omega^2 r_2^2}{2g} \qquad (2.30)$$

which is a parabola. The free surface is a paraboloid of revolution. An example illustrates the use of the above equations.

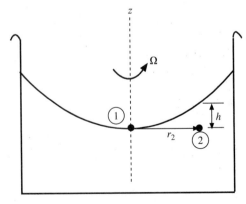

Figure 2.11   The free surface in a rotating container.

EXAMPLE 2.5 A 120-cm-long tank contains 80 cm of water and 20 cm of air maintained at 60 kPa above the water. The 60-cm-wide tank is accelerated at 10 m/s². After equilibrium is established, find the force acting on the bottom of the tank.

Solution: First, sketch the tank using the information given in the problem statement. It appears as in Fig. 2.12. The distance x can be related to y by using Eq. (2.26):

$$\tan \alpha = \frac{a_x}{g} = \frac{10}{9.81} = \frac{y}{x} \qquad \therefore y = 1.019x$$

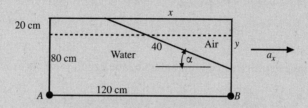

Figure 2.12

Equate the area of the air before and after to find either x or y:

$$120 \times 20 = \frac{1}{2}xy = \frac{1.019}{2}x^2 \qquad \therefore x = 68.63 \text{ cm} \quad \text{and} \quad y = 69.94 \text{ cm}$$

The pressure will remain unchanged in the air above the water since the air volume does not change. The pressures at A and B are then (use Eq. (2.25))

$$p_A = 60\,000 + 1000 \times 10 \times (1.20 - 0.6863) + 9810 \times 1.0 \text{ m} = 74\,900 \text{ Pa}$$
$$p_B = 60\,000 + 9810 \times (1.00 - 0.6994) = 62\,900 \text{ Pa}$$

The average pressure on the bottom is $(p_A + p_B)/2$. Multiply the average pressure by the area to find the force acting on the bottom:

$$F = \frac{p_A + p_B}{2}A = \frac{74\,900 + 62\,900}{2}(1.2 \times 0.6) = 49\,610 \text{ N}$$

EXAMPLE 2.6 The cylinder in Fig. 2.13 is rotated about the center axis as shown. What rotational speed is required so that the water just touches point A. Also, find the force on the bottom of the cylinder.

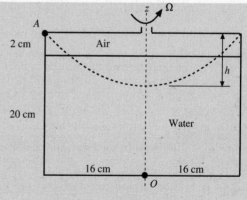

Figure 2.13

Solution: The volume of the air before and after must be the same. Recognizing that the volume of a paraboloid of revolution is half of the volume of a circular cylinder of the same radius and height, the height of the paraboloid of revolution is found:

$$\pi \times 0.16^2 \times 0.02 = \frac{1}{2}\pi \times 0.16^2 h \qquad \therefore h = 0.04 \text{ m}$$

Use Eq. (2.30) to find Ω:

$$0.04 = \frac{\Omega^2 \times 0.16^2}{2 \times 9.81} \qquad \Omega = 5.54 \, \text{rad/s}$$

The pressure on the bottom as a function of the radius r is $p(r)$, given by

$$p - p_0 = \frac{\rho \Omega^2}{2}(r^2 - r_1^2)$$

where $p_0 = 9810 \times (0.20 - 0.04) = 1570 \, \text{Pa}$. So,

$$p = \frac{1000 \times 5.54^2}{2} r^2 + 1570 = 15\,346 r^2 + 1570$$

The pressure is integrated over the area to find the force to be

$$\int_0^{0.16} (15\,346 r^2 + 1570) 2\pi r \, dr = 142.1 \, \text{N}$$

Solved Problems

2.1 Derive an expression for the density variation in a liquid assuming a constant bulk modulus and a constant temperature.

The density varies in a liquid according to Eq. (1.13), $B = \rho \, \Delta p / \Delta \rho|_T$. Over a small pressure difference, this can be written as, using Eq. (2.9),

$$dp = \frac{B}{\rho} d\rho = \rho g \, dh \quad \text{or} \quad \frac{d\rho}{\rho^2} = \frac{g}{B} dh$$

Assuming a constant value for B, set up an integration:

$$\int_{\rho_0}^{\rho} \frac{d\rho}{\rho^2} = \frac{g}{B} \int_0^h dh$$

Integrating gives the increase in density as

$$-\frac{1}{\rho} + \frac{1}{\rho_0} = \frac{gh}{B} \quad \text{or} \quad \rho = \frac{\rho_0}{1 - g\rho_0 h / B}$$

This could be used with $dp = \rho g \, dh$ to provide the pressure variation in the ocean.

2.2 A U-tube manometer measures the pressure in an air pipe to be 10 cm of water. Calculate the pressure in the pipe.

Refer to Fig. 2.3. Equation (2.15) provides the answer:

$$p_1 = \gamma_{\text{water}} H - \gamma_{\text{air}} h = 9810 \times 0.1 = 981 \, \text{Pa}$$

We have neglected the term $\gamma_{\text{air}} h$ since γ_{air} is small compared with γ_{water}.

2.3 Find the force P needed to hold the 2-m-wide gate in Fig. 2.14 in the position shown if $h = 1.2$ m.

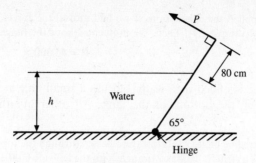

Figure 2.14

The force of the water on the gate is given by Eq. (2.17), using $\bar{h} = 0.6$ m, to be

$$F = \gamma \bar{h} A = 9810 \times 0.6 \times \left(\frac{1.2}{\sin 65°} \times 2 \right) = 15\,590 \text{ N}$$

The force F acts normal to the gate. Moments about the hinge gives

$$F d_1 = P d_2 \qquad 15\,590 \frac{0.6}{\sin 65°} = P \left(\frac{1.2}{\sin 65°} + 0.8 \right) \qquad \therefore P = 4860 \text{ N}$$

We have used d_1 as the distance to F and d_2 as the distance to P.

2.4 Find the force P needed to hold the 3-m-wide gate in the position shown in Fig. 2.15(a) if $r = 2$ m.

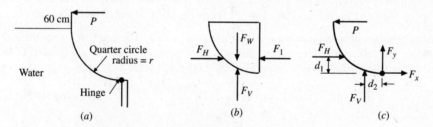

Figure 2.15

There are horizontal and vertical force components acting on the gate. The pressure distribution on the gate would be the same if water was above and to the right of the gate. So, only a free-body diagram of the water is shown in Fig. 2.15(b). The free-body diagram of the gate is shown in Fig. 2.15(c). The forces $F_1 = F_H$ and $F_W = F_V$ are

$$F_H = F_1 = \gamma \bar{h} A \qquad\qquad\qquad F_V = F_W = \gamma \Psi$$

$$= 9810 \times 1 \times (2 \times 3) = 58\,860 \text{ N} \qquad = 9810 \times \frac{1}{4} \pi \times 2^2 \times 3 = 92\,580 \text{ N}$$

The distances d_1 and d_2 (F_W acts through the centroid of the quarter circle) are

$$d_1 = \frac{1}{3} \times 2 = 0.667 \text{ m} \qquad\qquad d_2 = \frac{4r}{3\pi} = \frac{4 \times 2}{3 \times \pi} = 0.8488 \text{ m}$$

(The force F_1 is due to a triangular pressure distribution on the vertical rectangular area, so it must act through the centroid of that distribution: two-thirds the distance from the surface, or one-third the distance up from the hinge.) Moments about the hinge give

$$2.6P = d_1 F_H + d_2 F_V = 0.667 \times 58\,860 + 0.8488 \times 92\,580 \qquad \therefore P = 45\,300 \text{ N}$$

We could have simplified the calculations if we had moved the forces F_H and F_V to the center of the circular arc (review Example 2.4). Then, the moments about the hinge would have provided

$$2.6P = 2F_H \qquad \therefore \ P = 45\,300\,\text{N}$$

2.5   The tank of Example 2.5 is filled with water but has a small hole at the top of the very left. Now find the force acting on the bottom of the tank. All other quantities remain as stated in the example.

The constant-pressure line of zero gage pressure passes through the top left corner and extends below B at a distance z (make a sketch that has a triangle with the left side $(100 + z)$ cm high and the base 120 cm long), where z is found from

$$\tan \alpha = \frac{10}{9.81} = \frac{100 + z}{120} \qquad \therefore \ z = 22.3\,\text{cm}$$

Point B is 22.3 cm above the zero-pressure line so that the pressure at B is

$$p_B = -\gamma z = -9810 \times 0.223 = -2190\,\text{Pa}$$

The pressure at A and the average pressure on the bottom area are

$$p_A = 9810 \times 1.0 = 9810\,\text{Pa} \quad \text{and} \quad p_{\text{avg}} = \frac{p_A + p_B}{2} = \frac{9810 - 2190}{2} = 3810\,\text{Pa}$$

The force on the bottom is then

$$F = p_{\text{avg}} A = 3810(0.6 \times 1.2) = 2740\,\text{N}$$

2.6   A test tube is placed in a rotating device that gradually positions the tube to a horizontal position when it is rotating at a high enough rate. If that rate is 1000 rpm, estimate the pressure at the bottom of the relatively small-diameter test tube if the tube contains water and it is 12 cm long. The top of the tube is at a radius of 4 cm from the axis or rotation.

The paraboloid of revolution is a constant-pressure surface. The one that passes through the top of the rotating test tube is a surface of zero pressure. If we position point "1" on the axis of rotation and "2" on the bottom of the test tube, then Eq. (2.29) takes the form

$$p_2 - \cancel{p_1} = \frac{\rho \Omega^2}{2}(r_2 - r_1) \quad \text{or} \quad p_2 = \frac{1000(1000 \times 2\pi/60)^2}{2} 0.12 = 65\,800\,\text{Pa}$$

Supplementary Problems

Pressure Variation

2.7   Convert the following as indicated:

(a)   2 m of water to cm of mercury
(b)   20 kPa to mm of mercury
(c)   34 ft of water to kPa
(d)   760 mm of mercury to ft of water
(e)   250 kPa to psi
(f)   32 psi to kPa

2.8   Calculate the pressure difference from the top of a house to the ground if the distance is 10 m. Make the appropriate assumptions.

2.9   A weather person states that the barometric pressure is 29 in of mercury. Convert this pressure to (*a*) kPa, (*b*) psi, (*c*) ft of water, and (*d*) bars.

2.10   Determine the depth of a liquid needed to create a pressure difference of 225 kPa if the liquid is (*a*) water, (*b*) air at standard conditions, (*c*) mercury, and (*d*) oil with $S = 0.86$.

2.11   The specific gravity of a liquid is 0.75. What height of that liquid is needed to provide a pressure difference of 200 kPa?

2.12   Assume a pressure of 100 kPa absolute at ground level. What is the pressure at the top of a 3-m-high wall on the outside where the temperature is $-20°C$ and on the inside of a house where the temperature is $22°C$? (This difference results in infiltration even if no wind is present.)

2.13   Find an expression for the pressure variation in the ocean assuming $\rho_0 = 1030$ kg/m^3 for salt water using the bulk modulus to be 2100 MPa (see the solution to Solved Problem 2.1). Estimate the pressure at 2000 m using (*a*) the expression developed and (*b*) a constant density of 1030 kg/m^3. (*c*) Calculate the percent error in (*b*) assuming (*a*) is the accurate value.

2.14   From about 12 to 20 km, the temperature in the stratosphere is constant at 217 K. Assuming the pressure at 12 km to be 19.4 kPa, use Eq. (2.13) to approximate the pressure at 20 km. Calculate the error using Table C.3 in App. C to obtain the more accurate value.

2.15   Assume a temperature distribution of $T = 288 - 0.0065z$ K and integrate to find the pressure at 10 km in the atmosphere assuming $p = 101.3$ kPa at $z = 0$. Calculate the error.

Manometers

2.16   In Fig. 2.3, calculate the pressure in the water pipe if:

(*a*) $h = 10$ cm and $H = 20$ cm   (*b*) $h = 15$ cm and $H = 25$ cm
(*c*) $h = 20$ cm and $H = 30$ cm   (*d*) $h = 17$ cm and $H = 32$ cm

2.17   The pressure at the nose of a small airplane is given by $p = \frac{1}{2}\rho V^2$, where ρ is the density of air. A U-tube manometer measures 10 cm of water. Determine the airplane's speed if it is flying at an altitude of:

(*a*) 10 m   (*b*) 4000 m   (*c*) 6000 m

2.18   Calculate the pressure difference between the air pipe and the water pipe in Fig. 2.16 if H is:

(*a*) 5 cm   (*b*) 8 cm   (*c*) 10 cm

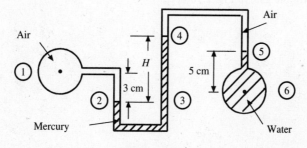

Figure 2.16

2.19   Replace the air between points 4 and 5 in Fig. 2.16 with oil having $S_{oil} = 0.86$ and let $z_4 - z_5 = 6$ cm. Calculate the pressure difference between the air pipe and water pipe if H is:

(*a*) 5 cm   (*b*) 8 cm   (*c*) 10 cm

2.20   If the manometer top in Fig. 2.17 is open, then the mercury level is 10 cm below the pressureless air pipe. The manometer top is sealed and the air pipe is pressurized. Estimate the reading for H for a pressure of 200 kPa in the air pipe. Assume an isothermal process for the air above the mercury.

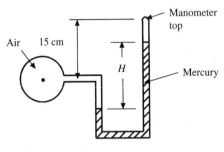

Figure 2.17

Forces on Plane and Curved Surfaces

2.21   A submersible has a viewing window that is 60 cm in diameter. Determine the pressure force of the water on the window if the center of the window is 30 m below the surface and the window is (*a*) horizontal, (*b*) vertical, and (*c*) on a 45° angle.

2.22   A concrete septic tank measures 2 m × 80 cm × 120 cm and has sides that are 8 cm thick. It is buried flush with the ground. If it is empty, how high would water saturating the soil have to rise on the outside of the tank to cause it to rise out of the ground? Assume $S_{concrete} = 2.4$.

2.23   In Solved Problem 2.3, calculate the force P if h is:

(*a*) 80 cm   (*b*) 2 m   (*c*) 2.4 m

2.24   The top of a 2-m-diamter vertical gate is 4 m below the water surface. It is hinged on the very bottom. What force, acting at the top of the gate, is needed to hold the gate closed?

2.25   Use Eq. (2.20) and show that the force on a plane rectangular surface at an angle β with the horizontal acts one-third up from the base provided the top of the rectangle is at the water's surface.

2.26   At what height H will the gate in Fig. 2.18 open if *h* is:

(*a*) 1.0 m
(*b*) 1.2 m
(*c*) 1.4 m
(*d*) 1.6 m

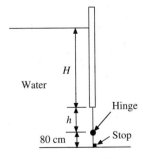

Figure 2.18

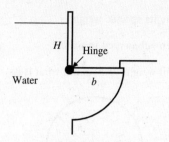

Figure 2.19

2.27 The gate shown in Fig. 2.19 will open automatically when the water level reaches a certain height above the hinge. Determine that height if b is:

    (*a*) 1.2 m   (*b*) 1.6 m   (*c*) 2.0 m

2.28 A pressure distribution exists under a concrete ($S = 2.4$) dam, as sketched in Fig. 2.20. Will the dam tend to topple (sum moments about the lower right-hand corner) if:

    (*a*) $H = 30$ m, $h = 4$ m
    (*b*) $H = 40$ m, $h = 6$ m
    (*c*) $H = 50$ m, $h = 8$ m

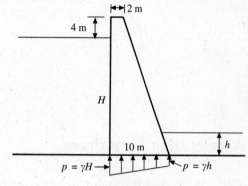

Figure 2.20

2.29 In Solved Problem 2.4, calculate the force P if r is:

    (*a*) 1.6 m   (*b*) 2.4 m   (*c*) 3 m

2.30 Consider the gate in Fig. 2.21 to be a quarter circle of radius 80 cm. Find the force P needed to just open the 1-m-wide gate if the hinge is:

    (*a*) 2 m below the surface.
    (*b*) 3 m below the surface.
    (*c*) 4 m below the surface.

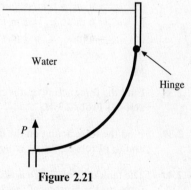

Figure 2.21

2.31 Calculate the force acting on the hinge of (*a*) Prob. 2.30*a*, (*b*) Prob. 2.30*b*, and (*c*) Prob. 2.30*c*.

2.32 Determine the force P needed to just open the 2-m-wide parabolic gate in Fig. 2.22 if the hinge is at the following y-position in the xy-plane:

    (*a*) 2 m
    (*b*) 8 m

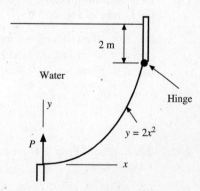

Figure 2.22

2.33   A body weighs 200 N in air and 125 N when submerged in water. Calculate its specific weight.

2.34   An object with a volume of 1200 cm³ weighs 20 N. What will it weigh when submerged in water?

2.35   A body lighter than water requires a force of 20 N to hold it under water. If it weighs 75 N in air, what is its density and specific gravity?

2.36   The cylinder shown in Fig. 2.23 pulls out the plug when the water depth reaches a certain height H. The circular plug and 2-m-long cylinder weigh 2000 N. Determine H if R is:

   (*a*)   20 cm   (*b*)   40 cm   (*c*)   60 cm

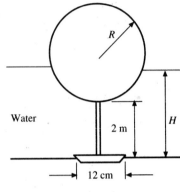

Figure 2.23

Accelerating Containers

2.37   The tank displayed in Fig. 2.24 is filled with water and accelerated with the two components shown. Calculate the pressures at A and B if:

   (*a*) $a_x = 6$ m/s², $a_z = 0$, and $h = 1.4$ m
   (*b*) $a_x = 0$, $a_z = 6$ m/s², and $h = 2.4$ m
   (*c*) $a_x = 6$ m/s², $a_z = 6$ m/s², and $h = 2$ m
   (*d*) $a_x = 6$ m/s², $a_z = 2$ m/s², and $h = 1.4$ m

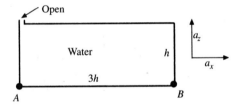

Figure 2.24

2.38   Find the force acting on the bottom of the 2-m-wide tank of (*a*) Prob. 2.37*a*, (*b*) Prob. 2.37*b*, (*c*) Prob. 2.37*c*, and (*d*) Prob. 2.37*d*.

2.39   Find the force acting on the left end of the 2-m-wide tank of (*a*) Prob. 2.37*a*, (*b*) Prob. 2.37*b*, (*c*) Prob. 2.37*c*, and (*d*) Prob. 2.37*d*.

2.40   The tank of Prob. 2.37*a* is accelerated to the left, rather than to the right. Calculate the pressure at A and the force on the bottom of the 2-m-wide tank.

2.41   Determine the pressures at points A and B in the water in the U-tube of Fig. 2.25 if:

   (*a*) $L = 40$ cm and $a_x = 6$ m/s²
   (*b*) $L = 60$ cm and $a_x = -10$ m/s²
   (*c*) $L = 50$ cm and $a_x = 4$ m/s²

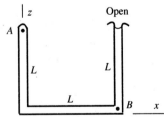

Figure 2.25

2.42   The U-tube of Prob. 2.41 is rotated about the right leg at 100 rpm. Calculate the pressures at A and B in the water if L is:

   (*a*)   40 cm   (*b*)   50 cm   (*c*)   60 cm

2.43   The U-tube of Prob. 2.41 is rotated about the left leg at 100 rpm. Calculate the pressures at A and B in the water if L is:

(*a*) 40 cm   (*b*) 50 cm   (*c*) 60 cm

2.44   The U-tube of Prob. 2.41 is rotated about the center of the horizontal part at 100 rpm. Calculate the pressures at A and B in the water if L is:

(*a*) 40 cm   (*b*) 50 cm   (*c*) 60 cm

2.45   Find the pressure at point A in the cylinder of Fig. 2.26 if $\Omega = 100$ rpm and R is:

(*a*) 40 cm   (*b*) 60 cm   (*c*) 80 cm

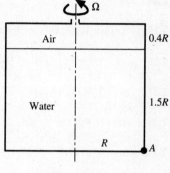

Figure 2.26

2.46   Determine the force on the bottom of the cylinder of (*a*) Prob. 2.45*a*, (*b*) Prob. 2.45*b*, and (*c*) Prob. 2.45*c*.

Answers to Supplementary Problems

2.7   (*a*) 4.7 cm   (*b*) 150 mm   (*c*) 101.7 kPa   (*d*) 33.9 ft   (*e*) 36.25 psi
(*f*) 221 kPa

2.8   120 Pa

2.9   (*a*) 15.23 kPa   (*b*) 2.21 psi   (*c*) 5.09 ft   (*d*) 0.1523 bars

2.10   (*a*) 22.9 m   (*b*) 18 650 m   (*c*) 1.686 m   (*d*) 26.7 m

2.11   27.2 m

2.12   99 959 Pa, 99 965 Pa

2.13   (*a*) 20.3 MPa   (*b*) 20.21 MPa   (*c*) -0.44%

2.14   5.50 kPa, -0.51%

2.15   26.3 kPa, -0.75%

2.16   (*a*) 25.7 kPa   (*b*) 31.9 kPa   (*c*) 38.1 kPa   (*d*) 41.0 kPa

2.17   (*a*) 40.1 m/s   (*b*) 49.1 m/s   (*c*) 54.5 m/s

2.18   (*a*) 6.18 kPa   (*b*) 10.2 kPa   (*c*) 12.8 kPa

2.19   (*a*) 5.67 kPa   (*b*) 9.68 kPa   (*c*) 12.34 kPa

2.20   30.7 cm

2.21   (*a*) 83.2 kN   (*b*) 83.2 kN   (*c*) 83.2 kN

2.22   54.2 cm

2.23   (*a*) 1212 N   (*b*) 10.6 kN   (*c*) 15.96 kN

2.24   24.4 kN

2.25   Answer is given in problem.

2.26   (*a*) 1.8 m   (*b*) 0.667 m   (*c*) 0.244 m   (*d*) 0 m

2.27   (*a*) 2.08 m   (*b*) 2.77 m   (*c*) 3.46 m

2.28   (*a*) It will tip   (*b*) It will tip   (*c*) It will not tip

2.29   (*a*) 17.1 kN   (*b*) 28.2 kN   (*c*) 36.8 kN

2.30   (*a*) 6500 N   (*b*) 7290 N   (*c*) 8070 N

2.31   (*a*) 4710 N   (*b*) 5490 N   (*c*) 6280 N

2.32   (*a*) 31.9 kN   (*b*) 91.4 kN

2.33   2.67, 0.00764 m^3

2.34   8.23 N

2.35   789 kg/m^3, 0.789

2.36   (*a*) 2.39 m   (*b*) 2.19 m

2.37   (*a*) 13.73 kPa, −11.47 kPa   (*b*) 37.9 kPa, 37.9 kPa   (*c*) 31.6 kPa, −4.38 kPa
         (*d*) 16.53 kPa, −8.67 kPa

2.38   (*a*) 9.49 kN   (*b*) 546 kN   (*c*) 327 kN   (*d*) 66 kN

2.39   (*a*) 19.22 kN   (*b*) 53.1 kN   (*c*) 44.3 kN   (*d*) 23.1 kN

2.40   13.73 kPa, 221 kN

2.41   (*a*) 2.40 kPa, 3.92 kPa   (*b*) 6.00 kPa, 3.92 kPa   (*c*) 2.00 kPa, 3.92 kPa

2.42   (*a*) 8.77 kPa, 3.92 kPa   (*b*) 13.7 kPa, 4.90 kPa   (*c*) 19.73 kPa, 5.89 kPa

2.43   (*a*) −8.77 kPa, 3.92 kPa   (*b*) −13.7 kPa, 4.90 kPa   (*c*) −19.73 kPa, 5.89 kPa

2.44   (*a*) 0 kPa, 3.92 kPa   (*b*) 0 kPa, 4.90 kPa   (*c*) 0 kPa, 5.89 kPa

2.45   (*a*) 10.98 kPa   (*b*) 21.3 kPa   (*c*) 39.5 kPa

2.46   (*a*) 3.31 kN   (*b*) 12.9 kN   (*c*) 44.1 kN

Chapter 3

Fluids in Motion

3.1 INTRODUCTION

This chapter introduces the general subject of the motion of fluid flows. Such motions are quite complex and require rather advanced mathematics to describe them if all details are to be included. With experience we can make simplifying assumptions to reduce the mathematics required, but even then the problems can get rather involved mathematically. To describe the motion of air around an airfoil, water around a ship, a tornado, a hurricane, the agitated motion in a washing machine, or even water passing through a valve, the mathematics becomes quite sophisticated and is beyond the scope of an introductory course. We will, however, derive the equations needed to describe such motions but will make simplifying assumptions that will allow a number of problems of interest to be solved. These problems will include flow in a pipe, through a channel, around rotating cylinders, and in a boundary layer near a flat wall. They will also include compressible flows involving simple geometries.

The assumptions that we will make include the nature of the geometry: pipes and channels are straight and possibly smooth, and walls are perfectly flat. Fluids are all viscous (viscosity causes fluid to stick to a boundary) but often we can ignore the viscous effects; however, if viscous effects are to be included we can demand that they behave in a linear fashion, a good assumption for water and air. Compressibility effects can also be ignored for low velocities such as those encountered in wind motions (including hurricanes) and flows around airfoils at speeds below about 100 m/s (220 mi/h) when flying near the ground.

In Sec. 3.2, we will describe fluid motion in general, the classification of different types of fluid motions will follow this, and then we will introduce the famous Bernoulli equation along with its numerous assumptions that make it applicable in only limited situations.

3.2 FLUID MOTION

3.2.1 Lagrangian and Eulerian Descriptions

The motion of a group of particles can be thought of in two basic ways: focus can be on an individual particle, such as following a particular car on a freeway jammed with cars (a police patrol car may do this while moving with traffic), or it can be at a particular location as the cars move by (a patrol car sitting along the freeway does this). When analyzed correctly, the solution to a problem would be the same using either approach (if you are speeding, you will get a ticket from either patrol car).

When solving a problem involving a single object, such as in a dynamics course, focus is always on the particular object. If there were several objects, we could establish the position $\mathbf{r}(x_0, y_0, z_0, t)$, velocity $\mathbf{V}(x_0, y_0, z_0, t)$, and acceleration $\mathbf{a}(x_0, y_0, z_0, t)$ of the object that occupied the position (x_0, y_0, z_0) at the starting time. The position (x_0, y_0, z_0) is the "name" of the object upon which attention is focused. This is the *Lagrangian description of motion*. It is quite difficult to use this description in a fluid flow where there are so many particles. Let us consider the second way to describe a fluid motion.

Let us now focus on a general point (x, y, z) in the flow with the fluid moving by the point having a velocity $\mathbf{V}(x, y, z, t)$. The rate of change of the velocity of the fluid as it passes the point is $\partial\mathbf{V}/\partial x, \partial\mathbf{V}/\partial y, \partial\mathbf{V}/\partial z$, and it may also change with time at the point: $\partial\mathbf{V}/\partial t$. We use partial derivatives here since the velocity is a function of all four variables. This is the *Eulerian description of motion*, the preferred description in our study of fluids. We have used rectangular coordinates here but other coordinate systems, such as cylindrical coordinates, can also be used. The region of interest is referred to as a *flow field* and the velocity in that flow field is often referred to as the *velocity field*. The flow field could be the inside of a pipe, the region around a turbine blade, or the water in a washing machine.

If the quantities of interest using a Eulerian description were not dependent on time t, we would have a *steady flow*; the flow variables would depend only on the space coordinates. For such a flow

$$\frac{\partial\mathbf{V}}{\partial t} = 0 \qquad \frac{\partial p}{\partial t} = 0 \qquad \frac{\partial \rho}{\partial t} = 0 \tag{3.1}$$

to list a few. In the above partial derivatives, it is assumed that the space coordinates remain fixed; we are observing the flow at a fixed point. If we followed a particular particle, as in a Lagrangian approach, the velocity of that particle would, in general, vary with time as it progressed through a flow field. Using the Eulerian description, as in Eq. (3.1), time would not appear in the expressions for quantities in a steady flow.

3.2.2   Pathlines, Streaklines, and Streamlines

There are three different lines in our description of a fluid flow. The locus of points traversed by a particular fluid particle is a *pathline*; it provides the history of the particle. A time exposure of an illuminated particle would show a pathline. A *streakline* is the line formed by all particles passing a given point in the flow; it would be a snapshot of illuminated particles passing a given point. A *streamline* is a line in a flow to which all velocity vectors are tangent at a given instant; we cannot actually photograph a streamline. The fact that the velocity is tangent to a streamline allows us to write

$$\mathbf{V} \times d\mathbf{r} = 0 \tag{3.2}$$

since $\mathbf{V}$ and $d\mathbf{r}$ are in the same direction, as shown in Fig. 3.1; recall that two vectors in the same direction have a cross product of 0.

In a steady flow, all three lines are coincident. So, if the flow is steady, we can photograph a pathline or a streakline and refer to such a line as a streamline. It is the streamline in which we have primary interest in our study of fluids.

A *streamtube* is the tube whose walls are streamlines. A pipe is a streamtube as is a channel. We often sketch a streamtube in the interior of a flow for derivation purposes.

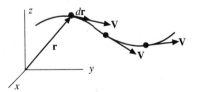

Figure 3.1   A streamline.

3.2.3 Acceleration

To make calculations for a fluid flow, such as pressures and forces, it is necessary to describe the motion in detail; the expression for the acceleration is needed assuming the velocity field is known. Consider a fluid particle having a velocity $\mathbf{V}(t)$ at an instant t, as shown in Fig. 3.2. At the next instant $t + \Delta t$ the particle will have velocity $\mathbf{V}(t + \Delta t)$, as shown. The acceleration of the particle is

$$\mathbf{a} = \frac{d\mathbf{V}}{dt} \tag{3.3}$$

where $d\mathbf{V}$ is shown in the figure. From the chain rule of calculus, we know that

$$d\mathbf{V} = \frac{\partial \mathbf{V}}{\partial x} dx + \frac{\partial \mathbf{V}}{\partial y} dy + \frac{\partial \mathbf{V}}{\partial z} dz + \frac{\partial \mathbf{V}}{\partial t} dt \tag{3.4}$$

since $\mathbf{V} = \mathbf{V}(x, y, z, \mathbf{t})$. This gives the acceleration as

$$\mathbf{a} = \frac{d\mathbf{V}}{dt} = \frac{\partial \mathbf{V}}{\partial x}\frac{dx}{dt} + \frac{\partial \mathbf{V}}{\partial y}\frac{dy}{dt} + \frac{\partial \mathbf{V}}{\partial z}\frac{dz}{dt} + \frac{\partial \mathbf{V}}{\partial t} \tag{3.5}$$

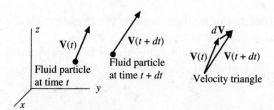

Figure 3.2   The velocity of a fluid particle.

Now, since $\mathbf{V}$ is the velocity of a particle at (x, y, z), we let

$$\mathbf{V} = u\mathbf{i} + v\mathbf{j} + w\mathbf{k} \tag{3.6}$$

where (u, v, w) are the velocity components of the particle in the x-, y-, and z-directions, respectively, and $\mathbf{i}$, $\mathbf{j}$, and $\mathbf{k}$ are the unit vectors. For the particle at the point of interest, we have

$$\frac{dx}{dt} = u \qquad \frac{dy}{dt} = v \qquad \frac{dz}{dt} = w \tag{3.7}$$

so that the acceleration can be expressed as

$$\mathbf{a} = u\frac{\partial \mathbf{V}}{\partial x} + v\frac{\partial \mathbf{V}}{\partial y} + w\frac{\partial \mathbf{V}}{\partial z} + \frac{\partial \mathbf{V}}{\partial t} \tag{3.8}$$

The time derivative of velocity represents the *local acceleration* and the other three terms represent the *convective acceleration*. In a pipe, local acceleration results if the velocity changes with time whereas convective acceleration results if velocity changes with position (as occurs at a bend or valve).

It is important to note that the expressions for the acceleration have assumed an inertial reference frame, i.e., the reference frame is not accelerating. It is assumed that a reference frame attached to the earth has negligible acceleration for problems of interest in this book. If a reference frame is attached to, say, a dishwasher spray arm, additional acceleration components enter the expressions for the acceleration vector.

The vector equation (3.8) can be written as the three scalar equations

$$a_x = u\frac{\partial u}{\partial x} + v\frac{\partial u}{\partial y} + w\frac{\partial u}{\partial z} + \frac{\partial u}{\partial t}$$

$$a_y = u\frac{\partial v}{\partial x} + v\frac{\partial v}{\partial y} + w\frac{\partial v}{\partial z} + \frac{\partial v}{\partial t} \qquad (3.9)$$

$$a_z = u\frac{\partial w}{\partial x} + v\frac{\partial w}{\partial y} + w\frac{\partial w}{\partial z} + \frac{\partial w}{\partial t}$$

We usually write Eq. (3.3) (and Eq. (3.8)) as

$$\mathbf{a} = \frac{D\mathbf{V}}{Dt} \qquad (3.10)$$

where D/Dt is called the *material,* or *substantial, derivative* since we have followed a material particle, or the substance, at an instant. In rectangular coordinates, the material derivative is

$$\frac{D}{Dt} = u\frac{\partial}{\partial x} + v\frac{\partial}{\partial y} + w\frac{\partial}{\partial z} + \frac{\partial}{\partial t} \qquad (3.11)$$

It can be used with other quantities of interest, such as the pressure: Dp/Dt would represent the rate of change of pressure of a fluid particle at some point (x, y, z).

The material derivative and acceleration components are presented for cylindrical and spherical coordinates in Table 3.1 at the end of this section.

3.2.4 Angular Velocity and Vorticity

Visualize a fluid flow as the motion of a collection of fluid particles that deform and rotate as they travel along. At some instant in time, we could think of all the particles that make up the flow as being little cubes. If the cubes simply deform and do not rotate, we refer to the flow, or a region of the flow, as an *irrotational flow*. Such flows are of particular interest in our study of fluids; they exist in tornados away from the "eye" and in the flow away from the surfaces of airfoils and automobiles. If the cubes do rotate, they possess vorticity. Let us derive the equations that allow us to determine if a flow is irrotational or if it possesses vorticity.

Consider the rectangular face of an infinitesimal volume shown in Fig. 3.3. The *angular velocity* Ω_z about the z-axis is the average of the angular velocity of segments AB and AC, counterclockwise taken as positive:

$$\Omega_z = \frac{\Omega_{AB} + \Omega_{AC}}{2} = \frac{1}{2}\left[\frac{v_B - v_A}{dx} + \frac{-(u_C - u_A)}{dy}\right]$$

$$= \frac{1}{2}\left[\frac{\frac{\partial v}{\partial x}dx}{dx} - \frac{\frac{\partial u}{\partial y}dy}{dy}\right] = \frac{1}{2}\left(\frac{\partial v}{\partial x} - \frac{\partial u}{\partial y}\right) \qquad (3.12)$$

If we select the other faces, we would find

$$\Omega_x = \frac{1}{2}\left(\frac{\partial w}{\partial y} - \frac{\partial v}{\partial z}\right) \qquad \Omega_y = \frac{1}{2}\left(\frac{\partial u}{\partial z} - \frac{\partial w}{\partial x}\right) \qquad (3.13)$$

These three components of the angular velocity components represent the rate at which a fluid particle rotates about each of the coordinate axes. The expression for Ω_z would predict the rate at which a cork would rotate in the xy-surface of the flow of water in a channel.

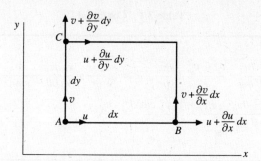

Figure 3.3   The rectangular face of a fluid element.

The *vorticity vector* $\boldsymbol{\omega}$ is defined as twice the angular velocity vector: $\boldsymbol{\omega} = 2\boldsymbol{\Omega}$. The vorticity components are

$$\omega_x = \frac{\partial w}{\partial y} - \frac{\partial v}{\partial z} \qquad \omega_y = \frac{\partial u}{\partial z} - \frac{\partial w}{\partial x} \qquad \omega_z = \frac{\partial v}{\partial x} - \frac{\partial u}{\partial y} \qquad\qquad (3.14)$$

The vorticity components in cylindrical coordinates are listed in Table 3.1. The vorticity and angular velocity components are 0 for an irrotational flow; the fluid particles do not rotate, they only deform.

Table 3.1   The Material Derivative, Acceleration, and Vorticity in Rectangular, Cylindrical, and Spherical Coordinates

---

Material derivative

Rectangular

$$\frac{D}{Dt} = u\frac{\partial}{\partial x} + v\frac{\partial}{\partial y} + w\frac{\partial}{\partial z} + \frac{\partial}{\partial t}$$

Cylindrical

$$\frac{D}{Dt} = v_r\frac{\partial}{\partial r} + \frac{v_\theta}{r}\frac{\partial}{\partial \theta} + v_z\frac{\partial}{\partial z} + \frac{\partial}{\partial t}$$

Spherical

$$\frac{D}{Dt} = v_r\frac{\partial}{\partial r} + \frac{v_\theta}{r}\frac{\partial}{\partial \theta} + \frac{v_\phi}{r\sin\theta}\frac{\partial}{\partial \phi} + \frac{\partial}{\partial t}$$

Acceleration

Rectangular

$$a_x = u\frac{\partial u}{\partial x} + v\frac{\partial u}{\partial y} + w\frac{\partial u}{\partial z} + \frac{\partial u}{\partial t} \qquad a_y = u\frac{\partial v}{\partial x} + v\frac{\partial v}{\partial y} + w\frac{\partial v}{\partial z} + \frac{\partial v}{\partial t} \qquad a_z = u\frac{\partial w}{\partial x} + v\frac{\partial w}{\partial y} + w\frac{\partial w}{\partial z} + \frac{\partial w}{\partial t}$$

Cylindrical

$$a_r = v_r\frac{\partial v_r}{\partial r} + \frac{v_\theta}{r}\frac{\partial v_r}{\partial \theta} + v_z\frac{\partial v_r}{\partial z} - \frac{v_\theta^2}{r} + \frac{\partial v_r}{\partial t} \qquad a_\theta = v_r\frac{\partial v_\theta}{\partial r} + \frac{v_\theta}{r}\frac{\partial v_\theta}{\partial \theta} + v_z\frac{\partial v_\theta}{\partial z} + \frac{v_r v_\theta}{r} + \frac{\partial v_\theta}{\partial t}$$

$$a_z = v_r\frac{\partial v_z}{\partial r} + \frac{v_\theta}{r}\frac{\partial v_z}{\partial \theta} + v_z\frac{\partial v_z}{\partial z} + \frac{\partial v_z}{\partial t}$$

Table 3.1   Continued

Spherical

$$a_r = v_r \frac{\partial v_r}{\partial r} + \frac{v_\theta}{r} \frac{\partial v_r}{\partial \theta} + \frac{v_\phi}{r \sin \theta} \frac{\partial v_r}{\partial \phi} - \frac{v_\theta^2 + v_\phi^2}{r} + \frac{\partial v_r}{\partial t} \qquad a_\theta = v_r \frac{\partial v_\theta}{\partial r} + \frac{v_\theta}{r} \frac{\partial v_\theta}{\partial \theta} + \frac{v_\phi}{r \sin \theta} \frac{\partial v_\theta}{\partial \phi} + \frac{v_r v_\theta - v_\phi^2 \cot \theta}{r} + \frac{\partial v_\theta}{\partial t}$$

$$a_\phi = v_r \frac{\partial v_\phi}{\partial r} + \frac{v_\theta}{r} \frac{\partial v_\phi}{\partial \theta} + \frac{v_\phi}{r \sin \theta} \frac{\partial v_\phi}{\partial \phi} + \frac{v_r v_\phi + v_\theta v_\phi \cot \theta}{r} + \frac{\partial v_\phi}{\partial t}$$

Vorticity

Rectangular

$$\omega_x = \frac{\partial w}{\partial y} - \frac{\partial v}{\partial z} \qquad \omega_y = \frac{\partial u}{\partial z} - \frac{\partial w}{\partial x} \qquad \omega_z = \frac{\partial v}{\partial x} - \frac{\partial u}{\partial y}$$

Cylindrical

$$\omega_r = \frac{1}{r} \frac{\partial v_z}{\partial \theta} - \frac{\partial v_\theta}{\partial z} \qquad \omega_\theta = \frac{\partial v_r}{\partial z} - \frac{\partial v_z}{\partial r} \qquad \omega_z = \frac{1}{r} \frac{\partial (r v_\theta)}{\partial r} - \frac{1}{r} \frac{\partial v_r}{\partial \theta}$$

It is the deformation of fluid particles that leads to the internal stresses in a flow. The study of the deformation of fluid particles leads to the rate-of-strain components and, with the use of constitutive equations that introduce the viscosity, to expressions for the normal and shear stresses. If Newton's second law is then applied to a particle, the famous Navier–Stokes equations result (see Chap. 5). We present these equations, along with the continuity equation (to be derived later), in Table 3.2 for completeness and consider their applications in later chapters.

Table 3.2   The Constitutive Equations, Continuity Equation, and Navier–Stokes Equations for an Incompressible Flow Using Rectangular Coordinates

Constitutive equations

$$\sigma_{xx} = -p + 2\mu \frac{\partial u}{\partial x} \qquad \tau_{xy} = \tau_{yx} = \mu \left(\frac{\partial u}{\partial y} + \frac{\partial v}{\partial x} \right) \qquad \tau_{xz} = \tau_{zx} = \mu \left(\frac{\partial u}{\partial z} + \frac{\partial w}{\partial x} \right)$$

$$\sigma_{yy} = -p + 2\mu \frac{\partial v}{\partial y} \qquad \tau_{yz} = \tau_{zy} = \mu \left(\frac{\partial v}{\partial z} + \frac{\partial w}{\partial y} \right)$$

$$\sigma_{zz} = -p + 2\mu \frac{\partial w}{\partial z}$$

Continuity equation

$$\frac{\partial u}{\partial x} + \frac{\partial v}{\partial y} + \frac{\partial w}{\partial z} = 0$$

Navier–Stokes equations

$$\rho \frac{Du}{Dt} = -\frac{\partial p}{\partial x} + \rho g_x + \mu \nabla^2 u \qquad \text{where} \qquad \frac{D}{Dt} = \frac{\partial}{\partial t} + u \frac{\partial}{\partial x} + v \frac{\partial}{\partial y} + w \frac{\partial}{\partial z}$$

$$\rho \frac{Dv}{Dt} = -\frac{\partial p}{\partial y} + \rho g_y + \mu \nabla^2 v \qquad \text{where} \qquad \nabla^2 = \frac{\partial^2}{\partial x^2} + \frac{\partial^2}{\partial y^2} + \frac{\partial^2}{\partial z^2}$$

$$\rho \frac{Dw}{Dt} = -\frac{\partial p}{\partial z} + \rho g_z + \mu \nabla^2 w$$

EXAMPLE 3.1 A velocity field in a plane flow is given by $\mathbf{V} = 2yt\mathbf{i} + x\mathbf{j}$. Find the equation of the streamline passing through (4, 2) at $t = 2$.

Solution: Equation (3.2) can be written in the form

$$(2yt\mathbf{i} + x\mathbf{j}) \times (dx\mathbf{i} + dy\mathbf{j}) = (2yt\,dy - x\,dx)\mathbf{k} = 0$$

This leads to the equation, at $t = 2$

$$4y\,dy = x\,dx$$

Integrate to obtain

$$2y^2 - \frac{x^2}{2} = C$$

The constant is evaluated at the point (4, 2) to be $C = 0$. So, the equation of the streamline is

$$x^2 = 4y^2$$

Distance is usually measured in meters and time in seconds so then velocity would have units of m/s.

EXAMPLE 3.2 For the velocity field $\mathbf{V} = 2xy\mathbf{i} + 4tz^2\mathbf{j} - yz\mathbf{k}$, find the acceleration, the angular velocity about the z-axis, and the vorticity vector at the point $(2, -1, 1)$ at $t = 2$.

Solution: The acceleration is found as follows:

$$\mathbf{a} = u\frac{\partial \mathbf{V}}{\partial x} + v\frac{\partial \mathbf{V}}{\partial y} + w\frac{\partial \mathbf{V}}{\partial z} + \frac{\partial \mathbf{V}}{\partial t}$$

$$= 2xy(2y\mathbf{i}) + 4tz^2(2x\mathbf{i} - z\mathbf{k}) - yz(8tz\mathbf{j} - y\mathbf{k}) + 4z^2\mathbf{j}$$

At the point $(2, -1, 1)$ and $t = 2$ there results

$$\mathbf{a} = 2(2)(-1)(-2\mathbf{i}) + 4(2)(1^2)(4\mathbf{i} - \mathbf{k}) - (-1)(1)(16\mathbf{j} + \mathbf{k}) + 4(1^2)\mathbf{j}$$

$$= 8\mathbf{i} + 32\mathbf{i} - 8\mathbf{k} + 16\mathbf{j} + \mathbf{k} + 4\mathbf{j}$$

$$= 40\mathbf{i} + 20\mathbf{j} - 7\mathbf{k}$$

The angular velocity component Ω_z is

$$\Omega_z = \frac{1}{2}\left(\frac{\partial v}{\partial x} - \frac{\partial u}{\partial y}\right) = \frac{1}{2}(0 - 2x) = x$$

At the point $(2, -1, 1)$ and $t = 2$ it is $\Omega_z = 2$.

The vorticity vector is

$$\boldsymbol{\omega} = \left(\frac{\partial w}{\partial y} - \frac{\partial v}{\partial z}\right)\mathbf{i} + \left(\frac{\partial u}{\partial z} - \frac{\partial w}{\partial x}\right)\mathbf{j} + \left(\frac{\partial v}{\partial x} - \frac{\partial u}{\partial y}\right)\mathbf{k}$$

$$= (-z - 8tz)\mathbf{i} + (0 - 0)\mathbf{j} + (0 - 2x)\mathbf{k}$$

At the point $(2, -1, 1)$ and $t = 2$ it is

$$\boldsymbol{\omega} = (-1 - 16)\mathbf{i} - 4\mathbf{k}$$

$$= -17\mathbf{i} - 4\mathbf{k}$$

Distance is usually measured in meters and time in seconds. Thus, angular velocity and vorticity would have units of m/(s·m) or rad/s.

3.3 CLASSIFICATION OF FLUID FLOWS

Fluid mechanics is a subject in which many rather complicated phenomena are encountered, so it is important that we understand some of the descriptions and simplifications of several special fluid flows. Such special flows will be studied in detail in later chapters. Here we will attempt to classify them in as much detail as possible.

3.3.1   Uniform, One-, Two-, and Three-Dimensional Flows

A dependent variable in our study of fluids depends, in general, on the three space coordinates and time, e.g., $\mathbf{V}(x, y, z, t)$. The flow that depends on three space coordinates is a *three-dimensional flow*; it could be a steady flow if time is not involved, such as would be the case in the flow near the intersection of a wing and the fuselage of an aircraft flying at a constant speed. The flow in a washing machine would be an unsteady, three-dimensional flow.

Certain flows can be approximated as two-dimensional flows; flows over a wide weir, in the entrance region of a pipe, and around a sphere are examples that are of special interest. In such *two-dimensional flows* the dependent variables depend on only two space variables, i.e., $p(r, \theta)$ or $\mathbf{V}(x, y, t)$. If the space coordinates are x and y, we refer to the flow as a *plane flow*.

One-dimensional flows are flows in which the velocity depends on only one space variable. They are of special interest in our introductory study since they include the flows in pipes and channels, the two most studied flows in an introductory course. For flow in a long pipe, the velocity depends on the radius r, and in a wide channel (parallel plates) it depends on y, as shown in Fig. 3.4.

Figure 3.4   One-dimensional flow. (*a*) Flow in a pipe; (*b*) flow in a wide channel.

The flows shown in Fig. 3.4 are also referred to as *developed flows*; the velocity profiles do not change with respect to the downstream coordinate. This demands that the pipe flow shown is many diameters downstream of any change in geometry, such as an entrance, a valve, an elbow, or a contraction or expansion. If the flow has not developed, the velocity field depends on more than one space coordinate, as is the case near a geometry change. The developed flow may be unsteady, i.e., it may depend on time, such as when a valve is being opened or closed.

Figure 3.5   A uniform flow in a pipe.

Finally, there is the *uniform flow*, as sketched in Fig. 3.5; the velocity profile, and other properties such as pressure, is uniform across the section of pipe. This profile is often assumed in pipe and channel flow problems since it approximates the more common turbulent flow so well. We will make this assumption in many of the problems of future chapters.

3.3.2   Viscous and Inviscid Flows

In an *inviscid flow* the effects of viscosity can be completely neglected with no significant effects on the solution to a problem involving the flow. All fluids have viscosity and if the viscous effects cannot be neglected, it is a *viscous flow*. Viscous effects are very important in pipe flows and many other kinds of flows inside conduits; they lead to losses and require pumps in long pipe lines. But, are there flows in which we can neglect the influence of viscosity? Certainly, we would not even consider inviscid flows if no such flows could be found in our engineering problems.

Consider an *external flow*, flow external to a body, such as the flow around an airfoil or a hydrofoil, as shown in Fig. 3.6. If the airfoil is moving relatively fast (faster than about 1 m/s), the flow away from a thin layer near the boundary, a *boundary layer*, can be assumed to have zero viscosity with no significant effect on the solution to the flow field (the velocity, pressure, temperature fields). All the viscous effects are concentrated inside the boundary layer and cause the velocity to be zero at the surface of the airfoil, the *no-slip condition*. Since inviscid flows are easier to solve than viscous flows, the recognition that the viscosity can be ignored in the flow away from the surface in many flows leads to much simpler solutions. This will be demonstrated in Chap. 8.

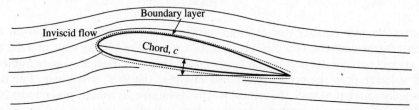

Figure 3.6   Flow around an airfoil.

3.3.3 Laminar and Turbulent Flows

A viscous flow is either a laminar flow or a turbulent flow. In a *turbulent flow* there is mixing of fluid particles so that the motion of a given particle is random and highly irregular; statistical averages are used to specify the velocity, the pressure, and other quantities of interest. Such an average may be "steady" in that it is independent of time, or it may be unsteady and depend on time. Figure 3.7 shows steady and unsteady turbulent flows. Notice the noisy turbulent flow from a faucet when you get a drink of water.

In a *laminar flow* there is negligible mixing of fluid particles; the motion is smooth and noiseless, like the slow water flow from a faucet. If a dye is injected into a laminar flow, it remains distinct for a relatively long period of time. The dye would be immediately diffused if the flow were turbulent. Figure 3.8 shows a steady and an unsteady laminar flow. A laminar flow could be made to appear turbulent by randomly controlling a valve in the flow of honey in a pipe so as to make the velocity appear as in Fig. 3.7. Yet, it would be a laminar flow since there would be no mixing of fluid particles. So, a simple

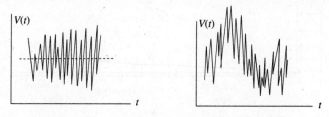

Figure 3.7   Steady and unsteady turbulent flows.

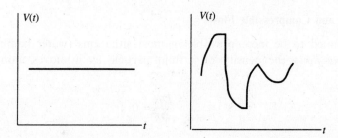

Figure 3.8   Steady and unsteady laminar flows.

display of $V(t)$ is not sufficient to decide if a particular flow is laminar or turbulent. To be turbulent, the motion has to be random, as in Fig. 3.7, but it also has to have mixing of fluid particles.

As a flow begins, as in a pipe, the flow starts out laminar, but as the average velocity increases, the laminar flow becomes unstable and turbulent flow ensues. In some cases, as in the flow between rotating cylinders, the unstable laminar flow develops into a secondary laminar flow of vortices, and then a third laminar flow, and finally a turbulent flow at higher speeds.

There is a quantity, called the *Reynolds number*, that is used to determine if a flow is laminar or turbulent. It is

$$\text{Re} = \frac{VL}{v} \qquad (3.15)$$

where V is a characteristic velocity (the average velocity in a pipe or the speed of an airfoil), L is a characteristic length (the diameter of a pipe or the distance from the leading edge of a flat plate), and v is the kinematic viscosity. If the Reynolds number is larger than a critical Reynolds number, the flow is turbulent; if it is lower than the critical Reynolds number, the flow is laminar. For flow in a pipe, assuming the usually rough pipe wall, the critical Reynolds number is usually taken to be 2000; if the wall is smooth and free of vibrations, and the entering flow is free of disturbances, the critical Reynolds number can be as high as 40 000. The critical Reynolds number is different for each geometry. For flow between parallel plates, it is taken as 1500 using the average velocity and the distance between the plates. For a boundary layer on a flat plate with a zero pressure gradient, it is between 3×10^5 and 10^6, using the distance from the leading edge.

We do not refer to an inviscid flow as laminar or turbulent. In an external flow, the inviscid flow is called a *free-stream flow*. A free stream has disturbances but the disturbances are not accompanied by shear stresses, another requirement of both laminar and turbulent flows; this will be discussed in a later chapter. The free stream can also be irrotational or it can possess vorticity.

A *boundary layer* is a thin layer of fluid that develops on a body due to the viscosity causing the fluid to stick to the boundary; it causes the velocity to be zero at the wall. The viscous effects in such a layer can actually burn up a satellite on reentry. Figure 3.9 shows the typical boundary layer on a flat plate. It is laminar near the leading edge and undergoes transition to a turbulent flow with sufficient length. For a smooth rigid plate with low free-stream fluctuation level, a laminar layer can exist up to $\text{Re} = 10^6$, where $\text{Re} = VL/v$, L being the length along the plate; for a rough plate, or a vibrating plate, or high free-stream fluctuations, a laminar flow exists up to about $\text{Re} = 3 \times 10^5$.

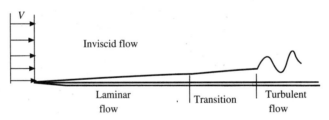

Figure 3.9   Boundary layer flow on a flat plate.

3.3.4   Incompressible and Compressible Flows

Liquid flows are assumed to be incompressible in most situations (water hammer is an exception). In such *incompressible flows* the density of a fluid particle as it moves along is assumed to be constant, i.e.,

$$\frac{D\rho}{Dt} = 0 \qquad (3.16)$$

This does not demand that the density of all the fluid particles be the same. For example, salt could be added to a water flow at some point in a pipe so that downstream of the point the density would be greater than at some upstream point. Atmospheric air at low speeds is incompressible but the density decreases with increased elevation, i.e., $\rho = \rho(z)$, where z is vertical. We usually assume a fluid to have constant density when we make the assumption of incompressibility, which is

$$\frac{\partial\rho}{\partial t} = 0 \qquad \frac{\partial\rho}{\partial x} = 0 \qquad \frac{\partial\rho}{\partial y} = 0 \qquad \frac{\partial\rho}{\partial z} = 0 \qquad (3.17)$$

The flow of air can be assumed to be incompressible if the velocity is sufficiently low. Air flow in conduits, around automobiles and small aircraft, and the takeoff and landing of commercial aircraft are all examples of incompressible airflows. The *Mach number* M where

$$M = \frac{V}{c} \tag{3.18}$$

is used to determine if a flow is compressible; V is the characteristic velocity and $c = \sqrt{kRT}$ is the speed of sound. If $M < 0.3$, we assume the flow to be incompressible. For air near sea level this is about 100 m/s (300 ft/sec) so many air flows can be assumed to be incompressible. Compressibility effects are considered in some detail in Chap. 9.

EXAMPLE 3.3   A river flowing through campus appears quite placid. A leaf floats by and we estimate the average velocity to be about 0.2 m/s. The depth is only 0.6 m. Is the flow laminar or turbulent?

Solution: We estimate the Reynolds number to be, assuming $T = 20°C$ (see Table C.1),

$$Re = \frac{Vh}{\nu} = \frac{0.2 \times 0.6}{10^{-6}} = 120\,000$$

This flow is highly turbulent at this Reynolds number, contrary to our observation of the placid flow. Most internal flows are turbulent, as observed when we drink from a drinking fountain. Laminar flows are of minimal importance to engineers when compared with turbulent flows; a lubrication problem is one exception.

3.4   BERNOULLI'S EQUATION

Bernoulli's equation may be the most often used equation in fluid mechanics but it is also the most often misused equation in fluid mechanics. In this section, that famous equation will be derived and the restrictions required for its derivation will be highlighted so that its misuse can be minimized. Before the equation is derived let us state the five assumptions required: negligible viscous effects, constant density, steady flow, the flow is along a streamline, and in an inertial reference frame. Now, let us derive the equation.

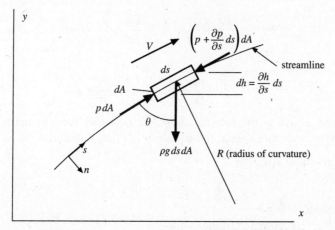

Figure 3.10   A particle moving along a streamline.

We apply Newton's second law to a cylindrical particle that is moving on a streamline, as shown in Fig. 3.10. A summation of infinitesimal forces acting on the particle is

$$p\,dA - \left(p + \frac{\partial p}{\partial s}\,ds\right)dA - \rho g\,ds\,dA\cos\theta = \rho\,ds\,dA\,a_s \tag{3.19}$$

where a_s is the s-component of the acceleration vector. It is given by Eq. (3.9a) where we think of the x-direction being in the s-direction so that $u = V$

$$a_s = V\frac{\partial V}{\partial s} + \frac{\partial V}{\partial t} \qquad (3.20)$$

where $\partial V/\partial t = 0$ assuming a steady flow. (This leads to the same acceleration expression as presented in physics or dynamics where $a_x = V\,dV/dx$ providing an inertial reference frame is used in which no Coriolis or other acceleration components are present.) Next, we observe that

$$dh = ds\cos\theta = \frac{\partial h}{\partial s}\,ds \qquad (3.21)$$

resulting in

$$\cos\theta = \frac{\partial h}{\partial s} \qquad (3.22)$$

Now, divide Eq. (3.19) by $ds\,dA$ and use the above expressions for a_s and $\cos\theta$ and rearrange. There results

$$-\frac{\partial p}{\partial s} - \rho g\frac{\partial h}{\partial s} = \rho V\frac{\partial V}{\partial s} \qquad (3.23)$$

If we assume that the density ρ is constant (this is more restrictive than incompressibility as we shall see later) so it can be moved after the partial derivative, and we recognize that $V\partial V/\partial s = \partial(V^2/2)/\partial s$, we can write our equation as

$$\frac{\partial}{\partial s}\left(\frac{V^2}{2g} + \frac{p}{\rho g} + h\right) = 0 \qquad (3.24)$$

This means that along a streamline the quantity in parentheses is constant, i.e.,

$$\frac{V^2}{2g} + \frac{p}{\rho g} + h = \text{const} \qquad (3.25)$$

where the constant may change from one streamline to the next; along a given streamline the sum of the three terms is constant. This is often written referring to two points on the same streamline as

$$\frac{V_1^2}{2g} + \frac{p_1}{\rho g} + h_1 = \frac{V_2^2}{2g} + \frac{p_2}{\rho g} + h_2 \qquad (3.26)$$

or

$$\frac{V_1^2}{2} + \frac{p_1}{\rho} + gh_1 = \frac{V_2^2}{2} + \frac{p_2}{\rho} + gh_2 \qquad (3.27)$$

Either of the two forms above is the famous *Bernoulli Equation* used in many applications. Let us highlight the assumptions once more since the equation is often misused:

- Inviscid flow (no shear stresses)
- Constant density
- Steady flow
- Along a streamline
- Applied in an inertial reference frame

The first three of these are the primary ones that are usually considered, but there are special applications where the last two must be taken into account; those special applications will not be presented in this book. Also, we often refer to a constant-density flow as an incompressible flow even though constant density is more restrictive (refer to the comments after Eq. (3.16)); this is because we do not typically make application to incompressible flows in which the density changes from one streamline to the next, such as in atmospheric flows.

Note that the units on all the terms in Eq. (3.26) are meters (feet when using English units). Consequently, $V^2/2g$ is called the *velocity head*, $p/\rho g$ is the *pressure head*, and h is simply the *head*. The sum of the three terms is often referred to as the *total head*. The pressure p is the *static pressure* and the sum $p + \rho V^2/2$ is the *total pressure* or *stagnation pressure* since it is the pressure at a *stagnation point*, a point where the fluid is brought to rest along a given streamline.

The difference in the pressures can be observed by considering the measuring probes sketched in Fig. 3.11. The probe in Fig. 3.11(*a*) is a *piezometer*; it measures the static pressure, or simply, the pressure at point 1. The *pitot tube* in Fig. 3.11(*b*) measures the total pressure, the pressure at a point where the velocity is 0, as at point 2. And, the *pitot-static tube*, which has a small opening in the side of the probe as shown in Fig. 3.11(*c*), is used to measure the difference between the total pressure and the static pressure, i.e., $\rho V^2/2$; this is used to measure the velocity. The expression for velocity is

$$V = \sqrt{\frac{2}{\rho}(p_2 - p_1)} \qquad (3.28)$$

where point 2 must be a stagnation point with $V_2 = 0$. So, if only the velocity is desired, we simply use the pitot-static probe sketched in Fig. 3.11(*c*).

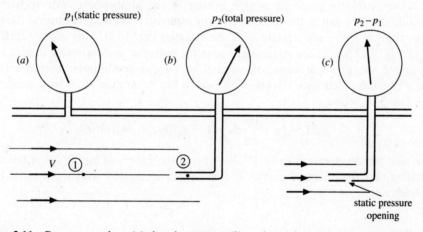

Figure 3.11 Pressure probes: (*a*) the piezometer, (*b*) a pitot tube, and (*c*) a pitot-static tube.

Bernoulli's equation is used in numerous fluid flows. It can be used in an internal flow in short reaches if the viscous effects can be neglected; such is the case in the well-rounded entrance to a pipe (see Fig. 3.12) or in a rather sudden contraction of a pipe. The velocity for such an entrance is approximated by Bernoulli's equation to be

$$V_2 = \sqrt{\frac{2}{\rho}(p_1 - p_2)} \qquad (3.29)$$

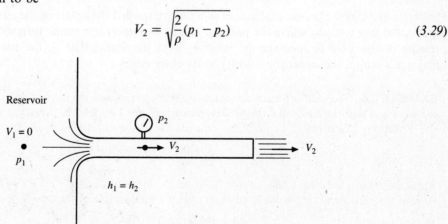

Figure 3.12 Flow from a reservoir through a pipe.

Another common application of the Bernoulli equation is from the free stream to the front area of a round object such as a sphere or a cylinder or an airfoil. A sketch is helpful as shown in Fig. 3.13. For many flow situations the flow separates from the surface, resulting in a separated flow, as sketched. If the flow approaching the object is uniform, the constant in Eq. (3.25) will be the same for all the streamlines and Bernoulli's equation can be applied from the free stream to the stagnation point at the front of the object and to points along the surface of the object up to the separation region.

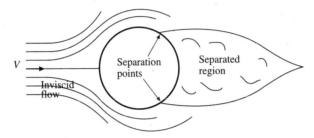

Figure 3.13   Flow around a sphere or a long cylinder.

We often solve problems involving a pipe exiting to the atmosphere. For such a situation the pressure just inside the pipe exit is the same as the atmospheric pressure just outside the pipe exit since the streamlines exiting the pipe are straight near the exit (see Fig. 3.12). This is quite different from the entrance flow of Fig. 3.12 where the streamlines near the entrance are extremely curved.

To approximate the pressure variation normal to curved streamlines, consider the particle of Fig. 3.10 to be a parallelepiped with thickness normal to the streamline of dn with area dA_s of the side with length ds. Use $\sum F_n = ma_n$:

$$p\,dA_s - \left(p + \frac{\partial p}{\partial n}\,dn\right)dA_s - \rho g\,dn\,dA_s = \rho\,dn\,dA_s\frac{V^2}{R} \tag{3.30}$$

where we have used the acceleration to be V^2/R, R being the radius of curvature in the assumed plane flow. If we assume that the effect of gravity is small when compared with the acceleration term, this equation simplifies to

$$-\frac{\partial p}{\partial n} = \rho\frac{V^2}{R} \tag{3.31}$$

Since we will use this equation to make estimations of pressure changes normal to a streamline, we approximate $\partial p/\partial n = \Delta p/\Delta n$ and arrive at the relationship

$$-\frac{\Delta p}{\Delta n} = \rho\frac{V^2}{R} \tag{3.32}$$

Hence, we see that the pressure decreases as we move toward the center of the curved streamlines; this is experienced in a tornado where the pressure can be extremely low in the tornado's "eye." This reduced pressure is also used to measure the intensity of a hurricane; that is, the lower the pressure in the hurricane's center, the larger the velocity at its outer edges.

EXAMPLE 3.4   The wind in a hurricane reaches 200 km/h. Estimate the force of the wind on a window facing the wind in a high-rise building if the window measures 1 m × 2 m. Use the density of the air to be 1.2 kg/m³.
   Solution: Use Bernoulli's equation to estimate the pressure on the window

$$p = \rho\frac{V^2}{2} = 1.2 \times \frac{(200 \times 1000/3600)^2}{2} = 1852\,\text{N/m}^2$$

where the velocity must have units of m/s. To check on the units, use kg = (N·s²)/m.
Assume the pressure to be essentially constant over the window so that the force is then

$$F = pA = 1852 \times 1 \times 2 = 3704\,\text{N} \quad\text{or}\quad 833\,\text{lb}$$

This force is large enough to break many windows, especially if they are not properly designed.

EXAMPLE 3.5   A piezometer is used to measure the pressure in a pipe to be 20 cm of water. A pitot tube measures the total pressure to be 33 cm of water at the same general location. Estimate the velocity of the water in the pipe.

Solution: The velocity using Eq. (3.27) is found to be

$$V = \sqrt{\frac{2}{\rho}(p_2 - p_1)} = \sqrt{2g(h_2 - h_1)} = \sqrt{2 \times 9.81 \times (0.33 - 0.20)} = 1.60 \, \text{m/s}$$

where we used the pressure relationship $p = \rho g h$.

Solved Problems

3.1   A velocity field in a plane flow is given by $\mathbf{V} = 2yt\mathbf{i} + x\mathbf{j}$ m/s, as in Example 3.1. Find the acceleration, the angular velocity, and the vorticity vector at the point $(4 \, \text{m}, 2 \, \text{m})$ at $t = 3$ s. (Note: the constants have units so that the velocity has units of m/s.)

The acceleration is given by

$$\mathbf{a} = \frac{\partial \mathbf{V}}{\partial t} + u\frac{\partial \mathbf{V}}{\partial x} + v\frac{\partial \mathbf{V}}{\partial y} + w\frac{\partial \mathbf{V}}{\partial z} = 2y\mathbf{i} + 2yt(\mathbf{j}) + x(2t\mathbf{i}) = 2(xt + y)\mathbf{i} + 2yt\mathbf{j}$$

At the point $(4, 2)$ and $t = 3$ s the acceleration is

$$\mathbf{a} = 2(4 \times 3 + 2)\mathbf{i} + 2 \times 2 \times 3t\mathbf{j} = 28\mathbf{i} + 12\mathbf{j} \, \text{m/s}^2$$

The angular velocity is

$$\Omega = \frac{1}{2}\left(\frac{\partial w}{\partial y} - \frac{\partial v}{\partial z}\right)\mathbf{i} + \frac{1}{2}\left(\frac{\partial u}{\partial z} - \frac{\partial w}{\partial x}\right)\mathbf{j} + \frac{1}{2}\left(\frac{\partial v}{\partial x} - \frac{\partial u}{\partial y}\right)\mathbf{k} = \frac{1}{2}(1 - 2t)\mathbf{k}$$

At $t = 3$ s, it is

$$\Omega_z = \frac{1}{2}(1 - 2 \times 3) = -\frac{5}{2} \, \text{rad/s}$$

The vorticity vector is twice the angular velocity vector so

$$\boldsymbol{\omega} = -5\mathbf{k} \, \text{rad/s}$$

3.2   Find the rate-of-change of the density in a stratified flow where $\rho = 1000(1 - 0.2z)$ and the velocity is $\mathbf{V} = 10(z - z^2)\mathbf{i}$.

The velocity is in the x-direction only and the density varies with z (usually the vertical direction). The material derivative provides the answer

$$\frac{D\rho}{Dt} = u\frac{\partial \rho}{\partial x} + v\frac{\partial \rho}{\partial y} + w\frac{\partial \rho}{\partial z} + \frac{\partial \rho}{\partial t} = 0$$

So, there is no density variation of a particular particle as that particle moves through the field of flow.

3.3   A velocity field is given in cylindrical coordinates as

$$v_r = \left(2 - \frac{8}{r^2}\right)\cos \theta \, \text{m/s} \qquad v_\theta = -\left(2 + \frac{8}{r^2}\right)\sin \theta \, \text{m/s} \qquad v_z = 0$$

What is the acceleration at the point $(3 \, \text{m}, 90°)$?

Table 3.1 provides the equations for the acceleration components. We have

$$a_r = v_r \frac{\partial v_r}{\partial r} + \frac{v_\theta}{r}\frac{\partial v_r}{\partial \theta} + v_z \frac{\partial \phi_r}{\partial z} - \frac{v_\theta^2}{r} + \frac{\partial \phi_r}{\partial t}$$

$$= \left(2 - \frac{8}{r^2}\right)\cos\theta \left(\frac{16}{r^3}\right)\cos\theta + \left(\frac{2}{r} + \frac{8}{r^3}\right)\sin\theta\left(2 - \frac{8}{r^2}\right)\sin\theta - \frac{1}{r}\left(2 + \frac{8}{r^2}\right)^2 \sin^2\theta$$

$$= 0 + \left(\frac{2}{3} + \frac{8}{27}\right)\left(2 - \frac{8}{9}\right) - \frac{1}{3}\left(2 + \frac{8}{9}\right)^2 = -1.712\,\text{m/s}^2$$

$$a_\theta = v_r \frac{\partial v_\theta}{\partial r} + \frac{v_\theta}{r}\frac{\partial v_\theta}{\partial \theta} + v_z \frac{\partial \phi_\theta}{\partial z} + \frac{v_r v_\theta}{r} + \frac{\partial \phi_\theta}{\partial t}$$

$$= \left(2 - \frac{8}{r^2}\right)\cos\theta\left(\frac{16}{r^3}\right)\sin\theta + \frac{1}{r}\left(2 + \frac{8}{r^2}\right)^2 \sin\theta\cos\theta - \left(\frac{2}{r} - \frac{8}{r^2}\right)\cos\theta\left(2 + \frac{8}{r^2}\right)\sin\theta$$

$$= 0$$

$$a_z = 0$$

Note that $\cos 90° = 0$ and $\sin 90° = 1$.

3.4 A laminar flow of 20°C water in an 8-mm diameter pipe is desired. A 2-L container, used to catch the water, is filled in 82 s. Is the flow laminar?

To make the determination, the Reynolds number must be calculated. First, determine the average velocity. It is

$$V = \frac{Q}{A} = \frac{2 \times 10^{-3}/82}{\pi \times 0.004^2} = 0.485$$

Using the kinematic viscosity of water to be about 10^{-6} m²/s (see Table C.1), the Reynolds number is

$$\text{Re} = \frac{Vh}{v} = \frac{0.485 \times 0.008 \times 0.5}{10^{-6}} = 3880$$

This is greater than 2000 so if the pipe is not smooth or the entrance is not well-rounded, the flow would be turbulent. It could, however, be laminar if care is taken to avoid building vibrations and water fluctuations with a smooth pipe.

3.5 The pitot and piezometer probes read the total and static pressures as shown in Fig. 3.14. Calculate the velocity V.

Bernoulli's equation provides

$$\frac{V_2^2}{2} + \frac{p_2}{\rho} + g h_2 = \frac{V_1^2}{2} + \frac{p_1}{\rho} + g h_1$$

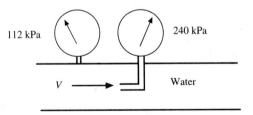

Figure 3.14

where point 2 is just inside the pitot tube. Using the information given, there results

$$\frac{240\,000}{1000} = \frac{V_1^2}{2} + \frac{112\,000}{1000} \qquad \therefore V_1 = 16\,\text{m/s}$$

Check the units on the first term of the above equation: $\dfrac{\text{N/m}^2}{\text{kg/m}^3} = \dfrac{(\text{kg·m/s}^2)/\text{m}^2}{\text{kg/m}^3} = \dfrac{\text{m}^2}{\text{s}^2}$.

3.6 A nozzle on a hose accelerates water from 4-cm diameter to 1-cm diameter. If the pressure is 400 kPa upstream of the nozzle, what is the maximum velocity exiting the nozzle?

The continuity equation relates the velocities

$$A_1 V_1 = A_2 V_2 \qquad \pi \times 2^2 \times V_1 = \pi \times 0.5^2 \times V_2 \qquad \therefore V_2 = 16 V_1$$

The Bernoulli equation provides

$$\frac{V_1^2}{2} + \frac{400\,000}{1000} + g\cancel{h_1} = \frac{256 V_1^2}{2} + \frac{100\,000}{1000} + g\cancel{h_2}$$

$$\therefore V_1 = 1.534\,\text{m/s} \quad \text{and} \quad V_2 = 24.5\,\text{m/s}$$

This represents the maximum since we have assumed no losses due to viscous effects and have assumed uniform velocity profiles.

3.7   Water flows through a long-sweep elbow on a 2-cm diameter pipe at an average velocity of 20 m/s. Estimate the increase in pressure from the inside of the pipe to the outside of the pipe midway through the elbow if the radius of curvature of the elbow averages 4 cm at the midway section.

Equation (3.32) provides the relationship between the pressure increase and the radius of curvature

$$-\frac{\Delta p}{\Delta n} = \rho\frac{V^2}{R} \qquad -\frac{\Delta p}{0.02} = 1000 \times \frac{20^2}{0.04} \qquad \therefore \Delta p = 200\,000\,\text{Pa} \ \ \text{or} \ \ 200\,\text{kPa}$$

This surprisingly high pressure difference can move the slow-moving water near the pipe wall (the water sticks to the wall due to viscosity) from the outside to the inside of the corner thereby creating a secondary flow as the water leaves the elbow. This secondary flow is eventually dissipated and accounts for a relatively large loss due to the elbow.

Supplementary Problems

Fluid Motion

3.8   The traffic in a large city is to be studied. Explain how it would be done using (a) the Lagrangian approach and (b) the Eulerian approach.

3.9   A light bulb and battery are attached to a large number of bars of soap that float. Explain how pathlines and streaklines would be photographed in a stream.

3.10   The light from a single car is photographed from a high vantage point with a time exposure. What is the line that is observed in the photograph? A long time passes as a large number of car lights are photographed instantaneously on the same road from the same high vantage point. What is the relation between the two photographs? Explain similarities and differences.

3.11   The parabolic velocity distribution in a channel flow is given by $u(y) = 0.2(1 - y^2)$ m/s with y measured in centimeters. What is the acceleration of a fluid particle on the centerline where $y = 0$? At a location where $y = 0.5$ cm?

3.12   Calculate the speed and acceleration of a fluid particle at the point $(2, 1, -3)$ when $t = 2$ s if the velocity field is given by (distances are in meters and the constants have the necessary units):

   (a) $\mathbf{V} = 2xy\mathbf{i} + y^2 t\mathbf{j} + yz\mathbf{k}$ m/s

   (b) $\mathbf{V} = 2(xy - z^2)\mathbf{i} + xyt\mathbf{j} + xzt\mathbf{k}$ m/s

3.13   Find the unit vector normal to the streamline at the point $(2, -1)$ when $t = 2$ s if the velocity field is given by:

   (a) $\mathbf{V} = 2xy\mathbf{i} + y^2 t\mathbf{j}$ m/s
   (b) $\mathbf{V} = 2y(x - y)\mathbf{i} + xyt\mathbf{j}$ m/s

3.14 What is the equation of the streamline that passes through the point $(2, -1)$ when $t = 2$ s if the velocity field is given by:

(a) $\mathbf{V} = 2xy\mathbf{i} + y^2t\mathbf{j}$ m/s
(b) $\mathbf{V} = 2y^2\mathbf{i} + xyt\mathbf{j}$ m/s

3.15 Determine the acceleration (vector and magnitude) of the fluid particle occupying the point $(-2, 1, 1)$ m when $t = 2$ s if the velocity field is given by:

(a) $\mathbf{V} = 2xy\mathbf{i} + xz\mathbf{j} + yz\mathbf{k}$ m/s
(b) $\mathbf{V} = 2y^2\mathbf{i} + (x - 2t)\mathbf{j} + z^2\mathbf{k}$ m/s
(c) $\mathbf{V} = 2yz\mathbf{i} + (x^2 - 2y^2)\mathbf{j} + z^2t\mathbf{k}$ m/s

3.16 Find the angular velocity and vorticity vectors at the point $(1, 2, 3)$ when $t = 3$ s for the velocity field of:

(a) Prob. 3.13a
(b) Prob. 3.13b
(c) Prob. 3.14a
(d) Prob. 3.14b

3.17 The velocity field in a fluid flow is given by $\mathbf{V} = 2y\mathbf{i} + x\mathbf{j} + t\mathbf{k}$. Determine the magnitudes of the acceleration, the angular velocity, and the vorticity at the point $(2, 1, -1)$ at $t = 4$ s.

3.18 The temperature field of a flow in which $\mathbf{V} = 2y\mathbf{i} + x\mathbf{j} + t\mathbf{k}$ is given by $T(x, y, z) = 20xy$ °C. Determine the rate of change of the temperature of a fluid particle in the flow at the point $(2, 1, -2)$ at $t = 2$ s.

3.19 A velocity field is given in cylindrical coordinates as

$$v_r = \left(4 - \frac{1}{r^2}\right)\sin\theta \text{ m/s} \qquad v_\theta = -\left(4 + \frac{1}{r^2}\right)\cos\theta \text{ m/s} \qquad v_z = 0$$

(a) What is the acceleration at the point $(0.6 \text{ m}, 90°)$?
(b) What is the vorticity at the point $(0.6 \text{ m}, 90°)$?

3.20 A velocity field is given in spherical coordinates as

$$v_r = \left(8 - \frac{1}{r^3}\right)\cos\theta \text{ m/s} \qquad v_\theta = -\left(8 + \frac{1}{r^3}\right)\sin\theta \text{ m/s} \qquad v_\phi = 0$$

What is the acceleration at the point $(0.6 \text{ m}, 90°)$?

Classification of Fluid Flows

3.21 Select the word: uniform, one-dimensional, two-dimensional, or three-dimensional, that best describes each of the following flows:

(a) Developed flow in a pipe
(b) Flow of water over a long weir
(c) Flow in a long, straight canal
(d) The flow of exhaust gases exiting a rocket
(e) Flow of blood in an artery
(f) Flow of air around a bullet
(g) Flow of blood in a vein
(h) Flow of air in a tornado

3.22 Select the flow in Prob. 3.21 that could be modeled as a plane flow.

3.23   Select the flow in Prob. 3.21 that would be modeled as an unsteady flow.

3.24   Select the flow in Prob. 3.21 that would have a stagnation point.

3.25   Which flows in Prob. 3.21 could be modeled as inviscid flows?

3.26   Which flow in Prob. 3.21 would be an external flow?

3.27   Which flows in Prob. 3.21 would be compressible flows?

3.28   Which flow in Prob. 3.21 would have a boundary layer?

3.29   Which flows in Prob. 3.21 would definitely be modeled as turbulent flows?

3.30   Water exits a 1-cm-diameter outlet of a faucet. Estimate the maximum speed that would result in a laminar flow if the water temperature is (a) 20°C, (b) 50°C, and (c) 100°C. Assume Re = 2000.

3.31   Air flows over and parallel to a flat plate at 2 m/s. How long is the laminar portion of the boundary layer if the air temperature is (a) 30°C, (b) 70°C, and (c) 200°C. Assume a high-fluctuation level on a smooth rigid plate.

3.32   Decide if each of the following can be modeled as an incompressible flow or a compressible flow:

   (a)   the take-off and landing of commercial airplanes
   (b)   the airflow around an automobile
   (c)   the flow of air in a hurricane
   (d)   the airflow around a baseball thrown at 100 mi/h

3.33   Write all the non-zero terms of $D\rho/Dt$ for a stratified flow in which:

   (a) $\rho = \rho(z)$ and $\mathbf{V} = z(2-z)\mathbf{i}$
   (b) $\rho = \rho(z)$ and $\mathbf{V} = f(x, z)\mathbf{i} + g(x, z)\mathbf{j}$

Bernoulli's equation

3.34   A pitot-static tube measures the total pressure p_T and the local pressure p in a uniform flow in a 4-cm-diameter water pipe. Calculate the flow rate if:

   (a) $p_T = 1500$ mm of mercury and $p = 150$ kPa
   (b) $p_T = 250$ kPa and $p = 800$ mm of mercury
   (c) $p_T = 900$ mm of mercury and $p = 110$ kPa
   (d) $p_T = 10$ in. of water and $p = 30$ lb/ft^2

3.35   Find an expression for the pressure distribution along the horizontal negative x-axis given the velocity field in Solved Problem 3.3 if $p(-\infty, 180°) = p_\infty$. Viscous effects are assumed to be negligible.

3.36   Determine v the velocity V in the pipe if the fluid in the pipe of Fig. 3.15 is:

   (a)   Atmospheric air and $h = 10$ cm of water
   (b)   Water and $h = 10$ cm of mercury
   (c)   Kerosene and $h = 20$ cm of mercury
   (d)   Gasoline and $h = 40$ cm of water

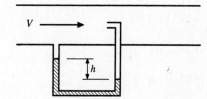

Figure 3.15

3.37

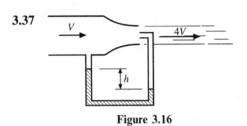

Figure 3.16

Determine the velocity V in the pipe if the fluid in the pipe of Fig. 3.16 is:

   (*a*) Atmospheric air and $h = 40$ cm of water
   (*b*) Water and $h = 20$ cm of mercury
   (*c*) Kerosene and $h = 30$ cm of mercury
   (*d*) Gasoline and $h = 80$ cm of water

Answers to Supplementary Problems

3.8   Ride in cars. Stand on corners.

3.9   A time exposure. An instantaneous picture.

3.10   A pathline. A streakline.

3.11   0, 0

3.12   (*a*) 5.385 m/s, $10\mathbf{i} + 9\mathbf{j} - 3\mathbf{k}\,\text{m/s}^2$ (*b*) 12.81 m/s, $-156\mathbf{i} - 10\mathbf{j} + 30\mathbf{k}\,\text{m/s}^2$

3.13   (*a*) $(\mathbf{i} - 2\mathbf{j})/\sqrt{5}$ (*b*) $(-2\mathbf{i} + 3\mathbf{j})/\sqrt{13}$

3.14   (*a*) $x = -2y$ (*b*) $x^2 - y^2 = 3$

3.15   (*a*) $-2\mathbf{i} - 3\mathbf{j}$ (*b*) $-24\mathbf{i} + 2\mathbf{k}$ (*c*) $-4\mathbf{i} - 8\mathbf{j} + 9\mathbf{k}$

3.16   (*a*) $6\mathbf{i} - \mathbf{k}, 12\mathbf{i} - 2\mathbf{k}$ (*b*) $-3\mathbf{i}/2 - 6\mathbf{k}, -3\mathbf{i} - 12\mathbf{k}$ (*c*) $6\mathbf{i} - \mathbf{k}, 12\mathbf{i} - 2\mathbf{k}$ (*d*) $-3\mathbf{i}/2 - 4\mathbf{k}, -3\mathbf{i} - 8\mathbf{k}$

3.17 $4.583\,\text{m/s}^2 - 0.5\,\text{rad/s} - 1.0\,\text{rad/s}$

3.18 $120\,°\text{C/s}$

3.19   (*a*) $a_r = 11.31\,\text{m/s}^2, a_\theta = 0$ (*b*) 0

3.20 $336.8\,\text{m/s}^2$

3.21   (*a*) 1-D   (*b*) 2-D   (*c*) 2-D   (*d*) 3-D   (*e*) 1-D   (*f*) 2-D   (*g*) 1-D   (*h*) 3-D

3.22   (*b*)

3.23   (*e*)

3.24   (*f*)

3.25   (*b*) (*h*)

3.26   (f)

3.27   (*d*) (f)

3.28   (*f*)

3.29　(*c*) (*d*)

3.30　(*a*) 0.201 m/s　　　(*b*) 0.111 m/s　　　(*c*) 0.0592 m/s

3.31　(*a*) 5.58 m　　　(*b*) 6.15 m　　　(*c*) 7.71 m

3.32　(*a*) incompressible　　　(*b*) incompressible　　　(*c*) incompressible　　　(*d*) incompressible

3.33　(*a*) none　　　(*b*) none

3.34　(*a*) 10.01 m/s　　　(*b*) 16.93 m/s　　　(*c*) 4.49 m/s　　　(*d*) 1.451 m/s

3.35　$2\rho\left(1 + \dfrac{8}{x^2} - \dfrac{16}{x^4}\right)$

3.36　(*a*) 39.9 m/s　　　(*b*) 4.97 m/s　　　(*c*) 7.88 m/s　　　(*d*) 1.925 m/s

3.37　(*a*) 79.8 m/s　　　(*b*) 7.03 m/s　　　(*c*) 9.65 m/s　　　(*d*) 2.72 m/s

The Integral Equations

4.1 INTRODUCTION

Fluid mechanics is encountered in almost every area of our physical lives. Many, if not most, of the quantities of interest are integral quantities; they are found by integrating some property of interest over an area or a volume. Many times the property is essentially constant so the integration is easily performed but other times, the property varies over the area or volume and the required integration may be quite difficult.

What are some of the integral quantities of interest? The rate of flow through a pipe, the force on the vertical surface of a dam, the kinetic energy in the wind approaching a wind machine, the power generated by the blade of a turbine, the force on the blade of a snowplow, and the drag on an airfoil, to mention a few. There are quantities that are not integral in nature, such as the minimum pressure on a body or the point of separation on an airfoil; quantities such as these will be considered in Chap. 5.

To perform an integration over an area or a volume, it is necessary that the integrand be known. The integrand must either be given or information must be available so that it can be approximated with an acceptable degree of accuracy. There are numerous integrands where acceptable approximations cannot be made requiring the solutions of differential equations to provide the required relationships; external flow calculations, such as the lift and drag on an airfoil, often fall into this category. Some relatively simple integrals requiring solutions to the differential equations will be included in Chap. 5. In this chapter, only those problems that involve integral quantities with integrands that are given or that can be approximated will be considered.

4.2 SYSTEM-TO-CONTROL-VOLUME TRANSFORMATION

The three basic laws that are of interest in fluid mechanics are often referred to as the conservation of mass, energy, and momentum. The last two are more specifically called the first law of thermodynamics and Newton's second law. Each of these laws is expressed using a Lagrangian description of motion; they apply to a specified mass of the fluid. They are stated as follows:

Mass: The mass of a system remains constant.
Energy: The rate of heat transfer to a system minus the work rate done by a system equals the rate of change of the energy E of the system.

Momentum: The resultant force acting on a system equals the rate of momentum change of the system.

Each of these laws will now be stated mathematically recognizing that the rate of change applies to a collection of fluid particles and the fact that the density, specific energy, and velocity can vary from point to point in the volume of interest. This requires the material derivative and integration over the volume

$$0 = \frac{D}{Dt} \int_{\text{sys}} \rho \, d\Psi \qquad \text{(mass)} \qquad (4.1)$$

$$\dot{Q} - \dot{W} = \frac{D}{Dt} \int_{\text{sys}} e\rho \, d\Psi \qquad \text{(energy)} \qquad (4.2)$$

$$\sum \mathbf{F} = \frac{D}{Dt} \int_{\text{sys}} \mathbf{V}\rho \, d\Psi \qquad \text{(momentum)} \qquad (4.3)$$

where the dot over Q and W signifies a time rate and e is the specific energy included in the parentheses of Eq. (1.29). It is very difficult to apply Eqs. (4.1) to (4.3) directly to a collection of fluid particles as the fluid moves along in either a simple pipe flow or the more complicated flow through a turbine. So, let us convert these integrals that are expressed using a Lagrangian description to integrals expressed using an Eulerian description (see Sec. 3.2.1). This is a rather tedious derivation but an important one.

In this derivation, it is necessary to differentiate between two volumes: a *control volume* that is a fixed volume in space and a *system* that is a specified collection of fluid particles. Figure 4.1 illustrates the difference between these two volumes. It represents a general fixed volume in space through which a fluid is flowing; the volumes are shown at time t and at a slightly later time $t + \Delta t$. Let us select the energy $E = \int_{\text{sys}} e\rho \, d\Psi$ with which to demonstrate the material derivative; lowercase e denotes the specific energy. We then write, assuming Δt to be a small quantity

$$\begin{aligned}
\frac{DE_{\text{sys}}}{Dt} &\cong \frac{E_{\text{sys}}(t + \Delta t) - E_{\text{sys}}(t)}{\Delta t} \\
&= \frac{E_3(t + \Delta t) + E_2(t + \Delta t) - E_1(t) - E_2(t)}{\Delta t} \\
&= \frac{E_2(t + \Delta t) + E_1(t + \Delta t) - E_2(t) - E_1(t)}{\Delta t} + \frac{E_3(t + \Delta t) - E_1(t + \Delta t)}{\Delta t} \qquad (4.4)
\end{aligned}$$

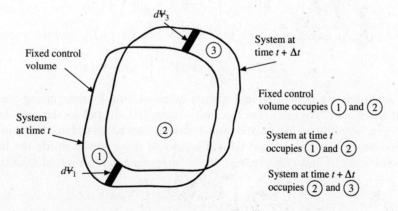

Figure 4.1   The system and the fixed control volume.

where we have simply added and subtracted $E_1(t + \Delta t)$ in the last line. Note that the first ratio in the last line above refers to the control volume so that

$$\frac{E_2(t + \Delta t) + E_1(t + \Delta t) - E_2(t) - E_1(t)}{\Delta t} \cong \frac{dE_{cv}}{dt} \qquad (4.5)$$

where an ordinary derivative is used since we are no longer following a specified fluid mass. Also, we have used "cv" to denote the control volume. The last ratio in Eq. (4.4) results from fluid flowing into volume 3 and out of volume 1. Consider the differential volumes shown in Fig. 4.1 and displayed with more detail in Fig. 4.2. Note that the area $A_1 + A_3$ completely surrounds the control volume so that

$$E_3(t + \Delta t) - E_1(t + \Delta t) = \int_{A_3} e\rho\, \hat{\mathbf{n}} \cdot \mathbf{V} \Delta t\, dA_3 + \int_{A_1} e\rho\, \hat{\mathbf{n}} \cdot \mathbf{V} \Delta t\, dA_1$$

$$= \int_{cs} e\rho\, \hat{\mathbf{n}} \cdot \mathbf{V} \Delta t\, dA \qquad (4.6)$$

where "cs" is the control surface that surrounds the control volume. Substituting Eqs. (4.5) and (4.6) into Eq. (4.4) results in the *Reynolds transfer theorem*, a system-to-control-volume transformation,

$$\frac{DE_{sys}}{Dt} = \frac{d}{dt}\int_{cv} e\rho\, d\Psi + \int_{cs} e\rho\, \hat{\mathbf{n}} \cdot \mathbf{V}\, dA \qquad (4.7)$$

where, in general, e would represent the specific property of E. Note that we could have taken the limit as $\Delta t \to 0$ to make the derivation more mathematically rigorous.

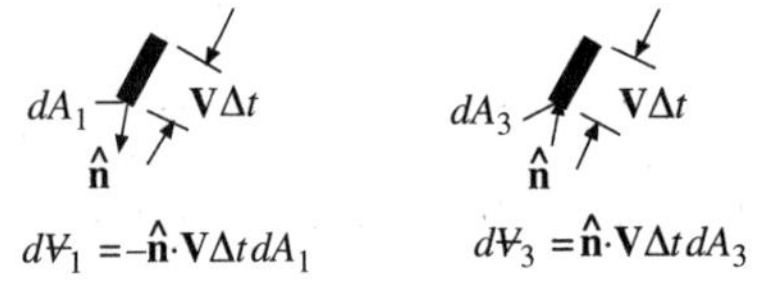

Figure 4.2   Differential volume elements from Fig. 4.1.

If we return to the energy equation of Eq. (4.2), we can now write it as

$$\dot{Q} - \dot{W} = \frac{d}{dt}\int_{cv} e\rho\, d\Psi + \int_{cs} e\rho\, \hat{\mathbf{n}} \cdot \mathbf{V}\, dA \qquad (4.8)$$

If we let $e = 1$ in Eq. (4.7) [see Eq. (4.1)], then the conservation of mass results. It is

$$0 = \frac{d}{dt}\int_{cv} \rho\, d\Psi + \int_{cs} \rho\, \hat{\mathbf{n}} \cdot \mathbf{V}\, dA \qquad (4.9)$$

And finally, if we replace e in Eq. (4.7) with the vector $\mathbf{V}$ [see Eq. (4.3)], Newton's second law results:

$$\sum \mathbf{F} = \frac{d}{dt}\int_{cv} \rho \mathbf{V}\, d\Psi + \int_{cs} \mathbf{V}\rho\, \hat{\mathbf{n}} \cdot \mathbf{V}\, dA \qquad (4.10)$$

These three equations can be written in a slightly different form by recognizing that a fixed control volume has been assumed. That means that the limits of the first integral on the right-hand side of each equation are independent of time. Hence, the time derivative can be moved inside the integral if desired; note that it would be written as a partial derivative should it be moved inside the integral since the integrand depends on $x, y, z,$ and t, in general. The momentum equation would take the form

$$\sum \mathbf{F} = \int_{cv} \frac{\partial}{\partial t}(\rho \mathbf{V})d\Psi + \int_{cs} \mathbf{V}\rho\, \hat{\mathbf{n}} \cdot \mathbf{V}\, dA \qquad (4.11)$$

The following three sections will apply these integral forms of the basic laws to problems in which the integrands are given or in which they can be assumed.

4.3 CONSERVATION OF MASS

The most general relationship for the conservation of mass using the Eulerian description that focuses on a fixed volume was developed in Sec. 4.2 and is

$$0 = \frac{d}{dt}\int_{cv} \rho \, d\mathcal{V} + \int_{cs} \rho \hat{\mathbf{n}} \cdot \mathbf{V} \, dA \tag{4.12}$$

Since the limits on the volume integral do not depend on time, this can be written as

$$0 = \int_{cv} \frac{\partial \rho}{\partial t} \, d\mathcal{V} + \int_{cs} \rho \hat{\mathbf{n}} \cdot \mathbf{V} \, dA \tag{4.13}$$

If the flow of interest can be assumed to be a steady flow so that time does not enter Eq. (4.13), the equation simplifies to

$$0 = \int_{cs} \rho \hat{\mathbf{n}} \cdot \mathbf{V} \, dA \tag{4.14}$$

Those flows in which the density ρ is uniform over an area are of particular interest in our study of fluids. Also, most applications have one entrance and one exit. For such a problem, Eq. (4.14) can then be written as

$$\rho_2 A_2 \overline{V}_2 = \rho_1 A_1 \overline{V}_1 \tag{4.15}$$

where an overbar denotes an average over an area, i.e., $\overline{V}A = \int V \, dA$. Note also that at an entrance, we use $\hat{\mathbf{n}} \cdot \mathbf{V}_1 = -V_1$ since the unit vector points out of the volume and the velocity is into the volume, but at an exit $\hat{\mathbf{n}} \cdot \mathbf{V}_2 = V_2$ since the two vectors are in the same direction.

For incompressible flows in which the density does not change[*] between the entrance and the exit and the velocity is uniform over each area, the conservation of mass takes the simplified form:

$$A_2 V_2 = A_1 V_1 \tag{4.16}$$

We refer to each of the above equations as the *continuity equation*. The one in Eq. (4.16) will be used quite often. These equations are used most often to relate the velocities between sections.

The quantity $\rho V A$ is the *mass flux* and has units of kg/s (slugs per second). The quantity VA is the *flow rate* (or *discharge*) and has units of m^3/s (ft^3/sec or cfs). The mass flux is usually used in a gas flow and the discharge in a liquid flow. They are defined by

$$\dot{m} = \rho A V$$
$$Q = AV \tag{4.17}$$

where V is the average velocity at a section of the flow.

EXAMPLE 4.1 Water flows in a 6-cm-diameter pipe with a flow rate of $0.06 \, m^3/s$. The pipe is reduced in diameter to 2.8 cm. Calculate the maximum velocity in the pipe. Also calculate the mass flux. Assume uniform velocity profiles.

Solution: The maximum velocity in the pipe will be where the diameter is the smallest. In the 2.8-cm-diameter section we have

$$Q = AV$$
$$0.02 = \pi \times 0.014^2 V_2 \qquad \therefore \ V_2 = 32.5 \, \text{m/s}$$

The mass flux is

$$\dot{m} = \rho Q = 1000 \times 0.02 = 20 \, \text{kg/s}$$

---

[*] Not all incompressible flows have constant density. Atmospheric and oceanic flows are examples as is salt water flowing in a canal where fresh water is also flowing.

EXAMPLE 4.2 Water flows into a volume that contains a sponge with a flow rate of 0.02 m³/s. It exits the volume through two tubes, one 2 cm in diameter and the other with a mass flux of 10 kg/s. If the velocity out the 2-cm-diameter tube is 15 m/s, determine the rate at which the mass is changing inside the volume.

Solution: The continuity equation (4.12) is used. It is written in the form

$$0 = \frac{dm_{vol}}{dt} + \dot{m}_2 + \rho A_3 V_3 - \rho Q_1$$

where $m_{vol} = \int \rho \, d\Psi$ and the two exits and entrance account for the other three terms. Expressing the derivative term as $\dot{m}_{vol}$, the continuity equation becomes

$$\dot{m}_{vol} = \rho Q_1 - \dot{m}_2 - \rho A_3 V_3$$
$$= 1000 \times 0.02 - 10 - 1000 \times \pi \times 0.01^2 \times 15 = 5.29 \text{ kg/s}$$

The sponge is soaking up water at the rate of 5.29 kg/s.

4.4 THE ENERGY EQUATION

The first law of thermodynamics, or simply, the energy equation, is of use whenever heat transfer or work is desired. If there is essentially no heat transfer and no external work from a pump or some other device, the energy equation allows us to relate the pressure, the velocity, and the elevation. Let us see how this develops. We begin with the energy equation (4.8) in its general form

$$\dot{Q} - \dot{W} = \frac{d}{dt} \int_{cv} e\rho \, d\Psi + \int_{cs} e\rho \hat{\mathbf{n}} \cdot \mathbf{V} \, dA \qquad (4.18)$$

Most applications allow us to simplify this equation by assuming a steady, uniform flow with one entrance and one exit. The energy equation simplifies to

$$\dot{Q} - \dot{W} = e_2 \rho_2 V_2 A_2 - e_1 \rho_1 V_1 A_1 \qquad (4.19)$$

where we have used $\hat{\mathbf{n}} \cdot \mathbf{V} = -V_1$ at the entrance. Using the continuity equation (4.15), this is written as

$$\dot{Q} - \dot{W} = \dot{m}(e_2 - e_1) \qquad (4.20)$$

The work rate term results from a force moving with a velocity: $\dot{W} = \mathbf{F} \cdot \mathbf{V}$. The force can be a pressure or a shear multiplied by an area. If the flow is in a conduit, e.g., a pipe or a channel, the walls do not move so there is no work done by the walls. If there is a moving belt, there could be an input of work due to the shear between the belt and the fluid. The most common work rate terms result from the pressure forces at the entrance and the exit (pressure is assumed to be uniform over each area) and from any device between the entrance and the exit. The work rate term is expressed as

$$\dot{W} = p_2 A_2 V_2 - p_1 A_1 V_1 + \dot{W}_S \qquad (4.21)$$

where power output is considered positive and $\dot{W}_S$ is the shaft power output from the control volume (a pump would be a negative power and a turbine would provide a positive power output). Using the expression for e given in Eq. (1.29), Eq. (4.20) takes the form

$$\dot{Q} - p_2 A_2 V_2 + p_1 A_1 V_1 - \dot{W}_S = \dot{m}\left(\frac{V_2^2}{2} + gz_2 + \tilde{u}_2 - \frac{V_1^2}{2} - gz_1 - \tilde{u}_1\right) \qquad (4.22)$$

The heat-transfer term and the internal energy terms form the losses in the flow (viscous effects result in heat transfer and/or an increase in internal energy). Divide Eq. (4.22) by $\dot{m}g$ and simplify[*]

$$-\frac{\dot{W}_S}{\dot{m}g} = \frac{V_2^2}{2g} + \frac{p_2}{\gamma_2} + z_2 - \frac{p_1}{\gamma_1} - \frac{V_1^2}{2g} - z_1 + h_L \qquad (4.23)$$

---

[*] We used $\dot{m} = \rho_2 A_2 V_2 = \rho_1 A_1 V_1$.

where we have included the loss term as h_L, called the *head loss*; it is $h_L = (\tilde{u}_2 - \tilde{u}_1)/g + \dot{Q}/\dot{m}g$. An incompressible flow occurs in many applications so that $\gamma_1 = \gamma_2$. Recall that γ for water is 9810 N/m^3 (62.4 lb/ft^3).

The head loss term is often expressed in terms of a *loss coefficient K*

$$h_L = K\frac{V^2}{2g} \tag{4.24}$$

where V is some characteristic velocity in the flow; if it is not obvious it will be specified. Some loss coefficients are listed in Table 7.2; in this chapter they will be given.

The term h_L is called the head loss because it has the dimension of length. We also refer to $V^2/2g$ as the *velocity head*, p/γ as the *pressure head*, and z as the *head*. The sum of these three terms is the *total head*.

The shaft-work term in Eq. (4.23) is usually due to either a pump or a turbine. If it is a pump, we can define the *pump head H_P* as

$$H_P = \frac{-\dot{W}_S}{\dot{m}g} = \frac{\eta_P \dot{W}_P}{\dot{m}g} \tag{4.25}$$

where $\dot{W}_P$ is the power input to the pump and η_P is the *pump efficiency*. For a turbine the *turbine head H_T* is

$$H_T = \frac{\dot{W}_S}{\dot{m}g} = \frac{\dot{W}_T}{\dot{m}g\,\eta_T} \tag{4.26}$$

where $\dot{W}_T$ is the power output of the turbine and η_T is the *turbine efficiency*. Power has units of watts [(ft-lb)/sec] or horsepower.

If the flow is not uniform over the entrance and the exit, an integration must be performed to obtain the kinetic energy. The rate at which the kinetic energy crosses an area is [see Eqs. (4.18) and (1.29)]

$$\text{Kinetic energy rate} = \int \frac{V^2}{2}\rho V\,dA = \frac{1}{2}\int \rho V^3\,dA \tag{4.27}$$

If the velocity distribution is known, the integration can be performed. A *kinetic-energy correction factor* α is defined as

$$\alpha = \frac{\int V^3\,dA}{\overline{V}^3 A} \tag{4.28}$$

The kinetic energy term can then be written as

$$\frac{1}{2}\rho\int V^3\,dA = \frac{1}{2}\rho\alpha\overline{V}^3 A \tag{4.29}$$

so that, for non-uniform flows, the energy equation takes the form

$$-\frac{\dot{W}_S}{\dot{m}g} = \alpha_2\frac{\overline{V}_2^2}{2g} + z_2 + \frac{p_2}{\gamma_2} - \alpha_1\frac{\overline{V}_1^2}{2g} - z_1 - \frac{p_1}{\gamma_1} + h_L \tag{4.30}$$

where $\overline{V}_1$ and $\overline{V}_2$ are the average velocities at sections 1 and 2, respectively. Equation (4.30) is used if the α's are known; for parabolic profiles, $\alpha = 2$ in a pipe and $\alpha = 1.5$ between parallel plates. For turbulent flows (most flows in engineering applications), $\alpha \cong 1$.

EXAMPLE 4.3 Water flows from a reservoir with elevation 30 m out of a 5-cm-diameter pipe that has a 2-cm-diameter nozzle attached to the end, as shown in Fig. 4.3. The loss coefficient for the entire pipe is given as $K = 1.2$. Estimate the flow rate of water through the pipe. Also, predict the pressure just upstream of the nozzle (the losses can be neglected through the nozzle). The nozzle is at an elevation of 10 m.

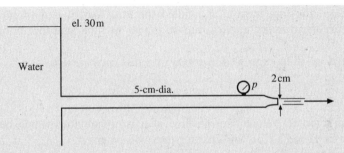

Figure 4.3

Solution: The energy equation is written in the form

$$\frac{\dot{W}_s}{\dot{m}g} = \frac{V_2^2}{2g} + z_2 + \frac{p_2}{\gamma_2} - \frac{V_1^2}{2g} - z_1 - \frac{p_1}{\gamma_1} + K\frac{V^2}{2g}$$

where the pressure is 0 at surface 1 and at the exit 2, the velocity is 0 at the surface, and there is no shaft work (there is no pump or turbine). The loss coefficient would be based on the characteristic velocity V in the pipe and not on the exit velocity V_2. Use the continuity equation to relate the velocities:

$$V = \frac{A_2}{A}V_2 = \frac{d_2^2}{d^2}V_2 = \frac{4}{25}V_2$$

The energy equation provides

$$0 = \frac{V_2^2}{2g} + 10 - 30 + 1.2\left(\frac{4}{25}\right)^2\frac{V_2^2}{2g} \qquad \therefore\ V_2 = 19.5\,\text{m/s}$$

The pressure just before the nozzle is found by applying the energy equation across the nozzle assuming no losses (Bernoulli's equation could also be used). It takes the form

$$-\frac{\dot{W}_s}{\dot{m}g} = \frac{V_2^2}{2g} + \frac{p_2}{\gamma} + z_2 - \frac{V^2}{2g} - \frac{p}{\gamma} - z$$

where area 2 is at the exit and p and V are upstream of the nozzle. The energy equation gives

$$0 = \frac{19.5^2}{2 \times 9.81} + \frac{p_2}{\gamma} + z_2 - \left(\frac{4}{25}\right)^2\frac{19.5^2}{2 \times 9.8} - \frac{p}{9810} - z \qquad \therefore\ p = 185\,300\,\text{Pa}\ \text{or}\ 185.3\,\text{kPa}$$

EXAMPLE 4.4 An energy conscious couple decides to dam up the creek flowing next to their cabin and estimates that a head of 4 m can be established above the exit to a turbine they bought on eBay. The creek is estimated to have a flow rate of 0.8 m³/s. What is the maximum power output of the turbine assuming no losses and a velocity at the turbine's exit of 3.6 m/s?

Solution: The energy equation is applied as follows:

$$-\frac{\dot{W}_T}{\dot{m}g} = \frac{V_2^2}{2g} + \frac{p_2}{\gamma} + z_2 - \frac{V_1^2}{2g} - \frac{p_1}{\gamma} - z_1 + h_L$$

It is only the head of the water above the turbine that provides the power; the exiting velocity subtracts from the power. There results, using $\dot{m} = \rho Q = 1000 \times 0.8 = 800\,\text{kg/s}$,

$$\dot{W}_T = \dot{m}gz_1 - \dot{m}\frac{V_2^2}{2}$$

$$= 800 \times 9.81 \times 4 - 800 \times \frac{3.6^2}{2} = 26\,200\,\text{J/s}\ \text{or}\ 26.2\,\text{kW}$$

Let us demonstrate that the units on $\dot{m}gz_1$ are J/s. The units on $\dot{m}gz_1$ are $\dfrac{\text{kg}}{\text{s}} \times \dfrac{\text{m}}{\text{s}^2} \times \text{m} = \dfrac{\text{kg·m}}{\text{s}^2} \times \dfrac{\text{m}}{\text{s}} = \dfrac{\text{N·m}}{\text{s}} = \text{J/s}$ where, from $F = ma$, we see that $\text{N} = \text{kg·m/s}^2$. If the proper units are included on the items in our equations, the units will come out as expected, i.e., the units on $\dot{W}_T$ must be J/s.

4.5 THE MOMENTUM EQUATION

When a force in involved in a calculation, it is often necessary to apply Newton's second law, or simply, the *momentum equation*, to the problem of interest. For some general volume, using the Eulerian description of motion, the momentum equation was presented in Eq. (4.10) in its most general form for a fixed control volume as

$$\sum \mathbf{F} = \frac{d}{dt} \int_{cv} \rho \mathbf{V} \, d\mathcal{V} + \int_{cs} \mathbf{V} \rho \hat{\mathbf{n}} \cdot \mathbf{V} \, dA \qquad (4.31)$$

When applying this equation to a control volume, we must be careful to include all forces acting on the control volume, so it is very important to sketch the control volume and place the forces on the sketched control volume. (The control volume takes the place of the free-body diagram utilized in courses in statics, dynamics, and solids.)

Most often, steady, uniform flows with one entrance and one outlet are encountered. For such flows, Eq. (4.31) reduces to

$$\sum \mathbf{F} = \rho_2 A_2 V_2 \mathbf{V}_2 - \rho_1 A_1 V_1 \mathbf{V}_1 \qquad (4.32)$$

Using continuity $\dot{m} = \rho_2 A_2 V_2 = \rho_1 A_1 V_1$, the momentum equation takes the simplified form

$$\sum \mathbf{F} = \dot{m}(\mathbf{V}_2 - \mathbf{V}_1) \qquad (4.33)$$

This is the form most often used when a force is involved in a calculation. It is a vector equation that contains the three scalar equations in a rectangular coordinate system

$$\sum F_x = \dot{m}(V_{2x} - V_{1x})$$
$$\sum F_y = \dot{m}(V_{2y} - V_{1y}) \qquad (4.34)$$
$$\sum F_z = \dot{m}(V_{2z} - V_{1z})$$

If the profiles at the entrance and exit are not uniform, Eq. (4.31) must be used and the integration performed or, if the *momentum-correction factor* β is known, it can be used. It is found from

$$\int_A V^2 \, dA = \beta \overline{V}^2 A \qquad (4.35)$$

The momentum equation for a steady flow with one entrance and one outlet then takes the form

$$\sum \mathbf{F} = \dot{m}(\beta_2 \mathbf{V}_2 - \beta_1 \mathbf{V}_1) \qquad (4.36)$$

where $\mathbf{V}_1$ and $\mathbf{V}_2$ represent the average velocity vectors over the two areas.

For parabolic profiles, $\beta = 1.33$ for a pipe and $\beta = 1.2$ for parallel plates. For turbulent flows (most flows in engineering applications), $\beta \cong 1$.

An important application of the momentum equation is to the deflectors (or vanes) of pumps, turbines, or compressors. The applications involve both stationary defectors and moving deflectors. The following assumptions are made for both:

- The frictional force between the fluid and the deflector is negligible.
- The pressure is assumed to be constant as the fluid moves over the deflector.
- The body force is assumed to be negligible.
- The effect of the lateral spreading of the fluid stream is neglected.

A sketch is made of a stationary deflector in Fig. 4.4. Bernoulli's equation predicts that the fluid velocity will not change ($V_2 = V_1$) as the fluid moves over the deflector since the pressure does not change, there is no friction, it is a steady flow, and the body forces are neglected. The component momentum equations appear as follows:

$$-R_x = \dot{m}(V_2 \cos \alpha - V_1) = \dot{m} V_1 (\cos \alpha - 1)$$
$$R_y = \dot{m} V_2 \sin \alpha = \dot{m} V_1 \sin \alpha \qquad (4.37)$$

Given the necessary information, the force components can be calculated.

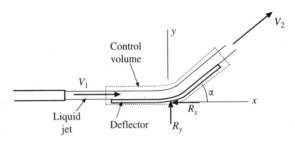

Figure 4.4   A stationary deflector.

The analysis of a moving deflector is more complicated. Is it a single deflector (a water scoop to slow a high-speed train) or is it a series of deflectors as in a turbine? First, let us consider a single deflector moving with speed V_B, as sketched in Fig. 4.5. The reference frame is attached to the deflector so the flow is steady from such a reference frame[*]. The deflector sees the velocity of the approaching fluid as the relative velocity V_{r1} and it is this relative velocity that Bernoulli's equation predicts will remain constant over the deflector, i.e., $V_{r2} = V_{r1}$. The velocity of the fluid exiting the fixed nozzle is V_1. The momentum equation then provides

$$-R_x = \dot{m}_r(V_1 - V_B)(\cos\alpha - 1)$$
$$R_y = \dot{m}_r(V_1 - V_B)\sin\alpha \qquad\qquad (4.38)$$

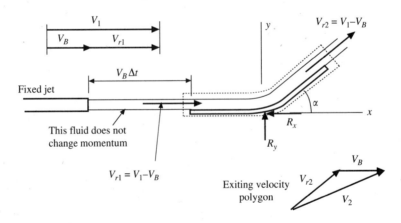

Figure 4.5   A single moving deflector.

where $\dot{m}_r$ is that part of the exiting fluid that has its momentum changed. As the deflector moves away from the nozzle, the fluid represented by the length $V_B\Delta t$ does not experience a change in momentum. The mass flux of fluid that experiences a momentum change is

$$\dot{m}_r = \rho A(V_1 - V_B) \qquad\qquad (4.39)$$

so it is that mass flux used in the expressions for the force components.

For a series of vanes, the nozzles are typically oriented such that the fluid enters the vanes from the side at an angle β_1 and leaves the vanes at an angle β_2, as shown in Fig. 4.6. The vanes are

---

[*] If the deflector is observed from the fixed jet, the deflector moves away from the jet and the flow is not a steady flow. It is steady if the flow is observed from the deflector.

designed so that the relative inlet velocity V_{r1} enters the vanes tangent to a vane (the relative velocity always leaves tangent to the vane) as shown in Fig. 4.7. It is the relative speed that remains constant in magnitude as the fluid moves over the vane, i.e., $V_{r2} = V_{r1}$. We also note that all of the fluid exiting the fixed jet has its momentum changed. So, the expression to determine the x-component of the force is

$$-R_x = \dot{m}(V_{2x} - V_{1x}) \qquad (4.40)$$

It is this x-component of the force that allows the power to be calculated; the y-component does no work and hence does not contribute to the power. The power is found from

$$\dot{W} = N R_x V_B \qquad (4.41)$$

where N is the number of jets in the device and we have observed that the force R_x moves with velocity V_B.

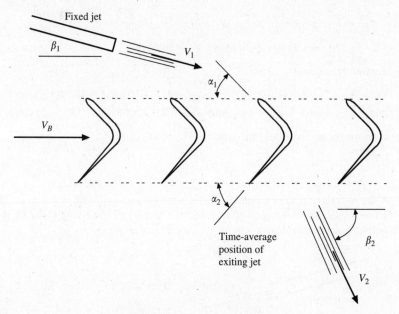

Figure 4.6   A series of vanes.

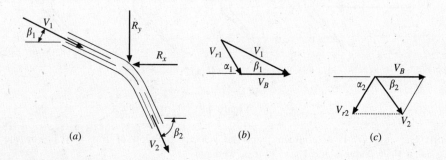

Figure 4.7   (*a*) Average position of the jet, (*b*) the entrance velocity polygon, and (*c*) the exit velocity polygon.

EXAMPLE 4.5 A 10-cm-diameter hose maintained at a pressure of 1600 kPa provides water from a tanker to a fire. There is a nozzle at the end of the hose that reduces the diameter to 2.5 cm. Estimate the force that the water exerts on the nozzle. The losses can be neglected in a short nozzle.

Solution: A sketch of the water contained in the nozzle is important so that the control volume is carefully identified. It is shown in Fig. 4.8. Note that $p_2 = 0$ and we expect that the force F_N of the nozzle on the water acts to the left. The velocities are needed upstream and at the exit of the nozzle. Continuity provides

$$A_2 V_2 = A_1 V_1 \qquad \therefore V_2 = \frac{10^2}{2.5^2} V_1 = 16 V_1$$

Figure 4.8

The energy equation requires

$$\frac{V_2^2}{2} + \frac{\cancel{p_2}}{\rho} + \cancel{g z_2} = \frac{V_1^2}{2} + \frac{p_1}{\rho} + \cancel{g z_1} + \cancel{h_L} \qquad 16^2 \frac{V_1^2}{2} = \frac{V_1^2}{2} + \frac{1\,600\,000}{1000}$$

$$\therefore V_1 = 3.54 \text{ m/s} \qquad \text{and} \qquad V_2 = 56.68 \text{ m/s}$$

The momentum equation then gives

$$p_1 A_1 - F_N = \dot{m}(V_2 - V_1) = \rho A_1 V_1 (V_2 - V_1) = 15 \rho A_1 V_1^2$$

$$1\,600\,000 \times \pi \times 0.05^2 - F_N = 15 \times 1000 \times \pi \times 0.05^2 \times 3.54^2 \qquad \therefore F_N = 12\,400 \text{ N}$$

The force of the water on the nozzle would be equal and opposite to F_N.

EXAMPLE 4.6 A steam turbine contains eight 4-cm-diameter nozzles each exiting steam at 200 m/s as shown in Fig. 4.9. The turbine blades are moving at 80 m/s and the density of the steam is 2.2 kg/m^3. Calculate the maximum power output assuming no losses.

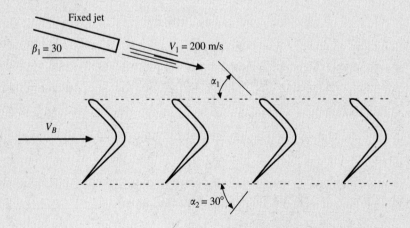

Figure 4.9

Solution: The angle α_1 is determined from the velocity polygon of Fig. 4.7(b). For the x- and y-components, using $V_1 = 200$ m/s and $V_B = 80$ m/s, we have

$$200 \sin 30° = V_{r1} \sin \alpha_1$$

$$200 \cos 30° = 80 + V_{r1} \cos \alpha_1$$

There are two unknowns in the above two equations: V_{r1} and α_1. A simultaneous solution provides

$$V_{r1} = 136.7 \, \text{m/s} \qquad \text{and} \qquad \alpha_1 = 47.0°$$

Neglecting losses allows $V_{r2} = V_{r1} = 136.7 \, \text{m/s}$ so the velocity polygon at the exit [Fig. 4.7(c)] provides

$$V_2 \sin \beta_2 = 136.7 \sin 30°$$
$$V_2 \cos \beta_2 = 80 - 136.7 \cos 30°$$

These two equations are solved to give

$$V_2 = 78.39 \, \text{m/s} \qquad \text{and} \qquad \beta_2 = 119.3°$$

Observe that the exiting velocity polygon appears as in Fig. 4.10.

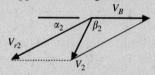

Figure 4.10

The force acting on the blades due to one nozzle is

$$-F = \dot{m}(V_{2x} - V_{1x})$$
$$= 2.2 \times \pi \times 0.02^2 \times 200(-78.39 \cos 60.7° - 200 \cos 30°) \qquad \therefore \ F = 11.7 \, \text{N}$$

The power output is then

$$\dot{W} = N \times F \times V_B = 8 \times 11.7 \times 80 = 7488 \, \text{W} \quad \text{or} \quad 10.04 \, \text{hp}$$

EXAMPLE 4.7   The relatively rapid flow of water in a horizontal rectangular channel can suddenly "jump" to a higher level (an obstruction downstream may be the cause). This is called a *hydraulic jump*. For the situation shown in Fig. 4.11, calculate the higher depth downstream. Assume uniform flow.

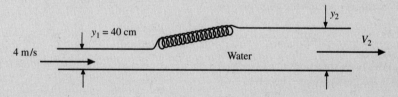

Figure 4.11

Solution: For a short section of water, the frictional force on the walls can be neglected. The forces acting on the water are F_1 acting to the right and F_2 acting to the left; they are (assume a width w)

$$F_1 = \gamma \bar{h}_1 A_1 = 9810 \times 0.20 \times 0.40 w = 785 w \qquad \text{and} \qquad F_2 = \gamma \bar{h}_2 A_2 = \gamma \frac{y_2}{2} \times y_2 w$$

Applying the momentum equation gives

$$\sum F_x = \dot{m}(V_2 - V_1) = \rho A_1 V_1 (V_2 - V_1)$$
$$785w - 4905 \times w y_2^2 = 1000 \times 0.4 w \times 4(V_2 - 4)$$

The width w divides out of this equation but there are two unknowns, y_2 and V_2. The continuity equation relates these two variables

$$A_2 V_2 = A_1 V_1$$
$$w y_2 V_2 = w \times 0.4 \times 4 \qquad \therefore \ V_2 = \frac{1.6}{y_2}$$

Substitute this into the momentum equation and obtain

$$785 - 4905y_2^2 = 1600\left(\frac{1.6}{y_2} - 4\right)$$

This equation is a cubic but with a little ingenuity it is a quadratic. Let us factor:

$$7^2(4 - 10y_2)(4 + 10y_2) = \frac{1600}{y_2}(1.6 - 4y_2) = \frac{1600}{2.5y_2}(4 - 10y_2)$$

The factor $(4 - 10y_2)$ divides out and a quadratic equation results

$$y_2^2 + 0.4y_2 - 1.306 = 0$$

It has two roots. The one of interest is

$$y_2 = 2.12\,\text{m}$$

This rather interesting effect is analogous to the shock wave that occurs in a supersonic gas flow. It is nature's way of moving from something traveling quite fast to something moving much slower while maintaining continuity and momentum. A significant amount of energy is lost when making this sudden change through the hydraulic jump; it can be found by using the energy equation.

Solved Problems

4.1   A balloon is being filled with water at an instant when the diameter is 50 cm. If the flow rate into the balloon is 200 gal/min, what is the rate of increase in the diameter?

The rate of increase in the volume of the balloon is

$$\frac{d\mathcal{V}}{dt} = \frac{d}{dt}\left(\frac{4}{3}\pi R^3\right) = 4\pi R^2 \frac{dR}{dt} = \frac{\pi}{2}D^2 \frac{dD}{dt}$$

Convert gallons per minute to m^3/s

$$200\,\frac{\text{gallons}}{\text{minute}} \times 0.003785\,\frac{\text{m}^3}{\text{gallon}} \times \frac{1}{60}\frac{\text{minute}}{\text{seconds}} = 0.01262\,\text{m}^3/\text{s}$$

The above two expressions must be equal if mass is conserved (in this case, volume is conserved because water is incompressible). This gives

$$\frac{\pi}{2} \times 0.50^2 \times \frac{dD}{dt} = 0.01262 \qquad \therefore \frac{dD}{dt} = 0.0321\,\text{m/s}$$

4.2   Air at 40°C and 250 kPa is flowing in a 32-cm-diameter pipe at 10 m/s. The pipe changes diameter to 20 cm and the density of the air changes to 3.5 kg/m^3. Calculate the velocity in the smaller diameter pipe.

The continuity equation (4.15) is used

$$\rho_1 A_1 V_1 = \rho_2 A_2 V_2 \qquad \therefore \frac{p_1}{RT_1}\pi\frac{d_1^2}{4}V_1 = \rho_2\pi\frac{d_2^2}{4}V_2 \qquad \therefore V_2 = \frac{d_1^2 p_1}{\rho_2 d_2^2 RT_1}V_1$$

Substitute the given information into the equation and

$$V_2 = \frac{d_1^2 p_1}{\rho_2 d_2^2 RT_1}V_1 = \frac{0.32^2 \times 350}{3.5 \times 0.20^2 \times 0.287 \times 313} \times 10 = 28.5\,\text{m/s}$$

Note: The pressure is assumed to be gauge pressure when given in a problem statement, so 100 kPa is added to convert it to absolute pressure. The pressure is used as kPa since the gas constant has units of kJ/(kg·K).

4.3   A liquid flows as a uniform flow in a 2 cm×4 cm rectangular conduit. It flows out a 2-cm-diameter pipe with a parabolic profile. If the maximum velocity in the pipe is 4 m/s, what is the velocity in the rectangular conduit?

The equation of the parabola for $u(r)$ must allow the velocity to be 4 m/s where $r = 0$ and 0 m/s where $r = 0.01$ m. The velocity profile that accomplishes this is

$$u(r) = 40\,000(0.01^2 - r^2)$$

The continuity equation of the incompressible flow (it is a liquid) takes the form

$$A_1 V_1 = \int_{A_2} u(r)2\pi r\,dr = \int_0^{0.01} 40\,000(0.01^2 - r^2)2\pi r\,dr$$

where $2\pi r\,dr$ in the integral is the differential area through which the fluid flows. The above equation provides

$$V_1 = \frac{40\,000 \times 2\pi}{0.02 \times 0.04}\left(0.01^2 \times \frac{0.01^2}{2} - \frac{0.01^4}{4}\right) = 0.785\,\text{m/s}$$

4.4   A turbine is designed to extract energy from a water source flowing through a 10-cm-diameter pipe at a pressure of 800 kPa with an average velocity of 10 m/s. If the turbine is 90 percent efficient, how much energy can be produced if the water is emitted to the atmosphere through a 20-cm-diameter pipe?

The flow rate and velocity at the exit are

$$Q = A_1 V_1 = \pi \times 0.05^2 \times 10 = 0.0854\,\text{m}^3/\text{s} \qquad V_2 = V_1\frac{d_1^2}{d_2^2} = 10 \times \frac{10^2}{20^2} = 2.5\,\text{m/s}$$

The pressure at the outlet is assumed to be atmospheric, i.e., $p_2 = 0$. The energy equation is applied between the inlet and the exit of the turbine

$$-\frac{\dot{W}_S}{\dot{m}g} = \frac{V_2^2}{2g} + \frac{p_2}{\gamma_2} + z_2 - \frac{p_1}{\gamma_1} - \frac{V_1^2}{2g} - z_1 + h_L$$

where the head loss term is omitted and included as an efficiency of the turbine. Substituting the appropriate information gives

$$-\frac{\dot{W}_S}{1000 \times 0.0854} = \frac{2.5^2 - 10^2}{2} - \frac{800\,000}{1000} \qquad \therefore\ \dot{W}_S = 72\,300\,\text{W}$$

This is the power extracted from the water. The power produced would be less than this due to the losses through the turbine measured by the efficiency, i.e.

$$\dot{W}_T = \eta_T \dot{W}_S = 0.9 \times 72.3 = 65.1\,\text{kW}$$

Check the units on the above equations to make sure they are consistent.

4.5   The flow rate in a pipe is determined by use of the *Venturi meter* shown in Fig. 4.12. Using the information given in the figure and $h = 4$ cm, calculate the flow rate assuming uniform flow and no losses (these assumptions are reasonable for highly turbulent flows).

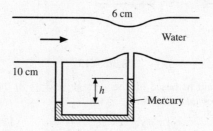

Figure 4.12

The manometer allows the pressures (measured at the centerline of the pipe) to be related by

$$p_1 + 9810 \times z + 0.04 \times 9810 = p_2 + 9810 \times z + 0.04 \times 13.6 \times 9810 \qquad \therefore \ p_1 - p_2 = 4944 \, \text{Pa}$$

where z is measured from the top of the mercury to the centerline. The continuity equation relates V_1 to V_2

$$V_2 = V_1 \frac{d_1^2}{d_2^2} = \frac{10^2}{6^2} V_1 = 2.778 V_1$$

The energy equation is then used to obtain

$$\frac{V_2^2}{2g} + \frac{p_2}{\gamma} + \cancel{z_2} = \frac{p_1}{\gamma} + \frac{V_1^2}{2g} + \cancel{z_1} + \cancel{h_L} \qquad \text{or} \qquad \frac{2.778^2 V_1^2 - V_1^2}{2 \times 9.81} = \frac{4944}{9810} \qquad \therefore \ V_1 = 1.213 \, \text{m/s}$$

The flow rate is

$$Q = A_1 V_1 = \pi \times 0.05^2 \times 1.213 = 0.00953 \, \text{m}^3/\text{s}$$

4.6   A dam is proposed on a remote stream that measures approximately 25-cm deep by 350-cm wide with an average velocity of 2.2 m/s. If the dam can be constructed so that the free surface above a turbine is 10 m, estimate the maximum power output of an 88 percent efficient turbine.

The flow rate of the water passing through the turbine is

$$Q = A_1 V_1 = 0.25 \times 3.5 \times 2.2 = 1.925 \, \text{m}^3/\text{s}$$

The energy equation is applied between the surface of the reservoir behind the dam, where $p_1 = 0$, $V_1 = 0$, and $z_1 = 10$ m and the outlet of the turbine where we assume, for maximum power output, that $V_2 \cong 0$, $p_2 \cong 0$, and $z_2 = 0$

$$\frac{\dot{W}_S}{\dot{m}g} = \frac{\cancel{V_2^2}}{2g} + \frac{\cancel{p_2}}{\cancel{\gamma_2}} + \cancel{z_2} - \frac{\cancel{p_1}}{\cancel{\gamma_1}} - \frac{\cancel{V_1^2}}{2g} - z_1 + \cancel{h_L}$$

or

$$\dot{W}_S = \dot{m}g z_1 = (1000 \times 1.925) \times 9.81 \times 10 = 189\,000 \, \text{W}$$

The turbine losses are included by the use of the efficiency. The maximum turbine output is

$$\dot{W}_T = \eta_T \dot{W}_S = 0.88 \times 189 = 166 \, \text{kW}$$

4.7   A pump is used to pump water from a reservoir to a water tank as shown in Fig. 4.13. Most pumps have a *pump curve* that relates the pump power requirement to the flow rate, like the one provided in the figure. Estimate the flow rate provided by the pump. The overall loss coefficient $K = 4$.

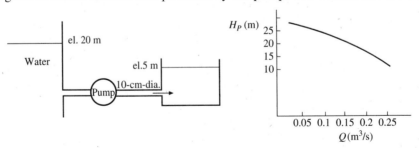

Figure 4.13

The loss coefficient would be based on the average velocity in the pipe. The energy equation applied between the two surfaces takes the form

$$\frac{\dot{W}_P}{\dot{m}g} = \frac{\cancel{V_2^2}}{2g} + \frac{\cancel{p_2}}{\cancel{\gamma}} + z_2 - \frac{\cancel{p_1}}{\cancel{\gamma}} - \frac{\cancel{V_1^2}}{2g} - z_1 + h_L = z_2 - z_1 + K \frac{V^2}{2g} = 15 + 4 \frac{Q^2}{\pi \times 0.05^2 \times 2g}$$

The energy equation is then [see Eq. (4.25)]

$$H_P = 15 + 26Q^2$$

This energy equation and the equation represented by the pump curve in the figure are solved simultaneously as follows:

$$\text{Try } Q = 0.1: \quad (H_P)_{\text{curve}} = 24\,\text{m} \quad \text{and} \quad (H_P)_{\text{energy}} = 15.3\,\text{m}$$
$$\text{Try } Q = 0.2: \quad (H_P)_{\text{curve}} = 17\,\text{m} \quad \text{and} \quad (H_P)_{\text{energy}} = 16\,\text{m}$$

The estimate is then $Q = 0.21$ m^3/s.

4.8 Integrate the appropriate velocity profile and calculate the kinetic energy transported by a flow of water that has a parabolic profile in a 4-cm-diameter pipe if the flow rate is 0.005 m^3/s.

The parabolic profile that has $u = 0$ at the wall where $r = 0.02$ m and $u = u_{\max}$ at the centerline is $u(r) = u_{\max}(1 - r^2/0.02^2)$. The flow rate is

$$Q = \int_A u(r)\,dA$$

$$0.005 = u_{\max} \int_0^{0.02} \left(\frac{1 - r^2}{0.02^2}\right) 2\pi r\,dr = 2\pi u_{\max}\left(\frac{0.02^2}{2} - \frac{0.02^2}{4}\right) \qquad \therefore u_{\max} \cong 8\,\text{m/s}$$

The rate of differential kinetic energy that passes through the differential area $2\pi r\,dr$ is $\frac{1}{2}\dot{m}v^2 = \frac{1}{2}(\rho 2\pi r\,dr \times v)v^2$. This is integrated to yield

$$\text{KE} = \int_A \frac{1}{2}(\rho 2\pi r\,dr \times v)v^2 = 1000\pi \int_0^{0.02} 8^3 \left(\frac{1 - r^2}{0.02^2}\right)^3 r\,dr$$

$$= -\frac{0.02^2}{2} \times 8^3 \times 1000\pi \frac{(1 - r^2/0.02^2)^4}{4}\bigg|_0^{0.02} = 80\,\text{J/s}$$

This can be checked using $\alpha = 2$, as noted after Eq. (4.30)

$$\frac{1}{2}\alpha \dot{m}\bar{V}^2 = \frac{1}{2} \times 2(0.005 \times 1000) \times 4^2 = 80\,\text{J/s}$$

where we have used the average velocity as half of the maximum velocity for a parabolic profile in a pipe.

4.9 A nozzle is attached to a 6-cm-diameter hose but the horizontal nozzle turns the water through an angle of 90°. The nozzle exit is 3 cm in diameter and the flow rate is 500 L/min. Determine the force components of the water on the nozzle and the magnitude of the resultant force. The pressure in the hose is 400 kPa and the water exits to the atmosphere.

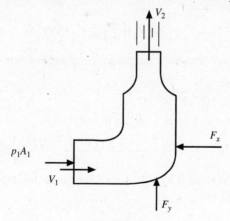

Figure 4.14

First, the control volume should be sketched since it is not given in the problem statement. It appears as shown in Fig. 4.14. The control volume shows the water with the force components of the nozzle on the water. The velocities are calculated to be

$$V_1 = \frac{Q}{A_1} = \frac{0.50/60}{\pi \times 0.03^2} = 2.95 \, \text{m/s} \qquad V_2 = 4 \times V_1 = 11.79 \, \text{m/s}$$

The pressure p_1 is found using the energy equation. The losses are neglected in the accelerated flow:

$$\frac{V_2^2}{2} + \frac{\cancel{p_2}}{\cancel{\rho}} + g\cancel{z_2} = \frac{V_1^2}{2} + \frac{p_1}{\rho} + g\cancel{z_1} \qquad p_1 = 1000\left(\frac{11.79^2 - 2.95^2}{2}\right) = 65\,150 \, \text{Pa}$$

The momentum equation provides the force components [see Eqs. (4.34)]

$$p_1 A_1 - F_x = \dot{m}(V_{2x} - V_{1x})$$
$$65\,150 \times \pi \times 0.03^2 - F_x = -(0.50/60) \times 1000 \times 2.95 \qquad \therefore F_x = 209 \, \text{N}$$

$$F_y - \cancel{p_2 A_{2y}} = \dot{m}(V_{2y} - \cancel{V_{2y}})$$
$$F_y = (0.50/60) \times 1000 \times 11.79 = 98.2 \, \text{N}$$

The magnitude of the resultant force is

$$F = \sqrt{F_x^2 + F_y^2} = \sqrt{209^2 + 98.2^2} = 231 \, \text{N}$$

The force of the water on the nozzle would be equal and opposite to F_x and F_y.

4.10 A fluid flows through a sudden expansion as shown in Fig. 4.15. The pressures before and after the expansion are p_1 and p_2, respectively. Find an expression for the head loss due to the expansion if uniform velocity profiles are assumed. *Note*: This problem requires the use of momentum, energy, and continuity.

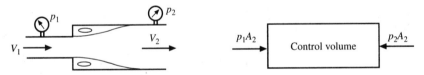

Figure 4.15

The control volume is shown from the expansion to the area downstream where the flow fills the area and the velocity is again uniform over the entire area A_2. Note that the pressure is p_1 over the area immediately after the expansion since the flow separates with parallel streamlines and then expands to fill the area. (The head loss is due to the energy needed to sustain the flow in the separated region.) The momentum equation provides

$$\sum F_x = \dot{m}(V_{2x} - V_{1x}) \qquad p_1 A_2 - p_2 A_2 = \rho A_2 V_2(V_2 - V_1) \qquad \therefore \frac{p_1 - p_2}{\rho} = V_2(V_2 - V_1)$$

The energy equation that introduces the head loss h_L, applied between sections 1 and 2, is

$$\frac{V_1^2}{2g} + \frac{p_1}{\gamma} + gz_1 = \frac{V_2^2}{2g} + \frac{p_2}{\gamma} + gz_2 + h_L \qquad \therefore h_L = \frac{p_2 - p_1}{\gamma} - \frac{V_2^2 - V_1^2}{2g}$$

Substituting the pressure difference from the momentum equation gives

$$h_L = \frac{2V_2(V_2 - V_1)}{2g} - \frac{V_2^2 - V_1^2}{2g} = \frac{(V_1 - V_2)^2}{2g}$$

The continuity equation requires $V_2 = V_1 A_1/A_2$. Substitute this into the above equation and obtain the expression for the head loss

$$h_L = \left(1 - \frac{A_1}{A_2}\right)^2 \frac{V_1^2}{2g}$$

The loss coefficient of Eq. (4.24) is $K = (1 - A_1/A_2)^2$ based on the inlet velocity V_1.

4.11   The blade on a snowplow turns the wet snow through an angle of $120°$ but off to one side at $30°$. If the snow has a density of 500 kg/m^3, what power is needed to move the blade at 40 mi/h if it scoops snow that is 15-cm deep and 3-m wide?

The momentum equation (4.37) is written to account for the component due to the side angle (the blade is stationary and the snow moves toward the blade)

$$R_x = -\dot{m}(V_2 \cos \alpha_1 \cos \theta - V_1) = \rho A V_1^2 (\cos \alpha_1 \cos \theta - 1)$$

$$= -500(0.15 \times 3)(40 \times 0.447)^2 (\cos 120° \cos 30° - 1) = 31\,150\,\text{N}$$

where 0.447 converts mi/h to m/s. We have neglected the friction generated by the snow moving over the blade, which would be small compared to the above force, so that the speed of the snow relative to the blade remains constant, i.e., $V_2 \cong V_1$. The power is then

$$31\,150(40 \times 0.447) = 557\,000\,\text{W} \quad \text{or} \quad 746\,\text{hp}$$

Supplementary Problems

4.12   What assumptions are needed on a flow to allow Eq. (4.3) to be simplified to $\sum \mathbf{F} = m\mathbf{a}$.

4.13   Sketch the three volumes V_1, V_2, and V_3, shown generally in Fig. 4.1, assuming a short time-increment Δt for the fixed control volume of

   (a)   A nozzle on the end of a hose.
   (b)   A balloon into which air is entering (the fixed volume is the balloon at time t).
   (c)   A balloon from which air is exiting (the fixed volume is the balloon at time t).
   (d)   A Tee in a pipe line.

4.14   Sketch the velocity vector $\mathbf{V}$ and the normal unit vector $\hat{\mathbf{n}}$ on each area.

   (a)   The free surface area of a water tank that is being drained.
   (b)   The inlet area of a turbine.
   (c)   The wall of a pipe.
   (d)   The bottom of a canal.
   (e)   The inlet area to a cylindrical screen around a drain.

4.15   A rectangle surrounds a two-dimensional, stationary airfoil. It is at a distance from the airfoil on all sides. Sketch the box containing the airfoil along with the velocity vector $\mathbf{V}$ and the normal unit vector $\hat{\mathbf{n}}$ on all four sides of the rectangle.

4.16   We used

$$\frac{d}{dt} \int_{cv} \rho e \, dV = \int_{cv} \frac{\partial}{\partial t} (\rho e) \, dV$$

in the derivation of the system-to-control-volume transformation. What constraint allows this equivalence? Why is it an ordinary derivative on the left but a partial derivative on the right?

Conservation of Mass

4.17   Apply Eq. (4.14) to a flow in a pipe that divides into two exiting areas with different densities at each area assuming uniform flows over all three areas.

4.18   Water flows in a 4-cm-diameter pipe at 20 m/s. The pipe enlarges to a diameter of 6 cm. Calculate the flow rate, the mass flux, and the velocity in the larger diameter section of pipe.

4.19   Water flows at a depth of 40 cm in a 100-cm-diameter storm sewer. Calculate the flow rate and the mass flux if the average velocity is 3 m/s.

4.20   Air at 25°C and 240 kPa flows in a 10-cm-diameter pipe at 40 m/s. What are the flow rate and the mass flux in the pipe? (Recall that pressures are always gauge pressures unless stated otherwise.)

4.21   Air flows in a 20-cm-diameter duct at 120°C and 120 kPa with a mass flux of 5 kg/s. The circular duct converts to a 20-cm square duct in which the temperature and pressure are 140°C and 140 kPa, respectively. Determine the velocities in both sections of the duct.

4.22   Air is exiting a 100-cm-diameter balloon out a 1-cm-diameter nozzle. If the pressure and temperature at the exit are 110 kPa and 22°C, respectively, and the exit velocity is 30 m/s, calculate the flow rate, the mass flux, and the rate at which the diameter is changing.

4.23   Water flows in a 4-cm-diameter pipe at 20 m/s. The pipe divides into two pipes, one 2 cm in diameter and the other 3 cm in diameter. If 10 kg/s flows from the 2-cm-diameter pipe, calculate the flow rate from the 3-cm-diameter pipe.

4.24   Water flows in a 2-cm-diameter pipe at 10 m/s vertically upward to the center of two horizontal circular disks separated by 8 mm. It flows out between the disks at a radius of 25 cm. Sketch the pipe/disk arrangement. Calculate average velocity of the water leaving the disks. Also, calculate the average velocity of the water between the disks at a position where the radius of the disks is 10 cm.

4.25   High-velocity air at 20°C and 100 kPa absolute flows in a conduit at 600 m/s. It undergoes a sudden change (a shock wave) to 263 m/s and 438°C with no change in conduit dimensions. Determine the mass flux and the downstream pressure if the conduit cross-sectional area is 500 cm^2.

4.26   Water flows in a 12-cm-diameter pipe with the velocity profiles shown in Fig. 4.16. The maximum velocity for each profile is 20 m/s. Calculate the mass flux, the flow rate, and the average velocity.

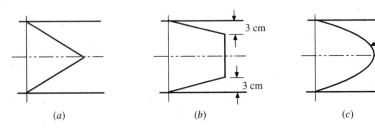

| (a) | (b) | (c) |

Figure 4.16

4.27   Water flows in a rectangular conduit 12-cm high and 60-cm wide having a maximum velocity of 20 m/s with the profiles shown in Fig. 4.16. Assume the profile exists across the entire cross section with negligible end effects. Calculate the mass flux, the flow rate, and the average velocity.

4.28   A sponge is contained in a volume that has one 4-cm-diameter inlet A_1 into which water flows and two outlets, A_2 and A_3. Determine dm/dt of the sponge if

   (a) $V_1 = 5$ m/s, $Q_2 = 0.002$ m^3/s, and $\dot{m}_3 = 2.5$ kg/s.

(b) $V_1 = 10$ m/s, $\dot{m}_2 = 1.5$ kg/s, and $Q_3 = 0.003$ m³/s.

(c) $\dot{m}_3 = 9.5$ kg/s, $Q_2 = 0.003$ m³/s, and $V_1 = 12$ m/s.

4.29   A sponge is contained in a volume that has one 4-cm-diameter inlet A_1 into which water flows and two 2-cm-diameter outlets, A_2 and A_3. The sponge is to have $dm/dt = 0$.

(a)   Find V_1 if $Q_2 = 0.002$ m³/s and $\dot{m}_3 = 2.5$ kg/s.

(b)   Find $\dot{m}_2$ if $V_1 = 10$ m/s and $Q_3 = 0.003$ m³/s.

(c)   Find Q_2 if $\dot{m}_1 = 4.5$ kg/s and $V_3 = 4$ m/s.

4.30   Atmospheric air flows over a flat plate as shown in Fig. 4.17. Viscosity makes the air stick to the surface creating a thin boundary layer. Estimate the mass flux $\dot{m}$ of the air across the surface that is 10 cm above the 120-cm wide plate if $u(y) = 800y$.

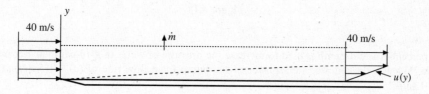

Figure 4.17

4.31   If a streamline is 5 cm above the flat plate of Fig. 4.17 at the leading edge, how far is it above the plate at the location where $u(y) = 800y$?

The Energy Equation

4.32   Water enters a horizontal nozzle with diameters d_1 and d_2 at 10 m/s and exits to the atmosphere. Estimate the pressure upstream of the nozzle if

(a) $d_1 = 8$ cm and $d_2 = 6$ cm

(b) $d_1 = 8$ cm and $d_2 = 4$ cm

(c) $d_1 = 10$ cm and $d_2 = 6$ cm

(d) $d_1 = 12$ cm and $d_2 = 5$ cm

4.33   Water is contained in a large tower that supplies a city. If the top of the water is 30 m above an outlet at the base of the tower, what maximum velocity can be expected at the outlet (to the atmosphere)? How does this maximum velocity compare with that of a rock dropped from the same height?

4.34   A high-speed jet is used to cut solid materials. Estimate the maximum pressure developed on the material if the velocity issuing from the water jet is (a) 100 m/s, (b) 120 m/s, and (c) 120 m/s.

4.35   Rework Solved Problem 4.5 with (a) $h = 5$ cm, (b) $h = 6$ cm, and (c) $h = 8$ cm.

4.36   Integrate the appropriate velocity profile and calculate the rate of kinetic energy transported by a flow of water that has a parabolic profile in a channel that measures 2 cm × 15 cm if the flow rate is 0.012 m³/s. Check your calculation using Eq. (4.30) with $\alpha = 1.5$.

4.37   The loss coefficient in Example 4.3 is increased to (a) 2.0, (b) 3.2, and (c) 6.0. Rework the problem. (The loss coefficient depends primarily on the pipe material, such as plastic, copper, wrought iron, so it can vary markedly.)

4.38 Water is transported from one reservoir with surface elevation of 135 m to a lower reservoir with surface elevation of 25 m through a 24-cm-diameter pipe. Estimate the flow rate and the mass flux through the pipe if the loss coefficient between the two surfaces is (a) 20, (b) 30, and (c) 40.

4.39 Assume uniform flow in the pipe of Fig. 4.18 and calculate the velocity in the larger pipe if the manometer reading h is (a) 30 cm, (b) 25 cm, and (c) 20 cm.

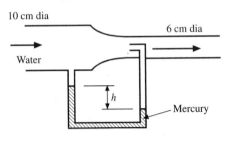

Figure 4.18

4.40 An 85 percent-efficient pump is used to increase the pressure of water in a 10-cm-diameter pipe from 120 to 800 kPa. What is the required horsepower of the pump for a flow rate of (a) 0.015 m^3/s, (b) 20 L/s, and (c) 4000 gal/h?

4.41 A 90 percent-efficient turbine accepts water at 400 kPa in a 16-cm-diameter pipe. What is the maximum power output if the flow rate is (a) 0.08 m^3/s, (b) 0.06 m^3/s, and (c) 0.04 m^3/s? The water is emitted to the atmosphere.

4.42 Air enters a compressor at 25°C and 10 kPa with negligible velocity. It exits through a 2-cm-diameter pipe at 400 kPa and 160°C with a velocity of 200 m/s. Determine the heat transfer if the power required is 18 kW.

4.43 Rework Solved Problem 4.7 if the overall loss coefficient K is (a) 2, (b) 8, and (c) 12.

The Momentum Equation

4.44 A strong wind at 30 m/s blows directly against a 120 cm × 300 cm window in a large building. Estimate the force of the wind on the window.

4.45 A 10-cm-diameter hose delivers 0.04 m^3/s of water through a 4-cm-diameter nozzle. What is the force of the water on the nozzle?

4.46 A 90°-nozzle with exit diameter d is attached to a hose of diameter $3d$ with pressure p. The nozzle changes the direction of the water flow from the hose through an angle of 90°. Calculate the magnitude of the force of the water on the nozzle if

    (a) $p = 200$ kPa, $d = 1$ cm

    (b) $p = 400$ kPa, $d = 6$ mm

    (c) $p = 300$ kPa, $d = 1.2$ cm

    (d) $p = 500$ kPa, $d = 2.2$ cm

4.47 A hydraulic jump, sketched in Fig. 4.19, can occur in a channel with no apparent cause, such as when a fast flowing stream flows from the mountains to the plains. (It is analogous to a shock wave that exists in a gas flow.) The momentum equation allows the height downstream to be calculated if the upstream height and velocity are known. Neglect any frictional force on the bottom and sidewalls and determine y_2 in the rectangular channel if

    (a) $V_1 = 10$ m/s and $y_1 = 50$ cm

    (b) $V_1 = 8$ m/s and $y_1 = 60$ cm

(c) $V_1 = 12$ m/s and $y_1 = 40$ cm

(d) $V_1 = 16$ m/s and $y_1 = 40$ cm

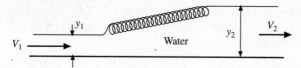

Figure 4.19

4.48 Determine the power lost in the hydraulic jump if the channel is 8-m wide in

  (a) Prob. 4.47(b)

  (b) Prob. 4.47(d)

4.49 It is desired to create a hydraulic jump, as in Fig. 4.19, in a 6-m wide, rectangular channel so that $V_2 = \dfrac{1}{4}V_1$. Calculate V_1 and the power lost if

  (a) $y_1 = 60$ cm

  (b) $y_1 = 40$ cm

4.50 A pipe transporting water undergoes a sudden expansion (Fig. 4.15). If the upstream pressure is 200 kPa and the mass flux is 40 kg/s, find the pressure downstream, where a uniform flow can be assumed, and the head lost due to the expansion. Use the following dimensions:

  (a) $d_1 = 4$ cm and $d_2 = 10$ cm

  (b) $d_1 = 4$ cm and $d_2 = 8$ cm

  (c) $d_1 = 6$ cm and $d_2 = 12$ cm

4.51 A 6-cm-diameter horizontal stationary water jet having a velocity of 40 m/s strikes a vertical plate. Determine the force needed to hold the plate if

  (a) it is stationary

  (b) it moves away from the jet at 20 m/s

  (c) it moves into the jet at 20 m/s

4.52 A 4-cm-diameter horizontal stationary water jet having a velocity of 50 m/s strikes a cone having an included angle at the apex of 60°. The water leaves the cone symmetrically. Determine the force needed to hold the cone if

  (a) it is stationary

  (b) it moves away from the jet at 20 m/s

  (c) it moves into the jet at 20 m/s

4.53 A jet boat traveling at 12 m/s takes in 0.08 m³/s of water and discharges it at 24 m/s faster than the boat's speed. Estimate the thrust produced and power required.

4.54 The deflector of Fig. 4.4 changes the direction of a 60 mm × 24 cm sheet of water with $V_1 = 30$ m/s such that $\alpha = 60°$. Calculate the force components of the water on the deflector if

  (a) it is stationary

  (b) it moves away from the jet at 20 m/s

  (c) it moves into the jet at 20 m/s

4.55 The blades of Fig. 4.6 deflect 10, 2-cm-diameter jets of water each having $V_1 = 40$ m/s. Determine the blade angle α_1 and the power output assuming no losses if

  (a) $\beta_1 = 30°$, $\alpha_2 = 45°$, and $V_B = 20$ m/s
  (b) $\beta_1 = 20°$, $\alpha_2 = 50°$, and $V_B = 15$ m/s

(c) $\beta_1 = 20°$, $\alpha_2 = 40°$, and $V_B = 20$ m/s
(d) $\beta_1 = 40°$, $\alpha_2 = 35°$, and $V_B = 20$ m/s

4.56   A rectangular jet strikes a stationary plate as shown in Fig. 4.20. Calculate the force F and the two mass fluxes if the velocity V_1 exiting the jet is (a) 20 m/s, (b) 40 m/s, and (c) 60 m/s. Neglect all frictional forces and any spreading of the stream.

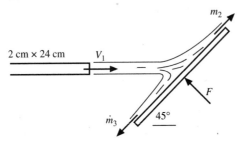

Figure 4.20

4.57   Estimate the drag force on the plate of Prob. 4.30 up to the position where the velocity profile is shown.

Answers to Supplementary Problems

4.12 $\rho = $ const, $V = V(t)$, inertial ref frame

4.16   Fixed cv

4.17 $\rho_1 A_1 V_1 = \rho_2 A_2 V_2 + \rho_3 A_3 V_3$

4.18   0.0251 m³/s, 25.1 kg/s, 8.89 m/s

4.19   1.182 m³/s, 1182 kg/s

4.20   0.314 m³/s, 1.25 kg/s

4.21   81.6 m/s, 61.7 m/s

4.22   0.236 m³/s, 0.585 kg/s, 0.075 m/s

4.23   0.01513 m³/s

4.24   0.25 m/s, 0.625 m/s

4.25   35.7 kg/s, 554 kPa

4.26   (a)   0.0754 m³/s, 75.4 kg/s, 6.67 m/s   (b)   0.1369 m³/s, 1369 kg/s, 12.1 m/s
          (c)   0.1131 m³/s, 1131 kg/s, 10 m/s

4.27   (a)   180 kg/s, 0.018 m³/s, 10 m/s   (b)   1080 kg/s, 1.08 m³/s, 11.25 m/s
          (c)   960 kg/s, 0.096 m³/s, 10 m/s

4.28   (a)   2.33 kg/s   (b)   8.07 kg/s   (c)   2.58 kg/s

4.29   (a)   3.58 m/s   (b)   9.57 kg/s   (c)   3.24 m³/s

4.30   2.20 kg/s

4.31   52.5 mm

4.32 (*a*) 17.78 m/s, 108 kPa   (*b*) 40 m/s, 750 kPa   (*c*) 27.8 m/s, 336 kPa
         (*d*) 57.6 m/s, 1609 kPa

4.33 24.3 m/s, same

4.34 (*a*) 5000 kPa   (*b*) 7200 kPa   (*c*) 11 250 kPa

4.35 (*a*) 0.01065 m³/s   (*b*) 0.01167 m³/s   (*c*) 0.01348 m³/s

4.36 144 J/s

4.37 (*a*) 181.9 kPa   (*b*) 176.6 kPa   (*c*) 165.7 kPa

4.38 (*a*) 0.470 m³/s, 470 kg/s   (*b*) 0.384 m³/s, 384 kg/s   (*c*) 0.332 m³/s, 332 kg/s

4.39 (*a*) 8.58 m/s   (*b*) 7.83 m/s   (*c*) 7.00 m/s

4.40 (*a*) 16.1 hp   (*b*) 21.4 hp   (*c*) 4.51 hp

4.41 (*a*) 28.8 kW   (*b*) 21.6 kW   (*c*) 14.4 kW

4.42 − 3140 J/s

4.43 (*a*) 0.22 m³/s   (*b*) 0.20 m³/s   (*c*) 0.19 m³/s

4.44 1980 N

4.45 2780 N

4.46 (*a*) 148 N   (*b*) 106.7 N   (*c*) 320 N   (*d*) 1795 N

4.47 (*a*) 2.95 m   (*b*) 2.51 m   (*c*) 3.23 m   (*d*) 4.37

4.48 (*a*) 439 kW   (*b*) 1230 kW

4.49 (*a*) 7.67 m/s, 272 kW   (*b*) 6.26 m/s, 99.5 kW

4.50 (*a*) 336 kPa, 36.4 m   (*b*) 390 kPa, 29.2 m   (*c*) 238 kPa, 5.73 m

4.51 (*a*) 4524 N   (*b*) 1131 N   (*c*) 10 180 N

4.52 (*a*) 421 N   (*b*) 151.5 N   (*c*) 825 N

4.53 30.9 hp

4.54 (*a*) 6480 N, 11 220 N   (*b*) 720 N, 1247 N   (*c*) 18 000 N, 31 200 N

4.55 (*a*) 53.79°, 108 hp   (*b*) 31.2°, 100 hp   (*c*) 37.87°, 17 hp
         (*d*) 67.5°, 113 hp

4.56 (*a*) 1358 N, 81.9 kg/s, 11.5 kg/s   (*b*) 5430 N, 163.9 kg/s, 22.9 kg/s
         (*c*) 12 220 N, 246 kg/s, 34.4 kg/s

4.57 9.9 N

Chapter 5

Differential Equations

5.1 INTRODUCTION

The differential equations introduced in this chapter are often omitted in an introductory course. The derivations in subsequent chapters will either not require these differential equations or there will be two methods to derive the equations: one using the differential equations and one utilizing differential elements. So, this chapter can be emitted with no loss of continuity.

In Chap. 4, problems were solved using integrals for which the integrands were known or could be approximated. Partial differential equations are needed in order to solve for those quantities in the integrands that are not known, such as the velocity distribution in a pipe or the pressure distribution on an airfoil. The partial differential equations may also contain information of interest, such as a point of separation of a fluid from a surface.

To solve a partial differential equation for the dependent variable, certain conditions are required, i.e., the dependent variable must be specified at certain values of the independent variables. If the independent variables are space coordinates (such as the velocity at the wall of a pipe), the conditions are called *boundary conditions*. If the independent variable is time, the conditions are called *initial conditions*. The general problem is usually referred to as a *boundary-value problem*.

The boundary conditions typically result from one or more of the following:

- The no-slip condition in a viscous flow. Viscosity causes any fluid, be it a gas or a liquid, to stick to the boundary so that the velocity of the fluid at a boundary takes on the velocity of the boundary. Most often the boundary is not moving.
- The normal component of the velocity in an inviscid flow. In an inviscid flow where the viscosity is neglected, the velocity vector is tangent to the boundary at the boundary, provided the boundary is not porous.
- The pressure at a free surface. For problems involving a free surface, a pressure condition is known at the free surface. This also applies to separated flows, where cavitation is present, and in wave motions.

For an unsteady flow, initial conditions are required, e.g., the initial velocity must be specified at some time, usually at $t = 0$. It would be a very difficult task to specify the three velocity components at $t = 0$ for most unsteady flows of interest. So, the problems requiring the solution of the partial differential equations derived in this chapter are those requiring boundary conditions.

The differential equations in this chapter will be derived using rectangular coordinates. It is often easier to solve problems using cylindrical or spherical coordinates; the differential equations using those two-coordinate systems are presented in Table 5.1.

The differential energy equation will not be derived in this book. It would be needed if there are temperature differences on the boundaries or if viscous effects are so large that temperature gradients are developed in the flow. A course in heat transfer would include such effects.

5.2   THE DIFFERENTIAL CONTINUITY EQUATION

To derive the differential continuity equation, the infinitesimal element of Fig. 5.1 is utilized. It is a small control volume into and from which the fluid flows. It is shown in the xy-plane with depth dz. Let us assume that the flow is only in the xy-plane so that no fluid flows in the z-direction. Since mass could be changing inside the element, the mass that flows into the element minus that which flows out must equal the change in mass inside the element. This is expressed as

$$\rho u \, dy \, dz - \left(\rho u + \frac{\partial(\rho u)}{\partial x} dx\right) dy \, dz + \rho v \, dx \, dz - \left(\rho v + \frac{\partial(\rho v)}{\partial y} dy\right) dx \, dz = \frac{\partial}{\partial t}(\rho \, dx \, dy \, dz) \qquad (5.1)$$

where the density ρ is allowed to change* across the element. Simplifying the above, recognizing that the elemental control volume is fixed, results in

$$\frac{\partial(\rho u)}{\partial x} + \frac{\partial(\rho v)}{\partial y} = -\frac{\partial \rho}{\partial t} \qquad (5.2)$$

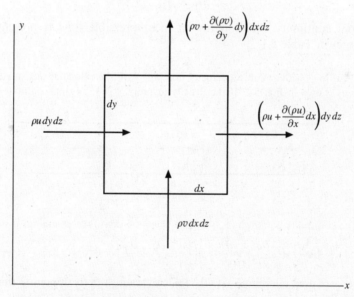

Figure 5.1   Infinitesimal control volume.

Differentiate the products and include the variation in the z-direction. Then the differential continuity equation can be put in the form

$$\frac{\partial \rho}{\partial t} + u \frac{\partial \rho}{\partial x} + v \frac{\partial \rho}{\partial y} + w \frac{\partial \rho}{\partial z} + \rho\left(\frac{\partial u}{\partial x} + \frac{\partial v}{\partial y} + \frac{\partial w}{\partial z}\right) = 0 \qquad (5.3)$$

---

* The product ρu could have been included as $\left(\rho + \frac{\partial \rho}{\partial x} dx\right)\left(u + \frac{\partial u}{\partial x} dx\right)$ on the right-hand side of the element, but the above is equivalent.

The first four terms form the material derivative [see Eq. (3.11)] so Eq. (5.3) becomes

$$\frac{D\rho}{Dt} + \rho\left(\frac{\partial u}{\partial x} + \frac{\partial v}{\partial y} + \frac{\partial w}{\partial z}\right) = 0 \tag{5.4}$$

providing the most general form of the *differential continuity equation* expressed using rectangular coordinates.

The differential continuity equation is often written using the vector operator

$$\mathbf{V} = \frac{\partial}{\partial x}\mathbf{i} + \frac{\partial}{\partial y}\mathbf{j} + \frac{\partial}{\partial z}\mathbf{k} \tag{5.5}$$

so that Eq. (5.4) takes the form

$$\frac{D\rho}{Dt} + \rho\mathbf{V}\cdot\mathbf{V} = 0 \tag{5.6}$$

where the velocity vector is $\mathbf{V} = u\mathbf{i} + v\mathbf{j} + w\mathbf{k}$. The scalar $\mathbf{V}\cdot\mathbf{V}$ is called the *divergence* of the velocity vector.

For an incompressible flow, the density of a fluid particle remains constant, i.e.,

$$\frac{D\rho}{Dt} = \frac{\partial\rho}{\partial t} + u\frac{\partial\rho}{\partial x} + v\frac{\partial\rho}{\partial y} + w\frac{\partial\rho}{\partial z} = 0 \tag{5.7}$$

so it is not necessary that the density be constant. If the density is constant, as it often is, then each term in Eq. (5.7) is 0. For an incompressible flow, Eqs. (5.4) and (5.6) also demand that

$$\frac{\partial u}{\partial x} + \frac{\partial v}{\partial y} + \frac{\partial w}{\partial z} = 0 \quad \text{or} \quad \mathbf{V}\cdot\mathbf{V} = 0 \tag{5.8}$$

The differential continuity equation for an incompressible flow is presented in cylindrical and spherical coordinates in Table 5.1.

EXAMPLE 5.1 Air flows with a uniform velocity in a pipe with the velocities measured along the centerline at 40-cm increments as shown in Fig. 5.2. If the density at point 2 is 1.2 kg/m³, estimate the density gradient at point 2.

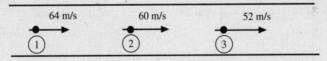

Figure 5.2

Solution: The continuity equation (5.3) is used since the density is changing. It is simplified as follows:

$$\frac{\partial \rho}{\partial t} + u\frac{\partial \rho}{\partial x} + v\frac{\partial \rho}{\partial y} + w\frac{\partial \rho}{\partial z} + \rho\left(\frac{\partial u}{\partial x} + \frac{\partial v}{\partial y} + \frac{\partial w}{\partial z}\right) = 0 \qquad \therefore u\frac{\partial \rho}{\partial x} = -\rho\frac{\partial u}{\partial x}$$

Central differences* are used to approximate the velocity gradient $\partial u\partial x$ at point 2 since information at three points is given as follows:

$$\frac{\partial u}{\partial x} \cong \frac{\Delta u}{\Delta x} = \frac{52 - 64}{0.80} = -15\,\text{m/(s·m)}$$

The best estimate of the density gradient, using the information given, is then

$$\frac{\partial \rho}{\partial x} = -\frac{\rho}{u}\frac{\partial u}{\partial x} = -\frac{1.2}{60}(-15) = 0.3\,\text{kg/(m}^4)$$

---

*A forward difference would give $\partial u/\partial x \cong (52-60)/0.40 = -20$. A backward difference would provide $\partial u/\partial x \cong (60-64)/0.40 = -10$. The central difference is the best approximation.

5.3 THE DIFFERENTIAL MOMENTUM EQUATION

The differential continuity equation derived in Sec. 5.2 contains the three velocity components as the dependent variables for an incompressible flow. If there is a flow of interest in which the velocity field and pressure field are not known, such as the flow around a turbine blade or over a weir, the differential momentum equation provides three additional equations since it is a vector equation containing three component equations. The four unknowns are then u, v, w, and p when using a rectangular coordinate system. The four equations provide us with the necessary equations and then the initial and boundary conditions allow a tractable problem. The problems of the turbine blade and the weir are quite difficult to solve, and their solutions will not be attempted in this book, but there are problems with simple geometries that will be solved.

So, let us go about deriving the differential momentum equations, a rather challenging task. First, stresses exist on the faces of an infinitesimal, rectangular fluid element, as shown in Fig. 5.3 for the xy-plane. Similar stress components act in the z-direction. The *normal stresses* are designated with σ and the *shear stresses* with τ. There are nine stress components: σ_{xx}, σ_{yy}, σ_{zz}, τ_{xy}, τ_{yx}, τ_{xz}, τ_{zx}, τ_{yz}, and τ_{zy}. If moments are taken about the x-axis, the y-axis, and the z-axis, respectively, they would show that

$$\tau_{yx} = \tau_{xy} \qquad \tau_{zx} = \tau_{xz} \qquad \tau_{zy} = \tau_{yz} \tag{5.9}$$

So, there are six stress components that must be related to the pressure and velocity components. Such relationships are called constitutive equations; they are equations that are not derived but are found using observations in the laboratory.

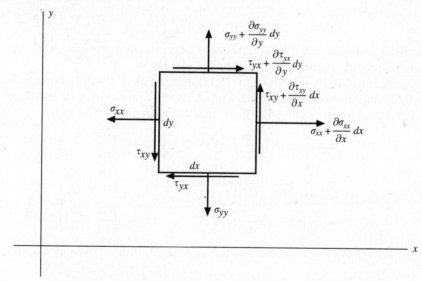

Figure 5.3   Rectangular stress components on a fluid element.

Next, apply Newton's second law to the element of Fig. 5.3, assuming that no shear stresses act in the z-direction (we will simply add those in later) and that gravity acts in the z-direction only:

$$\left(\sigma_{xx} + \frac{\partial \sigma_{xx}}{\partial x}dx\right)dy\,dz - \sigma_{xx}\,dy\,dz + \left(\tau_{xy} + \frac{\partial \tau_{xy}}{\partial y}dy\right)dx\,dz - \tau_{xy}\,dx\,dz = \rho\,dx\,dy\,dz\frac{Du}{Dt}$$

$$\left(\sigma_{yy} + \frac{\partial \sigma_{yy}}{\partial y}dy\right)dx\,dz - \sigma_{yy}\,dx\,dz + \left(\tau_{xy} + \frac{\partial \tau_{xy}}{\partial x}dx\right)dy\,dz - \tau_{xy}\,dy\,dz = \rho\,dx\,dy\,dz\frac{Dv}{Dt} \tag{5.10}$$

These are simplified to

$$\frac{\partial \sigma_{xx}}{\partial x} + \frac{\partial \tau_{xy}}{\partial y} = \rho\frac{Du}{Dt}$$

$$\frac{\partial \sigma_{yy}}{\partial y} + \frac{\partial \tau_{xy}}{\partial x} = \rho\frac{Dv}{Dt} \tag{5.11}$$

If the z-direction components are included, the differential equations become

$$\frac{\partial \sigma_{xx}}{\partial x} + \frac{\partial \tau_{xy}}{\partial y} + \frac{\partial \tau_{xz}}{\partial z} = \rho \frac{Du}{Dt}$$

$$\frac{\partial \sigma_{yy}}{\partial y} + \frac{\partial \tau_{xy}}{\partial x} + \frac{\partial \tau_{yz}}{\partial z} = \rho \frac{Dv}{Dt} \qquad (5.12)$$

$$\frac{\partial \sigma_{zz}}{\partial z} + \frac{\partial \tau_{xz}}{\partial x} + \frac{\partial \tau_{yz}}{\partial y} - \rho g = \rho \frac{Dw}{Dt}$$

assuming that the gravity term, $\rho g \, dx \, dy \, dz$, acts in the negative z-direction.

In many flows, the viscous effects that lead to the shear stresses can be neglected and the normal stresses are the negative of the pressure. For such inviscid flows, Eqs. (5.12) take the form

$$\rho \frac{Du}{Dt} = -\frac{\partial p}{\partial x}$$

$$\rho \frac{Dv}{Dt} = -\frac{\partial p}{\partial y} \qquad (5.13)$$

$$\rho \frac{Dw}{Dt} = -\frac{\partial p}{\partial z} - \rho g$$

In the vector form [see Eq. (5.5)], they become the famous *Euler's equation*

$$\rho \frac{D\mathbf{V}}{Dt} = -\nabla p - \rho g \hat{\mathbf{k}} \qquad (5.14)$$

which is applicable to inviscid flows. For a constant-density, steady flow, Eq. (5.14) can be integrated along a streamline to provide Bernoulli's equation [Eq. (3.25)].

If viscosity significantly affects the flow, Eqs. (5.12) must be used. Constitutive equations[*] relate the stresses to the velocity and pressure fields; they are not derived but are written by making observations in the laboratory. For a Newtonian,[†] isotropic[‡] fluid, they have been observed to be

$$\sigma_{xx} = -p + 2\mu \frac{\partial u}{\partial x} + \lambda \nabla \cdot \mathbf{V} \qquad \tau_{xy} = \mu \left(\frac{\partial u}{\partial y} + \frac{\partial v}{\partial x} \right)$$

$$\sigma_{yy} = -p + 2\mu \frac{\partial v}{\partial y} + \lambda \nabla \cdot \mathbf{V} \qquad \tau_{xz} = \mu \left(\frac{\partial u}{\partial z} + \frac{\partial w}{\partial x} \right) \qquad (5.15)$$

$$\sigma_{zz} = -p + 2\mu \frac{\partial w}{\partial z} + \lambda \nabla \cdot \mathbf{V} \qquad \tau_{yz} = \mu \left(\frac{\partial v}{\partial z} + \frac{\partial w}{\partial y} \right)$$

For most gases, *Stokes hypothesis* can be used so that $\lambda = -2\mu/3$. If the above normal stresses are added, there results

$$p = -\frac{1}{3}(\sigma_{xx} + \sigma_{yy} + \sigma_{zz}) \qquad (5.16)$$

showing that the pressure is the negative average of the three normal stresses in most gases, including air, and in all liquids in which $\nabla \cdot \mathbf{V} = 0$.

---

[*] The constitutive equations for cylindrical and spherical coordinates are displayed in Table 5.1.
[†] A Newtonian fluid has a linear stress–strain rate relationship.
[‡] An isotropic fluid has properties that are independent of direction at a point.

Table 5.1   The Differential Continuity, Momentum Equations, and Stresses for Incompressible Flows in Cylindrical and Spherical Coordinates

Continuity

Cylindrical

$$\frac{1}{r}\frac{\partial}{\partial r}(rv_r) + \frac{1}{r}\frac{\partial v_\theta}{\partial \theta} + \frac{\partial v_z}{\partial z} = 0$$

Spherical

$$\frac{1}{r^2}\frac{\partial}{\partial r}\left(r^2 v_r\right) + \frac{1}{r\sin\theta}\frac{\partial}{\partial \theta}(v_\theta \sin\theta) + \frac{1}{r\sin\theta}\frac{\partial v_\phi}{\partial \phi} = 0$$

Momentum

Cylindrical

$$\rho\frac{Dv_r}{Dt} - \frac{v_\theta^2}{r} = -\frac{\partial p}{\partial r} + \rho g_r + \mu\left(\nabla^2 v_r - \frac{v_r}{r^2} - \frac{2}{r^2}\frac{\partial v_\theta}{\partial \theta}\right)$$

$$\rho\frac{Dv_\theta}{Dt} + \frac{v_\theta v_r}{r} = -\frac{1}{r}\frac{\partial p}{\partial \theta} + \rho g_\theta + \mu\left(\nabla^2 v_\theta - \frac{v_\theta}{r^2} + \frac{2}{r^2}\frac{\partial v_r}{\partial \theta}\right)$$

$$\rho\frac{Dv_z}{Dt} = -\frac{\partial p}{\partial z} + \rho g_z + \mu\nabla^2 v_z$$

$$\frac{D}{Dt} = v_r\frac{\partial}{\partial r} + \frac{v_\theta}{r}\frac{\partial}{\partial \theta} + v_z\frac{\partial}{\partial z} + \frac{\partial}{\partial t}$$

$$\nabla^2 = \frac{\partial^2}{\partial r^2} + \frac{1}{r}\frac{\partial}{\partial r} + \frac{1}{r^2}\frac{\partial^2}{\partial \theta^2} + \frac{\partial^2}{\partial z^2}$$

Spherical

$$\rho\frac{Dv_r}{Dt} - \rho\frac{v_\theta^2 + v_\phi^2}{r} = -\frac{\partial p}{\partial r} + \rho g_r$$

$$+ \mu\left(\nabla^2 v_r - \frac{2v_r}{r^2} - \frac{2}{r^2}\frac{\partial v_\theta}{\partial \theta} - \frac{2v_\theta \cot\theta}{r^2} - \frac{2}{r^2\sin\theta}\frac{\partial v_\phi}{\partial \phi}\right)$$

$$\rho\frac{Dv_\theta}{Dt} + \rho\frac{v_r v_\theta + v_\phi^2 \cot\theta}{r} = -\frac{1}{r}\frac{\partial p}{\partial \theta} + \rho g_\theta$$

$$+ \mu\left(\nabla^2 v_\theta + \frac{2}{r^2}\frac{\partial v_r}{\partial \theta} - \frac{v_\theta}{r^2\sin\theta} - \frac{2\cos\theta}{r^2\sin^2\theta}\frac{\partial v_\phi}{\partial \phi}\right)$$

$$\rho\frac{Dv_\phi}{Dt} - \rho\frac{v_r v_\phi + v_\theta v_\phi \cot\theta}{r} = -\frac{1}{r\sin\theta}\frac{\partial p}{\partial \phi} + \rho g_\phi$$

$$+ \mu\left(\nabla^2 v_\phi - \frac{v_\phi}{r^2\sin^2\theta} + \frac{2}{r^2\sin^2\theta}\frac{\partial v_r}{\partial \phi} + \frac{2\cos\theta}{r^2\sin^2\theta}\frac{\partial v_\theta}{\partial \phi}\right)$$

$$\frac{D}{Dt} = v_r\frac{\partial}{\partial r} + \frac{v_\theta}{r}\frac{\partial}{\partial \theta} + \frac{v_\phi}{r\sin\theta}\frac{\partial}{\partial \phi} + \frac{\partial}{\partial t}$$

$$\nabla^2 = \frac{1}{r^2}\frac{\partial}{\partial r}\left(r^2\frac{\partial}{\partial r}\right) + \frac{1}{r^2\sin\theta}\frac{\partial}{\partial \theta}\left(\sin\theta\frac{\partial}{\partial \theta}\right) + \frac{1}{r^2\sin\theta}\frac{\partial^2}{\partial \phi^2}$$

Stresses

Cylindrical

$$\sigma_{rr} = -p + 2\mu\frac{\partial v_r}{\partial y} \qquad \tau_{r\theta} = \mu\left(r\frac{\partial(v_\theta/r)}{\partial r} + \frac{1}{r}\frac{\partial v_z}{\partial \theta}\right)$$

$$\sigma_{\theta\theta} = -p + 2\mu\left(\frac{1}{r}\frac{\partial v_\theta}{\partial \theta} + \frac{v_r}{r}\right) \qquad \tau_{\theta z} = \mu\left(\frac{\partial v_\theta}{\partial z} + \frac{1}{r}\frac{\partial v_r}{\partial \theta}\right)$$

$$\sigma_{zz} = -p + 2\mu\frac{\partial v_z}{\partial z} \qquad \tau_{rz} = \mu\left(\frac{\partial v_r}{\partial z} + \frac{\partial v_z}{\partial r}\right)$$

Spherical

$$\sigma_{rr} = -p + 2\mu\frac{\partial v_r}{\partial r}$$

$$\sigma_{\theta\theta} = -p + 2\mu\left(\frac{1}{r\sin\theta}\frac{\partial v_\theta}{\partial \theta} + \frac{v_r}{r}\right)$$

$$\sigma_{\phi\phi} = -p + 2\mu\left(\frac{1}{r\sin\theta}\frac{\partial v_\phi}{\partial \phi} + \frac{v_r}{r} + \frac{v_\phi \cot\theta}{r}\right)$$

$$\tau_{r\theta} = \mu\left[r\frac{\partial}{\partial r}\left(\frac{v_\theta}{r}\right) + \frac{1}{r}\frac{\partial v_r}{\partial \theta}\right]$$

$$\tau_{\theta\phi} = \mu\left[\frac{\sin\theta}{r}\frac{\partial}{\partial \theta}\left(\frac{v_\phi}{\sin\theta}\right) + \frac{1}{r\sin\theta}\frac{\partial v_\theta}{\partial \phi}\right]$$

$$\tau_{r\phi} = \mu\left[\frac{1}{r\sin\theta}\frac{\partial v_r}{\partial \phi} + r\frac{\partial}{\partial r}\left(\frac{v_\phi}{r}\right)\right]$$

If Eqs. (5.15) are substituted into Eqs. (5.12) using $\lambda = -2\mu/3$, there results

$$\rho\frac{Du}{Dt} = -\frac{\partial p}{\partial x} + \mu\left(\frac{\partial^2 u}{\partial x^2} + \frac{\partial^2 u}{\partial y^2} + \frac{\partial^2 u}{\partial z^2}\right) + \frac{\mu}{3}\frac{\partial}{\partial x}\left(\frac{\partial u}{\partial x} + \frac{\partial v}{\partial y} + \frac{\partial w}{\partial z}\right)$$

$$\rho\frac{Dv}{Dt} = -\frac{\partial p}{\partial y} + \mu\left(\frac{\partial^2 v}{\partial x^2} + \frac{\partial^2 v}{\partial y^2} + \frac{\partial^2 v}{\partial z^2}\right) + \frac{\mu}{3}\frac{\partial}{\partial y}\left(\frac{\partial u}{\partial x} + \frac{\partial v}{\partial y} + \frac{\partial w}{\partial z}\right) \qquad (5.17)$$

$$\rho\frac{Dw}{Dt} = -\frac{\partial p}{\partial z} + \mu\left(\frac{\partial^2 w}{\partial x^2} + \frac{\partial^2 w}{\partial y^2} + \frac{\partial^2 w}{\partial z^2}\right) + \frac{\mu}{3}\frac{\partial}{\partial z}\left(\frac{\partial u}{\partial x} + \frac{\partial v}{\partial y} + \frac{\partial w}{\partial z}\right) - \rho g$$

where gravity acts in the negative z-direction and a *homogeneous** *fluid* has been assumed so that, for example, $\partial\mu/\partial x = 0$.

Finally, if an incompressible flow is assumed so that $\mathbf{\nabla}\cdot\mathbf{V} = 0$, the *Navier–Stokes equations* result

$$\rho\frac{Du}{Dt} = -\frac{\partial p}{\partial x} + \mu\left(\frac{\partial^2 u}{\partial x^2} + \frac{\partial^2 u}{\partial y^2} + \frac{\partial^2 u}{\partial z^2}\right)$$

$$\rho\frac{Dv}{Dt} = -\frac{\partial p}{\partial y} + \mu\left(\frac{\partial^2 v}{\partial x^2} + \frac{\partial^2 v}{\partial y^2} + \frac{\partial^2 v}{\partial z^2}\right) \qquad (5.18)$$

$$\rho\frac{Dw}{Dt} = -\frac{\partial p}{\partial z} + \mu\left(\frac{\partial^2 w}{\partial x^2} + \frac{\partial^2 w}{\partial y^2} + \frac{\partial^2 w}{\partial z^2}\right) - \rho g$$

where the z-direction is vertical.

If we introduce the scalar operator called the *Laplacian*, defined by

$$\nabla^2 = \frac{\partial^2}{\partial x^2} + \frac{\partial^2}{\partial y^2} + \frac{\partial^2}{\partial z^2} \qquad (5.19)$$

and review the steps leading from Eq. (5.13) to Eq. (5.14), the Navier–Stokes equations can be written in vector form as

$$\rho\frac{D\mathbf{V}}{Dt} = -\mathbf{\nabla}p + \mu\nabla^2\mathbf{V} + \rho\mathbf{g} \qquad (5.20)$$

The Navier–Stokes equations expressed in cylindrical and spherical coordinates are presented in Table 5.1.

The three scalar Navier–Stokes equations and the continuity equation constitute the four equations that can be used to find the four variables u, v, w, and p, provided there are appropriate initial and boundary conditions. The equations are nonlinear due to the acceleration terms, such as $u\partial v/\partial y$ on the left-hand side; consequently, the solution to these equations may not be unique, i.e., the solution that is determined from the above equations may not be the one observed in the laboratory. For example, the flow between two rotating cylinders can be solved using the Navier–Stokes equations to be a relatively simple flow with circular streamlines; it could also be a flow with streamlines that are like a spring wound around the cylinders as a torus, and there are even more complex flows that are also solutions to the Navier–Stokes equations, all satisfying the identical boundary conditions.

The differential momentum equations (the Navier–Stokes equations) can be solved with relative ease for some simple geometries. But the equations cannot be solved for a turbulent flow even for the simplest of examples; a turbulent flow is highly unsteady and three-dimensional and thus requires that the three velocity components be specified at all points in a region of interest at some initial time, say $t = 0$. Such information would be nearly impossible to obtain, even for the simplest geometry. Consequently, the solutions of turbulent flows are left to the experimentalist and are not attempted by solving the equations.

---

* A homogeneous fluid has properties that are independent of position.

EXAMPLE 5.2 Water flows from a reservoir in between two closely aligned parallel plates, as shown in Fig. 5.4. Write the simplified equations needed to find the steady-state velocity and pressure distributions between the two plates. Neglect any z variation of the distributions and any gravity effects. Do not neglect $v(x, y)$.

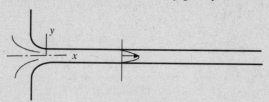

Figure 5.4

Solution: The continuity equation is simplified, for the incompressible water flow, to

$$\frac{\partial u}{\partial x} + \frac{\partial v}{\partial y} + \cancel{\frac{\partial w}{\partial z}} = 0$$

The differential momentum equations recognizing that

$$\frac{D}{Dt} = u\frac{\partial}{\partial x} + v\frac{\partial}{\partial y} + w\cancel{\frac{\partial}{\partial z}} + \cancel{\frac{\partial}{\partial t}}$$

are simplified as follows:

$$\rho\left(u\frac{\partial u}{\partial x} + v\frac{\partial u}{\partial y}\right) = -\frac{\partial p}{\partial x} + \mu\left(\frac{\partial^2 u}{\partial x^2} + \frac{\partial^2 u}{\partial y^2} + \cancel{\frac{\partial^2 u}{\partial z^2}}\right)$$

$$\rho\left(u\frac{\partial v}{\partial x} + v\frac{\partial v}{\partial y}\right) = -\cancel{\frac{\partial p}{\partial y}} + \mu\left(\frac{\partial^2 v}{\partial x^2} + \frac{\partial^2 v}{\partial y^2} + \cancel{\frac{\partial^2 v}{\partial z^2}}\right)$$

neglecting pressure variation in the y-direction since the plates are assumed to be a relatively small distance apart. So, the three equations that contain the three variables u, v, and p are

$$\frac{\partial u}{\partial x} + \frac{\partial v}{\partial y} = 0$$

$$\rho\left(u\frac{\partial u}{\partial x} + v\frac{\partial u}{\partial y}\right) = -\frac{\partial p}{\partial x} + \mu\left(\frac{\partial^2 u}{\partial x^2} + \frac{\partial^2 u}{\partial y^2}\right)$$

$$\rho\left(u\frac{\partial v}{\partial x} + v\frac{\partial v}{\partial y}\right) = \mu\left(\frac{\partial^2 v}{\partial x^2} + \frac{\partial^2 v}{\partial y^2}\right)$$

To find a solution to these equations for the three variables, it would be necessary to use the no-slip conditions on the two plates and assumed boundary conditions at the entrance, which would include $u(0, y)$ and $v(0, y)$. Even for this rather simple geometry, the solution to this entrance flow problem appears, and is, quite difficult. A numerical solution could be attempted.

EXAMPLE 5.3 Integrate Euler's equation (5.14) along a streamline as shown in Fig. 5.5 for a steady, constant-density flow and show that Bernoulli's equation (3.25) results.

Solution: First, sketch a general streamline and show the selected coordinates normal to and along the streamline so that the velocity vector can be written as $V\hat{s}$, as we did in Fig. 3.10. First, let us express DV/Dt in these coordinates.

$$\frac{D\mathbf{V}}{Dt} = \cancel{\frac{\partial \mathbf{V}}{\partial t}} + V\frac{\partial(V\hat{s})}{\partial s} + \cancel{V_n}\frac{\partial \mathbf{V}}{\partial n} = V\hat{s}\frac{\partial V}{\partial s} + V^2\frac{\partial \hat{s}}{\partial s}$$

where $\partial \hat{s}/\partial s$ is nonzero since $\hat{s}$ can change direction from point to point on the streamline; it is a vector quantity in the $\hat{n}$ direction. Applying Euler's equation along a streamline (in the s-direction) allows us to write

$$\rho V\frac{\partial V}{\partial s} = -\frac{\partial p}{\partial s} - \rho g\frac{\partial z}{\partial s}$$

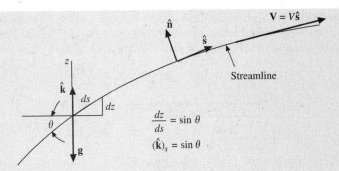

$$\frac{dz}{ds} = \sin\theta$$

$$(\hat{\mathbf{k}})_s = \sin\theta$$

Figure 5.5

where we have referred to Fig. 5.5 to write $(\hat{\mathbf{k}})_s = \partial z/\partial s$. Partial derivatives are necessary because quantities can vary in the normal direction. The above equation is then written as

$$\frac{\partial}{\partial s}\left(\rho\frac{V^2}{2} + p + \rho gz\right) = 0$$

provided the density ρ is constant. This means that along a streamline,

$$\frac{V^2}{2} + \frac{p}{\rho} + gz = \text{const}$$

This is Bernoulli's equation requiring the same conditions as it did when it was derived in Chap. 3.

5.4   THE DIFFERENTIAL ENERGY EQUATION

Most problems in an introductory fluid mechanics course involve isothermal fluid flows in which temperature gradients do not exist. So, the differential energy equation is not of interest. The study of flows in which there are temperature gradients are included in a course on heat transfer. For completeness, the differential energy equation is presented here without derivation. In general, it is

$$\rho\frac{Dh}{Dt} = K\nabla^2 T + \frac{Dp}{Dt} \tag{5.21}$$

where K is the *thermal conductivity*. For an incompressible ideal gas flow, it becomes

$$\rho c_p\frac{DT}{Dt} = K\nabla^2 T \tag{5.22}$$

and for a liquid flow it takes the form

$$\frac{DT}{Dt} = \alpha\nabla^2 T \tag{5.23}$$

where α is the *thermal diffusivity* defined by $\alpha = K/\rho c_p$.

Solved Problems

5.1   The x-component of the velocity in a certain plane flow depends only on y by the relationship $u(y) = Ay$. Determine the y-component $v(x, y)$ of the velocity if $v(x, 0) = 0$.

The continuity equation for this plane flow (in a plane flow there are only two velocity components that depend on two space variables) demands that

$$\frac{\partial u}{\partial x} = -\frac{\partial v}{\partial y} \qquad \therefore \frac{\partial v}{\partial y} = -\frac{\partial u}{\partial x} = -\frac{\partial(Ay)}{\partial x} = 0$$

The solution to $\partial v/\partial y = 0$ is $v(x, y) = f(x)$. But $v(x, 0) = 0$, as given, so that $f(x) = 0$ and $v(x, y) = 0$. The only way for $v(x, y)$ to be nonzero would be for $v(x, 0)$ to be nonzero.

5.2   Does the velocity field

$$v_r = 4\left(1 - \frac{1}{r^2}\right)\cos\theta \qquad v_\theta = -4\left(1 + \frac{1}{r^2}\right)\sin\theta \qquad v_z = 0$$

represent a possible incompressible flow?

The (r, θ, z) coordinates are cylindrical coordinates. So, Table 5.1 provides the continuity equation to be used:

$$\frac{1}{r}\frac{\partial}{\partial r}(rv_r) + \frac{1}{r}\frac{\partial v_\theta}{\partial \theta} + \frac{\partial v_z}{\partial z} = 0.$$

Substitute the velocity components into this equation and find

$$\frac{4\cos\theta}{r}\frac{\partial}{\partial r}\left(r - \frac{1}{r}\right) + \frac{-4}{r}\left(1 + \frac{1}{r^2}\right)\frac{\partial}{\partial \theta}(\sin\theta) + \frac{\partial v_z}{\partial z}\overset{?}{=} 0$$

Differentiate and find

$$\frac{4\cos\theta}{r}\left(1 + \frac{1}{r^2}\right) - \frac{4}{r}\left(1 + \frac{1}{r^2}\right)\cos\theta = 0$$

Continuity is satisfied, so the velocity field is a possible incompressible flow.

5.3   Use the differential momentum equations for an incompressible uniform flow that moves toward a flat plate, e.g., the wind hitting a vertical wall, and find an expression for the gradient of the pressure. Assume a plane flow in which only the x- and y-components are nonzero and viscous and gravity effects are negligible.

The Eqs. (5.18) are simplified as follows:

$$\rho\left(\frac{\partial u}{\partial t} + u\frac{\partial u}{\partial x} + v\frac{\partial u}{\partial y} + w\frac{\partial u}{\partial z}\right) = -\frac{\partial p}{\partial x} + \mu\left(\frac{\partial^2 u}{\partial x^2} + \frac{\partial^2 u}{\partial y^2} + \frac{\partial^2 u}{\partial z^2}\right) + \rho g_x$$

$$\rho\left(\frac{\partial v}{\partial t} + u\frac{\partial v}{\partial x} + v\frac{\partial v}{\partial y} + w\frac{\partial v}{\partial z}\right) = -\frac{\partial p}{\partial y} + \mu\left(\frac{\partial^2 v}{\partial x^2} + \frac{\partial^2 v}{\partial y^2} + \frac{\partial^2 v}{\partial z^2}\right) + \rho g_y$$

This provides the pressure gradient to be related to the velocity field by

$$\nabla p = \frac{\partial p}{\partial x}\mathbf{i} + \frac{\partial p}{\partial y}\mathbf{j} = -\left(u\frac{\partial u}{\partial x} + v\frac{\partial u}{\partial y}\right)\mathbf{i} - \left(u\frac{\partial v}{\partial x} + v\frac{\partial v}{\partial y}\right)\mathbf{j}$$

5.4   Show that Du/Dt can be written as $\mathbf{V}\cdot\nabla u$ for a steady flow. Then write an expression for $D\mathbf{V}/Dt$.

Expand Du/Dt for a steady flow as

$$\frac{Du}{Dt} = \frac{\partial u}{\partial t} + u\frac{\partial u}{\partial x} + v\frac{\partial u}{\partial y} + w\frac{\partial u}{\partial z} = \left(u\frac{\partial}{\partial x} + v\frac{\partial}{\partial y} + w\frac{\partial}{\partial z}\right)u = \mathbf{V}\cdot\nabla u$$

where we have used

$$\mathbf{V}\cdot\nabla = (u\mathbf{i} + v\mathbf{j} + w\mathbf{k})\cdot\left(\frac{\partial}{\partial x}\mathbf{i} + \frac{\partial}{\partial y}\mathbf{j} + \frac{\partial}{\partial z}\mathbf{k}\right) = u\frac{\partial}{\partial x} + v\frac{\partial}{\partial y} + w\frac{\partial}{\partial z}$$

Finally, we observe that

$$\frac{D\mathbf{V}}{Dt} = \frac{Du}{Dt}\mathbf{i} + \frac{Dv}{Dt}\mathbf{j} + \frac{Dw}{Dt}\mathbf{k} = \mathbf{V}\cdot\nabla u\mathbf{i} + \mathbf{V}\cdot\nabla v\mathbf{j} + \mathbf{V}\cdot\nabla w\mathbf{k}$$

$$= \mathbf{V}\cdot\nabla(u\mathbf{i} + v\mathbf{j} + w\mathbf{k})$$

$$= (\mathbf{V}\cdot\nabla)\mathbf{V}$$

Supplementary Problems

The Differential Continuity Equation

5.5   Refer to the first footnote and include $\left(\rho + \frac{\partial \rho}{\partial x}dx\right)\left(u + \frac{\partial u}{\partial x}dx\right)$ from the right-hand side of the element and $\left(\rho + \frac{\partial \rho}{\partial y}dy\right)\left(v + \frac{\partial v}{\partial y}dy\right)$ from the top area of the element and show that Eq. (5.2) results.

5.6   The *divergence theorem*, also called *Gauss' theorem*, is written in vector form as

$$\int_A \mathbf{B}\cdot\hat{\mathbf{n}}\,dA = \int_{V} \mathbf{\nabla}\cdot\mathbf{B}\,dV$$

where **B** represents any vector and the surface area A surrounds the volume V. Apply this theorem to the integral continuity equation of Eq. (4.13) for a steady flow and derive Eq. (5.6).

5.7   A compressible flow of a gas occurs in a pipeline. Assume uniform flow with the x-direction along the pipe axis and state the simplified continuity equation.

5.8   An incompressible steady flow of a fluid, such as a stratified flow of salt water (as in the isthmus between a fresh body of water and a body of salt water), flows in a channel with a sudden change in the height of the channel bottom (this allows for nonzero u and v). Assume no variation in the z-direction and write the two equations that result from the continuity equation. (Experiments show that a stagnant region of fluid exists in front of a sudden increase in the height of the bottom of a channel in a stratified flow. This phenomenon causes the buildup of smog in Los Angeles when air flows toward the city, but substantial smog does not appear in the more densely populated New York City. There are mountains east of Los Angeles but not west of New York.)

5.9   An isothermal flow occurs in a conduit. Show that the continuity equation can be written as $Dp = -p\mathbf{\nabla}\cdot\mathbf{V}$ for an ideal gas.

5.10 An incompressible fluid flows radially (no θ- or ϕ-component) into a small circular drain. How must the radial component of velocity vary with radius as demanded by continuity?

5.11 If the x-component of the velocity vector is constant in a plane flow, what is true of the y-component of the velocity vector?

5.12 Calculate the density gradient in Example 5.1 if (*a*) forward differences were used and (*b*) if backward differences were used. What is the percent error for each, assuming the answer in Example 5.1 is correct.

5.13 The x-component of the velocity vector is measured at three locations 8 mm apart on the centerline of a symmetrical contraction. At points A, B, and C, the measurements produce 8.2, 9.4, and 11.1 m/s, respectively. Estimate the y-component of the velocity 2 mm above point B in this steady, plane, incompressible flow.

5.14 If, in a plane flow, the two velocity components are given by

$$u(x, y) = 8(x^2 + y^2) \qquad v(x, y) = 8xy$$

What is $D\rho/Dt$ at $(1, 2)$ m if at that point $\rho = 2$ kg/m³?

5.15 The velocity field for a particular plane flow ($w = 0$) of air is given by

$$u(x, y) = \frac{4y}{x^2 + y^2} \qquad v(x, y) = -\frac{4x}{x^2 + y^2}$$

Show that this is an incompressible flow.

5.16   If $u(x, y) = 4 + 2x/(x^2 + y^2)$ in a plane incompressible flow, what is $v(x, y)$ if $v(x, 0) = 0$?

5.17   If $v(x, y) = 8 + 4y/(x^2 + y^2)$ in a plane incompressible flow, what is $u(x, y)$ if $u(0, y) = 0$?

5.18   The velocity component $v_\theta = -(25 + 1/r^2)\cos\theta$ in a plane incompressible flow. Find $v_r(r, \theta)$ if $v_r(r, 0) = 0$.

5.19   The velocity component $v_\theta = -25(1 + 1/r^2)\sin\theta + 50/r^2$ in a plane incompressible flow. Find $v_r(r, \theta)$ if $v_r(r, 90°) = 0$.

The Differential Momentum Equation

5.20   Draw a rectangular element similar to the one in Fig. 5.3 in the xz-plane. Assume that no shear stresses act in the y-direction and that gravity acts in the z-direction. Apply Newton's second law to the element in the z-direction and write an equation similar to those of Eqs. (5.11).

5.21   If a steady flow of fluid occurs around a long cylinder, what three equations would be needed to find the velocity and pressure fields if viscous effects are significant but gravity effects are not significant? What boundary conditions would exist on the cylinder? Express the equations in cylindrical coordinates. Refer to Table 5.1.

5.22   If a steady flow of fluid occurs around a sphere, what three equations would be needed to find the velocity and pressure fields if viscous effects are significant but gravity effects are not significant? What boundary conditions would exist on the sphere? Express the equations in spherical coordinates. Refer to Table 5.1.

5.23   Verify that $\dfrac{D\mathbf{V}}{Dt} = (\mathbf{V} \cdot \nabla)\mathbf{V}$ using rectangular coordinates assuming a steady flow.

5.24   Find the pressure gradient ∇p for the incompressible flow of Prob. 5.15, assuming an inviscid flow with negligible gravity effects.

5.25   Find the pressure gradient ∇p for the incompressible flow of Solved Problem 5.2, assuming an inviscid flow with negligible gravity effects.

5.26   Simplify the appropriate Navier–Stokes equation for the flow between parallel plates assuming $u = u(y)$ and gravity in the z-direction. The streamlines are assumed to be parallel to the plates so that $v = w = 0$.

5.27   Simplify the Navier–Stokes equation for flow in a pipe assuming $v_z = v_z(r)$ and gravity in the z-direction. The streamlines are assumed to be parallel to the pipe wall so that $v_\theta = v_r = 0$.

5.28   The inner cylinder of two concentric cylinders rotates resulting such that $v_\theta = v_\theta(r)$ and $v_r = 0$. What equations are needed to find the velocity profile assuming vertical cylinders?

5.29   Substitute the constitutive equations (5.15) into the momentum equations (5.12) and show that the Navier–Stokes equations (5.18) result, assuming a homogeneous incompressible fluid.

5.30   Assume that a flow is not homogeneous, e.g., there is a temperature gradient in the flow such that the viscosity is not constant, and write the x-component differential momentum equations for an incompressible flow using the constitutive equations (5.15).

5.31   Let the negative average of the three normal stresses be denoted by $\bar{p}$ in a gas flow in which Stokes hypothesis is not applicable. Find an expression for $(p - \bar{p})$.

Answers to Supplementary Problems

5.5   See problem statement

5.6   See problem statement

5.7 $u\dfrac{\partial \rho}{\partial x} + \rho\dfrac{\partial u}{\partial x} = 0$

5.8 $u\dfrac{\partial \rho}{\partial x} = -v\dfrac{\partial \rho}{\partial y} \qquad \dfrac{\partial u}{\partial x} = -\dfrac{\partial v}{\partial y}$

5.9   See problem statement

5.10 $v_r = C$

5.11 $v = f(x)$

5.12 $0.4\,\text{kg/m}^4$, 33.3%

5.13 0.36 m/s

5.14 $-32\,\text{kg/(m}^3\text{·s)}$

5.15 $\dfrac{\partial u}{\partial x} + \dfrac{\partial v}{\partial y} = 0$

5.16 $-2y/(x^2 + y^2)$

5.17 $-4y/(x^2 + y^2)$

5.18 $-(25 - 1^2)\sin\theta$

5.19 $(25 - 1^2)\cos\theta$

5.20 $\dfrac{\partial \sigma_{xx}}{\partial x} + \dfrac{\partial \tau_{xz}}{\partial z} = \rho\dfrac{Du}{Dt}$ and $\dfrac{\partial \sigma_{zz}}{\partial z} + \dfrac{\partial \tau_{zx}}{\partial x} - \gamma = \rho\dfrac{Dw}{Dt}$

5.23   See problem statement

5.24 $\dfrac{16}{\rho(x^2 + y^2)^2}(x\mathbf{i} + y\mathbf{j})$

5.26 $\dfrac{\partial p}{\partial x} = \mu\dfrac{\partial^2 u}{\partial y^2}$

5.27 $\therefore \dfrac{\partial p}{\partial z} = \rho g_z + \mu\dfrac{\partial^2 v_z}{\partial r^2} + \dfrac{1}{r}\dfrac{\partial v_z}{\partial r}$

5.28 $\therefore \dfrac{1}{r}\dfrac{\partial p}{\partial \theta} = \mu\left(\dfrac{\partial^2 v_\theta}{\partial r^2} + \dfrac{1}{r}\dfrac{\partial v_\theta}{\partial r} - \dfrac{v_\theta}{r^2}\right)$

5.31 $-(\lambda + 2\mu/3)\mathbf{V}\cdot\mathbf{V}$

Chapter 6

Dimensional Analysis and Similitude

6.1 INTRODUCTION

Many problems of interest in fluid mechanics cannot be solved using the integral and/or differential equations. Wind motions around a football stadium, the flow of water through a large hydroturbine, the airflow around the deflector on a semitruck, the wave motion around a pier or a ship, and airflow around aircraft are all examples of problems that are studied in the laboratory with the use of models. A laboratory study with the use of models is very expensive, however, and to minimize the cost, dimensionless parameters are used. In fact, such parameters are also used in numerical studies for the same reason.

Dimensionless parameters are obtained using a method called *dimensional analysis*, to be presented in Sec. 6.2. It is based on the idea of *dimensional homogeneity*: all terms in an equation must have the same dimensions. Simply using this idea, we can minimize the number of parameters needed in an experimental or analytical analysis, as will be shown. Any equation can be expressed in terms of dimensionless parameters simply by dividing each term by one of the other terms. For example, consider Bernoulli's equation

$$\frac{V_2^2}{2} + \frac{p_2}{\rho} + gz_2 = \frac{V_1^2}{2} + \frac{p_1}{\rho} + gz_1 \tag{6.1}$$

Now, divide both sides by gz_2. The equation can then be written as

$$\frac{V_2^2}{2gz_2} + \frac{p_2}{\gamma z_2} + 1 = \left(\frac{V_1^2}{2gz_1} + \frac{p_1}{\gamma z_1} + 1\right)\frac{z_1}{z_2} \tag{6.2}$$

Note the dimensionless parameters, V^2/gz and $p/\gamma z$.

Once an analysis is performed on a model in the laboratory and all quantities of interest are measured, it is necessary to predict those same quantities on the prototype, such as the power generated by a large wind machine from the measurements on a much smaller model. *Similitude* is the study that allows us to predict the quantities to be expected on a prototype from the measurements on a model. This will be done after our study of dimensional analysis that guides the model study.

6.2 DIMENSIONAL ANALYSIS

An example will be used to demonstrate the usefulness of dimensional analysis. Suppose the drag force F_D is desired on an object with a spherical front that is shaped as shown in Fig. 6.1. A study could be

97

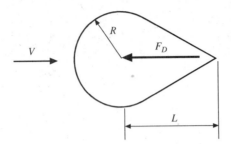

Figure 6.1   Flow around an object.

performed, the drag force measured for a particular radius R and length L in a fluid with velocity V, viscosity μ, and density ρ. Gravity is expected to not influence the force. This dependence of the drag force on the other variables would be written as

$$F_D = f(R, L, V, \mu, \rho) \qquad (6.3)$$

To present the results of an experimental study, the drag force could be plotted as a function of V for various values of the radius R holding all other variables fixed. Then a second plot could show the drag force for various values of L holding all other variables fixed, and so forth. The plots may resemble those of Fig. 6.2. To vary the viscosity holding the density fixed and then the density holding the viscosity fixed would require a variety of fluids leading to a very complicated study, possibly an impossible study.

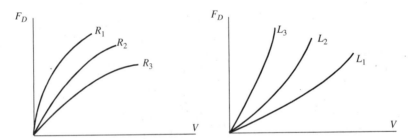

Figure 6.2   Drag force versus velocity. (a) L, μ, ρ fixed and (b) R, μ, ρ fixed.

The actual relationship that would relate the drag force to the other variables could be expressed as a set of dimensionless parameters, much like those of Eq. (6.2), as

$$\frac{F_D}{\rho V^2 R^2} = f\left(\frac{\rho V R}{\mu}, \frac{R}{L}\right) \qquad (6.4)$$

(The procedure to do this will be presented next). The results of a study using the above relationship would be much more organized than the study suggested by the curves of Fig. 6.2. An experimental study would require only several different models, each with different R/L ratios, and only one fluid, either air or water. Varying the velocity of the fluid approaching the model, a rather simple task, could vary the other two dimensionless parameters. A plot of $F_D/(\rho V^2 R^2)$ versus $\rho V R/\mu$ for the several values of R/L would then provide the results of the study.

Before we present the details of forming the dimensionless parameters of Eq. (6.4), let us review the dimensions on quantities of interest in fluid mechanics. Many quantities have obvious dimensions, but for some the dimensions are not so obvious. There are only three basic dimensions since Newton's second law can be used to relate the basic dimensions. Using F, M, L, and T as the dimensions on force, mass, length, and time, we see that $F = ma$ demands that the dimensions are related by

$$F = M\frac{L}{T^2} \qquad (6.5)$$

We choose to select the $M - L - T$ system[*] and use Eq. (6.5) to relate F to $M, L,$ and T. If temperature is needed, as with the flow of a compressible gas, an equation of state, such as

$$p = \rho RT \tag{6.6}$$

could be expressed dimensionally as

$$[RT] = [p/\rho] = \frac{F}{L^2}\frac{L^3}{M} = \frac{ML/T^2}{L^2}\frac{L^3}{M} = \frac{L^2}{T^2} \tag{6.7}$$

where the brackets mean "the dimensions of." The product RT does not introduce additional dimensions.

Table 6.1 has been included to aid in selecting the proper dimensions for the quantities of interest. It will simplify the creation of the dimensionless parameters. The dimensions are displayed for the $M - L - T$ system only, since that will be what is used in the solution to the problems in this chapter. The same results would be obtained using the $F - L - T$ system, should that system be selected.

The *Buckingham π-theorem* is used to create the dimensionless parameters, given a functional relationship such as that of Eq. (6.3). Write the primary variable of interest as a general function, such as

$$x_1 = f(x_2, x_3, x_4, \ldots, x_n) \tag{6.8}$$

Table 6.1   Symbols and Dimensions of Quantities of Interest Using the $M-L-T$ System

Quantity	Symbol	Dimensions
Length	l	L
Mass	m	M
Time	t	T
Velocity	V	L/T
Acceleration	a	L/T^2
Angular velocity	Ω	T^{-1}
Force	F	F
Gravity	g	L/T^2
Flow rate	Q	L^3/T
Mass flux	$\dot{m}$	M/T
Pressure	p	M/LT^2
Stress	τ	M/LT^2
Density	ρ	M/L^3
Specific weight	γ	M/L^2T^2
Work	W	ML^2/T^2
Viscosity	μ	M/LT
Kinematic viscosity	v	L^2/T
Power	$\dot{W}$	ML^2/T^3
Heat flux	$\dot{Q}$	ML^2/T^3
Surface tension	σ	M/T^2
Bulk modulus	B	M/LT^2

---

[*] The $F-L-T$ system could have been used. It is simply our choice to use the $M-L-T$ system.

where n is the total number of variables. If m is the number of basic dimensions, usually 3, the Buckingham π-theorem demands that $(n - m)$ dimensionless groups of variables, the π-terms, are related by

$$\pi_1 = f_1(\pi_2, \pi_3, \ldots, \pi_{n-m}) \qquad (6.9)$$

where π_1 is selected to contain the dependent variable [it would be F_D of Eq. (6.3)] and the remaining π-terms contain the independent variables. It should be noted that a functional relationship cannot contain a particular dimension in only one variable; for example, in the relationship $v = f(d, t, \rho)$, the density ρ cannot occur since it is the only variable that contains the dimension M, so M would not have the possibility of canceling out to form a dimensionless π-term.

The steps that are followed when applying the Buckingham π-theorem are:

1. Write the dependent variable as a function of the $(n - 1)$ independent variables. This step requires knowledge of the phenomenon being studied. All variables that influence the dependent variable must be included and all variables that do not influence the dependent variable should not be included. In most problems, this relationship will be given.
2. Identify m variables, the *repeating variables* that are combined with the remaining variables to form the π-terms. The m variables must include all the basic dimensions present in the n variables of the functional relationship, but they must not form a dimensionless π-term by themselves. Note that an angle is dimensionless, so it is not a candidate to be a repeating variable.
3. Combine each of the $(n - m)$ variables with the repeating variables to form the π-terms.
4. Write the π-term containing the dependent variable as a function of the remaining π-terms.

Step 3 is carried out by either inspection or an algebraic procedure. The method of inspection will be used in an example. To demonstrate the algebraic procedure, let us form a π-term of the variables V, R, ρ, and μ. This is written as

$$\pi = V^a R^b \rho^c \mu^d \qquad (6.10)$$

In terms of dimensions, this is

$$M^0 L^0 T^0 = \left(\frac{L}{T}\right)^a L^b \left(\frac{M}{L^3}\right)^c \left(\frac{M}{LT}\right)^d \qquad (6.11)$$

Equating exponents on each of the basic dimensions provides the system of equations

$$
\begin{aligned}
M: &\quad 0 = c + d \\
L: &\quad 0 = a + b - 3c - d \\
T: &\quad 0 = -a - d
\end{aligned}
\qquad (6.12)
$$

The solution is

$$c = -d \qquad a = -d \qquad b = -d \qquad (6.13)$$

The π-term is then written as

$$\pi = \left(\frac{\mu}{VR\rho}\right)^d \qquad (6.14)$$

This π-term is dimensionless regardless of the value of d. If we desire V to be in the denominator, select $d = 1$; if we desire V to be in the numerator, select $d = -1$. Select $d = -1$ so that

$$\pi = \frac{VR\rho}{\mu} \qquad (6.15)$$

Suppose that only one π-term results from an analysis. That π-term would be equal to a constant which could be determined by an experiment.

Finally, consider a very general functional relationship between a pressure change Δp, length l, velocity V, gravity g, viscosity μ, density ρ, speed of sound c, surface tension σ, and an angular velocity Ω.

All of these variables may not influence a particular problem, but it is interesting to observe the final relationship of dimensionless terms. Dimensional analysis using V, l, and ρ as the repeating variables provides the relationship

$$\frac{\Delta p}{\rho V^2} = f\left(\frac{V^2}{lg}, \frac{\rho V l}{\mu}, \frac{V}{c}, \frac{\rho l V^2}{\sigma}, \frac{\Omega l}{V}\right) \tag{6.16}$$

Each term that appears in this relationship is an important parameter in certain flow situations. The dimensionless term with its common name is listed as follows:

$$\frac{\Delta p}{\rho V^2} = \text{Eu} \qquad \text{Euler number}$$

$$\frac{V}{\sqrt{lg}} = \text{Fr} \qquad \text{Froude number}$$

$$\frac{\rho V l}{\mu} = \text{Re} \qquad \text{Reynolds number}$$

$$\frac{V}{c} = M \qquad \text{Mach number} \tag{6.17}$$

$$\frac{\rho l V^2}{\sigma} = \text{We} \qquad \text{Weber number}$$

$$\frac{\Omega l}{V} = \text{St} \qquad \text{Strouhal number}$$

Not all of the above numbers would be of interest in a particular flow; it is highly unlikely that both compressibility effects and surface tension would influence the same flow. These are, however, the primary dimensionless parameters in our study of fluid mechanics. The Euler number is of interest in most flows, the Froude number in flows with free surfaces in which gravity is significant (e.g., wave motion), the Reynolds number in flows in which viscous effects are important, the Mach number in compressible flows, the Weber number in flows affected by surface tension (e.g., sprays with droplets), and the Strouhal number in flows in which rotation or a periodic motion plays a role. Each of these numbers, with the exception of the Weber number (surface tension effects are of little engineering importance), will appear in flows studied in subsequent chapters. Note: The Froude number is often defined as V^2/lg; this would not influence the solution to problems.

EXAMPLE 6.1 The pressure drop Δp over a length L of pipe is assumed to depend on the average velocity V, the pipe's diameter D, the average height e of the roughness elements of the pipe wall, the fluid density ρ, and the fluid viscosity μ. Write a relationship between the pressure drop and the other variables.

Solution: First, select the repeating variables. Do not select Δp since that is the dependent variable. Select only one D, L, and e since they all have the dimensions of length. Select the variables that are thought[*] to most influence the pressure drop: V, D, and ρ. Now, list the dimensions on each variable (Table 6.1):

$$[\Delta p] = \frac{M}{LT^2} \qquad [L] = L \qquad [V] = \frac{L}{T} \qquad [D] = L \qquad [e] = L \qquad [\rho] = \frac{M}{L^3} \qquad [\mu] = \frac{M}{LT}$$

First, combine Δp, V, D, and μ into a π-term. Since only Δp and ρ have M as a dimension, they must occur as a ratio $\Delta p/\rho$. That places T in the denominator so that V must be in the numerator so the T's cancel out. Finally, check out the L's: there is L in the numerator, so D must be in the denominator providing

$$\pi_1 = \frac{\Delta p}{\rho V^2 D^2}$$

---

[*] This is often debatable. Either D or L could be selected, whichever is considered to be most influential.

The second π-term is found by combining L with the three repeating variables V, D, and ρ. Since both L and D have the dimension of length, the second π-term is

$$\pi_2 = \frac{L}{D}$$

The third π-term results from combining e with the repeating variables. It has the dimension of length so the third π-term is

$$\pi_3 = \frac{e}{D}$$

The last π-term is found by combining μ with V, D, and ρ. Both μ and ρ contain the dimension M demanding that they form the ratio ρ/μ. This puts T in the numerator demanding that V goes in the numerator. This puts L in the denominator so that D must appear in the numerator. The last π-term is then

$$\pi_4 = \frac{\rho V D}{\mu}$$

The final expression relates the π-terms as

$$\pi_1 = f(\pi_2, \pi_3, \pi_4)$$

or using the variables

$$\frac{\Delta p}{\rho V^2 D^2} = f\left(\frac{L}{D}, \frac{e}{D}, \frac{\rho V D}{\mu}\right)$$

If L had been chosen as a repeating variable, it would simply change places with D since it has the same dimension.

6.3 SIMILITUDE

After the dimensionless parameters have been identified and a study on a model has been accomplished in a laboratory, similitude allows us to predict the behavior of a prototype from the measurements made on the model. The measurements on the model of a ship in a towing basin or on the model of an aircraft in a wind tunnel are used to predict the performance of the ship or the aircraft.

The application of similitude is based on three types of similarity. First, a model must look like the prototype, i.e., the length ratio must be constant between corresponding points on the model and prototype. For example, if the ratio of the lengths of the model and prototype is λ, then every other length ratio is also λ. Hence, the area ratio would be λ^2 and the volume ratio λ^3. This is *geometric similarity*.

The second is *dynamic similarity*: all force ratios acting on corresponding mass elements in the model flow and the prototype flow are the same. This results by equating the appropriate dimensionless numbers of Eqs. (6.17). If viscous effects are important, the Reynolds numbers are equated; if compressibility is significant, Mach numbers are equated; if gravity influences the flows, Froude numbers are equated; if an angular velocity influences the flow, the Strouhal numbers are equated, and if surface tension affects the flow, the Weber numbers are equated. All of these numbers can be shown to be ratios of forces, so equating the numbers in a particular flow is equivalent to equating the force ratios in that flow.

The third type of similarity is *kinematic similarity*: the velocity ratio is the same between corresponding points in the flow around the model and the prototype. This can be shown by considering the ratio of inertial forces, using the inertial force as

$$F_I = mV\frac{dV}{ds} \approx m\frac{V^2}{l} \approx \rho l^3 \frac{V^2}{l} = \rho l^2 V^2 \qquad (6.18)$$

where the acceleration[*] $a = V\, dV/ds$ has been used. The ratio of forces between model and prototype is then

$$\frac{(F_I)_m}{(F_I)_p} = \frac{V_m^2 l_m^2}{V_p^2 l_p^2} = \text{const} \qquad (6.19)$$

---

[*] Recall $a = dV/dt$ and $V = ds/dt$ so that $a = V\,dV/ds$.

showing that the velocity ratio is a constant between corresponding points if the length ratio is a constant, i.e., if geometric similarity exists (we assume the density ratio ρ_m/ρ_p to be constant between corresponding points in the two flows).

Assuming complete similarity between model and prototype, quantities of interest can now be predicted. For example, if a drag force is measured on flow around a model in which viscous effects play an important role, the ratio of the forces [see Eq. (6.18)] would be

$$\frac{(F_D)_m}{(F_D)_p} = \frac{\rho_m V_m^2 l_m^2}{\rho_p V_p^2 l_p^2} \tag{6.20}$$

The velocity ratio would be found by equating the Reynolds numbers.

$$\mathrm{Re}_m = \mathrm{Re}_p \qquad \frac{\rho_m V_m l_m}{\mu_m} = \frac{\rho_p V_p l_p}{\mu_p} \tag{6.21}$$

If the length ratio, the *scale*, is given and the same fluid is used in model and prototype, the force acting on the prototype can be found. It would be

$$(F_D)_p = (F_D)_m \left(\frac{V_p}{V_m}\right)^2 \left(\frac{l_p}{l_m}\right)^2 = (F_D)_m \left(\frac{l_m}{l_p}\right)^2 \left(\frac{l_p}{l_m}\right)^2 = (F_D)_m \tag{6.22}$$

showing that, if the Reynolds number governs the model study and the same fluid is used in the model and prototype, the force on the model is the same as the force of the prototype. Note that the velocity in the model study is the velocity in the prototype multiplied by the length ratio so that the model velocity could be quite large.

If the Froude number governed the study, we would have

$$\mathrm{Fr}_m = \mathrm{Fr}_p \qquad \frac{V_m^2}{l_m g_m} = \frac{V_p^2}{l_p g_p} \tag{6.23}$$

The drag force on the prototype, with $g_m = g_p$, would then be

$$(F_D)_p = (F_D)_m \left(\frac{V_p}{V_m}\right)^2 \left(\frac{l_p}{l_m}\right)^2 = (F_D)_m \left(\frac{l_p}{l_m}\right)\left(\frac{l_p}{l_m}\right)^2 = (F_D)_m \left(\frac{l_p}{l_m}\right)^3 \tag{6.24}$$

This is the situation for the model study of a ship. The Reynolds number is not used even though the viscous drag force acting on the ship cannot be neglected. We cannot satisfy both the Reynolds number and the Froude number in a study if the same fluid is used for the model study as that exists in the prototype flow; the model study of a ship always uses water as the fluid. To account for the viscous drag, the results of the model study based on the Froude number are adapted using industrial modifiers not included in this book.

EXAMPLE 6.2 A clever design of the front of a ship is to be tested in a water basin. A drag of 12.2 N is measured on the 1:20 scale model when towed at a speed of 3.6 m/s. Determine the corresponding speed of the prototype ship and the drag to be expected.

Solution: The Froude number guides the model study of a ship since gravity effects (wave motions) are more significant than the viscous effects. Consequently,

$$\mathrm{Fr}_p = \mathrm{Fr}_m \quad \text{or} \quad \frac{V_p}{\sqrt{l_p g_p}} = \frac{V_m}{\sqrt{l_m g_m}}$$

Since gravity does not vary significantly on the earth, there results

$$V_p = V_m \sqrt{\frac{l_p}{l_m}} = 3.6 \times \sqrt{20} = 16.1 \,\mathrm{m/s}$$

To find the drag on the prototype, the drag ratio is equated to the gravity force ratio (the inertial force ratio could be used but not the viscous force ratio since viscous forces have been ignored).

$$\frac{(F_D)_p}{(F_D)_m} = \frac{\rho_p V_p^2 l_p^2}{\rho_m V_m^2 l_m^2} \qquad \therefore (F_D)_p = (F_D)_m \frac{V_p^2 l_p^2}{V_m^2 l_m^2} = 12.2 \times \frac{16.1^2}{3.6^2} \times 20^2 = 41\,000\,\text{N}$$

where we used $\rho_p \cong \rho_m$ since salt water and fresh water have nearly the same density. The above results would be modified based on the established factors to account for the viscous drag on the ship.

EXAMPLE 6.3 A large pump delivering 1.2 m³/s of water with a pressure rise of 400 kPa is needed for a particular hydroelectric power plant. A proposed design change is tested on a smaller 1:4 scale pump. Estimate the flow rate and pressure rise which would be expected in the model study. If the power needed to operate the model pump is measured to be 8000 kW, what power would be expected to operate the prototype pump?

   Solution: For this internal flow problem, Reynolds number would be equated

$$\text{Re}_p = \text{Re}_m \quad \text{or} \quad \frac{V_p d_p}{\nu_p} = \frac{V_m d_m}{\nu_m} \quad \text{or} \quad \frac{V_p}{V_m} = \frac{d_m}{d_p}$$

assuming $\nu_p \cong \nu_m$ for the water in the model and prototype. The ratio of flow rates is

$$\frac{Q_p}{Q_m} = \frac{A_p V_p}{A_m V_m} = \frac{l_p^2 V_p}{l_m^2 V_m} = 4^2 \times \frac{1}{4} = 4 \qquad \therefore Q_m = \frac{Q_p}{4} = \frac{1.2}{4} = 0.3\,\text{m}^3/\text{s}$$

The power ratio is found using power as force times velocity; this provides

$$\frac{\dot{W}_p}{\dot{W}_m} = \frac{\rho_p V_p^2 l_p^2}{\rho_m V_m^2 l_m^2}\frac{V_p}{V_m} \qquad \therefore \dot{W}_p = \dot{W}_m \left(\frac{d_m}{d_p}\right)^2 \left(\frac{d_p}{d_m}\right)^2 \frac{d_m}{d_p} = \frac{8000}{4} = 500\,\text{kW}$$

This is an unexpected result. When using the Reynolds number to guide a model study, the power measured on the model exceeds the power needed to operate the prototype since the pressures are so much larger on the model. Note that in this example the Euler number would be used to provide the model pressure rise as

$$\frac{\Delta p_p}{\Delta p_m} = \frac{\rho_p V_p^2}{\rho_m V_m^2} \qquad \therefore \Delta p_m = \Delta p_p \left(\frac{d_p}{d_m}\right)^2 = 400 \times 4^2 = 6400\,\text{kPa}$$

For this reason and the observation that the velocity is much larger on the model, model studies are not common for situations (e.g., flow around an automobile) in which the Reynolds number is the guiding parameter.

EXAMPLE 6.4 The pressure rise from free stream to a certain location on the surface of the model of a rocket is measured to be 22 kPa at an air speed of 1200 km/h. The wind tunnel is maintained at 90 kPa absolute and 15°C. What would be the speed and pressure rise on a rocket prototype at an elevation of 15 km?

   Solution: The Mach number governs the model study. Thus,

$$M_m = M_p \qquad \frac{V_m}{c_m} = \frac{V_p}{c_p} \qquad \frac{V_m}{\sqrt{kRT_m}} = \frac{V_p}{\sqrt{kRT_p}}$$

Using the temperature from Table B.3, the velocity is

$$V_p = V_m \sqrt{\frac{T_p}{T_m}} = 1200\sqrt{\frac{216.7}{288}} = 1041\,\text{km/h}$$

A pressure force is $\Delta p A \approx \Delta p l^2$ so that the ratio to the inertial force of Eq. (6.18) is the Euler number, $\Delta p / \rho V^2$. Equating the Euler numbers gives the pressure rise as

$$\Delta p_p = \Delta p_m \frac{\rho_p V_p^2}{\rho_m V_m^2} = 22 \times \frac{p_p T_m V_p^2}{p_m T_p V_m^2} = 22 \times \frac{12.3 \times 288 \times 1041^2}{90 \times 216.7 \times 1200^2} = 3.01\,\text{kPa}$$

Solved Problems

6.1 Write the dimensions of the kinetic energy term $\frac{1}{2}mV^2$ using the $F-L-T$ system of units.

The dimensions on mV^2 are

$$[mV^2] = M\frac{L^2}{T^2} = F\frac{T^2}{L}\frac{L^2}{T^2} = FL$$

where $M = FT^2/L$ comes from Newton's second law written as $m = F/a$. The units on FL would be N·m in the SI system, as expected. Using the $M-L-T$ system the units would be (kg·m^2)/s^2, which are equivalent to N·m.

6.2 The speed V of a weight when it hits the floor is assumed to depend on gravity g, the height h from which it was dropped, and the density ρ of the weight. Use dimensional analysis and write a relationship between the variables.

The dimensions of each variable are listed as

$$[V] = \frac{L}{T} \qquad [g] = \frac{L}{T^2} \qquad [h] = L \qquad [\rho] = \frac{M}{L^3}$$

Since M occurs in only one variable, that variable ρ cannot be included in the relationship. The remaining three terms are combined to form a single π-term; it is formed by observing that T occurs in only two of the variables, thus V^2 is in the numerator and g is in the denominator. The length dimension is then canceled by placing h in the denominator. The single π-term is

$$\pi_1 = \frac{V^2}{gh}$$

Since this π-term depends on all other π-terms and there are none, it must be at most a constant. Hence, we conclude that

$$V = C\sqrt{gh}$$

A simple experiment would show that $C = \sqrt{2}$. We see that dimensional analysis rules out the possibility that the speed of free fall, neglecting viscous effects (e.g., drag), depends on the density of the material (or the weight).

6.3 A new design is proposed for an automobile. It is suggested that a 1:5 scale model study be done to access the proposed design for a speed of 90 km/h. What speed should be selected for the model study and what drag force would be expected on the prototype if a force of 80 N were measured on the model?

The Reynolds number would be the controlling parameter. It requires

$$\frac{V_m l_m}{\nu_m} = \frac{V_p l_p}{\nu_p} . \qquad \therefore V_m = V_p \frac{l_p}{l_m} = 90 \times 5 = 450 \, \text{km/h}$$

This high speed would introduce compressibility effects. Hence, either a larger model would have to be selected or a lower prototype speed would be required.

For the speed calculated above, the drag force would be found using Eq. (6.22)

$$(F_D)_p = (F_D)_m \left(\frac{V_p}{V_m}\right)^2 \left(\frac{l_p}{l_m}\right)^2 = (F_D)_m \left(\frac{l_m}{l_p}\right)^2 \left(\frac{l_p}{l_m}\right)^2 = (F_D)_m = 80 \, \text{N}$$

It should be noted that for high Reynolds number flows, the flow around blunt objects often becomes independent of Reynolds number, as observed in Fig. 8.2 for flow around a sphere for $\text{Re} > 4 \times 10^5$. This would probably be the case for flow around an automobile. As long as $(\text{Re})_m > 5 \times 10^5$ any velocity could be selected for the model study. If the model were 40 cm wide, then a velocity of 100 km/h could be selected; at that velocity the Reynolds number, based on the width, would be

$\text{Re} = V_m l_m / \nu_m = (100\,000/3600) \times 0.4/1.6 \times 10^{-5} = 7 \times 10^5$. This would undoubtedly be an acceptable velocity. It is obvious that knowledge and experience is required for such studies.

Supplementary Problems

6.4   Divide Eq. (6.1) by V_1^2 thereby expressing Bernoulli's equation (6.1) as a group of dimensionless terms. Identify the dimensionless parameters introduced.

6.5   If the $F-L-T$ system is used, select the dimensions on each of the following: (*a*) mass flux, (*b*) pressure, (*c*) density, (*d*) viscosity, and (*e*) power.

Dimensional Analysis

6.6   Combine each of the following groups of variables into a single dimensionless group, a π-term.

    (*a*)   Velocity V, length l, gravity g, and density ρ
    (*b*)   Velocity V, diameter D, density ρ, and viscosity μ
    (*c*)   Velocity V, density ρ, diameter D, and kinematic viscosity ν
    (*d*)   Angular velocity Ω, gravity g, diameter d, and viscosity μ
    (*e*)   Angular velocity Ω, viscosity μ, distance b, and density ρ
    (*f*)   Power $\dot{W}$, diameter d, velocity V, and pressure rise Δp

6.7   What variable could not influence the velocity if it is proposed that the velocity depends on a diameter, a length, gravity, rotational speed, and viscosity?

6.8   An object falls freely in a viscous fluid. Relate the terminal velocity V to its width w, its length l, gravity g, and the fluid density ρ and viscosity μ. Relate the terminal velocity to the other variables. Select (*a*) w, g, and ρ as the repeating variables and (*b*) l, g, and ρ as the repeating variables. Show that the relationship for (*a*) is equivalent to that of (*b*).

6.9   It is proposed that the velocity V issuing from a hole in the side of an open tank depends on the density ρ of the fluid, the distance H from the surface, and gravity g. What expression relates the variables?

6.10   Include the viscosity μ in the list of variables in Prob. 6.9. Find the expression that relates the variables.

6.11   Include the diameter d of the hole and the viscosity μ in the list of variables in Prob. 6.9. Find an expression that relates the variables.

6.12   The pressure drop Δp over a horizontal section of pipe of diameter d depends on the average velocity, the viscosity, the fluid density, the average height of the surface roughness elements, and the length of the pipe section. Write an expression that relates the pressure drop to the other variables.

6.13   Assume a vertical pipe and include gravity in the list of variables in Prob. 6.12 and find an expression for the pressure drop.

6.14   The drag force on a sphere depends on the sphere's diameter and velocity, the fluid's viscosity and density, and gravity. Find an expression for the drag force.

6.15   The drag force on a cylinder is studied in a wind tunnel. If wall effects are negligible, relate the drag force to the wind's speed, density and kinematic viscosity, and the cylinder's diameter and length.

6.16   The distance of the flight of a golf ball is assumed to depend on the initial velocity of the ball, the angle of the ball from the club, the viscosity and density of the air, the number of dimples on the ball and its diameter, and gravity. Write an expression for the flight distance. How would the temperature of the air influence the flight distance?

6.17   The flow rate Q of water in an open channel is assumed to depend on the height h of the water and width w and slope S of the channel, the wall roughness height e, and gravity g. Relate the flow rate to the other variables.

6.18   The lift F_L on an airfoil is related to its velocity V, its length L, its chord length c, its angle of attack α, and the density ρ of the air. Viscous effects are assumed negligible. Relate the lift to the other variables.

6.19   The drag F_D on an airfoil is related to its velocity V, its length L, its chord length c, its angle of attack α, and the density ρ and viscosity μ of the air. Relate the drag to the other variables.

6.20   Find an expression for the torque required to rotate a disk of diameter d, a distance t from a flat plate at a rotational speed Ω, a liquid fills the space between the disk and the plate.

6.21   The power $\dot{W}_P$ required for a pump depends on the impeller rotational speed Ω, the impeller diameter d, the number N of impeller blades, the fluid viscosity and density, and the pressure difference Δp. What expression relates the power to the other variables?

6.22   Write an expression for the torque required to rotate the cylinder surrounded by a fluid as shown in Fig. 6.3. (*a*) Neglect the effects of h. (*b*) Include the effects of h.

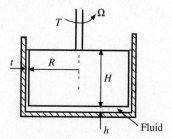

Figure 6.3   Similitude.

6.23   After a model study has been performed, quantities of interest are often predicted for the prototype. Using an average velocity V, a characteristic dimension l, and the fluid density ρ, write the ratio of prototype to model of (*a*) drag force F_D, (*b*) flow rate Q, (*c*) pressure drop Δp, and (*d*) torque T.

6.24   A model of a golf ball is to be studied to determine the effects of the dimples. A sphere 10 times larger than an actual golf ball is used in the wind tunnel study. What speed should be selected for the model to simulate a prototype speed of 50 m/s?

6.25   A proposed pier design is studied in a water channel to simulate forces due to hurricanes. Using a 1:10 scale model, what velocity should be selected in the model study to simulate a water speed of 12 m/s?

6.26   A proposed model study of a low-speed aircraft is to be performed using a 1:10 scale model. If the prototype is to travel at 25 m/s, what speed should be selected for a wind tunnel model? Is such a test advisable? Would it be better to test a 40:1 scale model in a water channel?

6.27   A towing force of 15 N is measured on a 1:40 scale model of a ship in a water channel. What velocity should be used to simulate a prototype speed of 10 m/s? What would be the predicted force on the ship at that speed?

6.28   A 1:20 scale model of an aircraft is studied in a 20°C supersonic wind tunnel at sea level. If a lift of 20 N at a speed of 250 m/s is measured in the wind tunnel, what velocity and lift does that simulate for the prototype? Assume the prototype is at (*a*) sea level, (*b*) 3000 m, and (*c*) 10 000 m.

6.29   The force on a weir is to be predicted by studying the flow of water over a 1:10 scale model. If 1.8 m³/s is expected over the weir, what flow rate should be used in the model study? What force should be expected on the weir if 20 N is measured on the model?

Answers to Supplementary Problems

6.4 $\dfrac{gz}{V^2}$ and $\dfrac{p}{\rho V^2}$

6.5   (*a*) FT/L (*b*) F/L^2 (*c*) FT^2/L^4 (*d*) FT/L^2 (*e*) LF/T

6.6   (*a*) V^2/lg (*b*) $V\rho D/\mu$ (*c*) VD/ν (*d*) $\Omega^2 d/g$ (*e*) $\Omega\rho b^2/\mu$ (*f*) $\dot{W}/\Delta p V^2 d$

6.7   Viscosity

6.8   (*a*) $\dfrac{V^2}{gw}=f\left(\dfrac{w}{l},\dfrac{\rho\sqrt{g}w^{3/2}}{\mu}\right)$ (*b*) $\dfrac{V^2}{gl}=f\left(\dfrac{l}{w},\dfrac{\rho\sqrt{g}l^{3/2}}{\mu}\right)$

6.9 $V=C\sqrt{gH}$

6.10 $\dfrac{V}{\sqrt{gH}}=f\left(\dfrac{\rho\sqrt{gH^3}}{\mu}\right)$

6.11 $\dfrac{V}{\sqrt{gH}}=f\left(\dfrac{H}{d},\dfrac{\rho\sqrt{gH^3}}{\mu}\right)$

6.12 $\dfrac{\Delta p}{\rho V^2}=f\left(\dfrac{e}{d},\dfrac{L}{d},\dfrac{\rho V d}{\mu}\right)$

6.13 $\dfrac{\Delta p}{\rho V^2}=f\left(\dfrac{e}{d},\dfrac{L}{d},\dfrac{\rho V d}{\mu},\dfrac{V^2}{dg}\right)$

6.14 $\dfrac{F_D}{\rho V^2 d^2}=f\left(\dfrac{\rho V d}{\mu},\dfrac{V^2}{dg}\right)$

6.15 $\dfrac{F_D}{\rho V^2 d^2}=f\left(\dfrac{Vd}{\nu},\dfrac{d}{l}\right)$

6.16 $\dfrac{L}{d}=f\left(\alpha,\dfrac{\rho V d}{\mu},N,\dfrac{V^2}{dg}\right)$

6.17 $\dfrac{Q}{\sqrt{gh^5}}=f\left(\dfrac{h}{w},S,\dfrac{h}{e}\right)$

6.18 $\dfrac{F_L}{\rho V^2 c^2}=f\left(\dfrac{c}{L},\alpha\right)$

6.19 $\dfrac{F_D}{\rho V^2 c^2} = f\left(\dfrac{c}{L}, \alpha, \dfrac{\rho V c}{\mu}\right)$

6.20 $\dfrac{T}{\rho \Omega^2 d^5} = f\left(\dfrac{d}{t}, \dfrac{\rho \Omega d^2}{\mu}\right)$

6.21 $\dfrac{\dot{W}_P}{\rho \Omega^3 d^5} = f\left(\dfrac{\rho \Omega d^2}{\mu}, N, \dfrac{\Delta p}{\rho \Omega^2 d^2}\right)$

6.22 $\dfrac{T}{\rho \Omega^2 d^5} = f\left(\dfrac{R}{t}, \dfrac{R}{H}, \dfrac{R}{h}, \dfrac{\rho \Omega d^2}{\mu}\right)$

6.23   (a) $\dfrac{F_{D,p}}{F_{D,m}} = \dfrac{\rho_p l_p^2 V_p^2}{\rho_m l_m^2 V_m^2}$ (b) $\dfrac{Q_p}{Q_m} = \dfrac{V_p l_p^2}{V_m l_m^2}$ (c) $\dfrac{\Delta p_p}{\Delta p_m} = \dfrac{\rho_p V_p^2}{\rho_m V_m^2}$ (d) $\dfrac{T_p}{T_m} = \dfrac{\rho_p V_p^2 l_p^3}{\rho_m V_m^2 l_m^3}$

6.24   5 m/s

6.25   3.79 m/s

6.26   500 m/s, 133 m/s. No model study feasible

6.27   1.58 m/s, 60 kN

6.28   (a)   250 m/s, 8000 N   (b)   258 m/s, 6350 N   (c)   283 m/s, 3460 N

6.29   56.9 m³/s, 20 kN

Chapter 7

Internal Flows

7.1 INTRODUCTION

The material in this chapter is focused on the influence of viscosity on the flows internal to boundaries, such as flow in a pipe or between rotating cylinders. Chapter 8 will focus on flows that are external to a boundary, such as an airfoil. The parameter that is of primary interest in an internal flow is the Reynolds number:

$$\mathrm{Re} = \frac{\rho V L}{\mu} \qquad (7.1)$$

where L is the primary characteristic length (e.g., the diameter of a pipe) in the problem of interest and V is usually the average velocity in a flow.

   If viscous effects dominate the flow (this requires a relatively large wall area), as in a long length of pipe, the Reynolds number is important; if inertial effects dominate, as in a sudden bend or a pipe entrance, then the viscous effects can often be ignored since they do not have a sufficiently large area upon which to act thereby making the Reynolds number less influential.

   We will consider internal flows in pipes, between parallel plates and rotating cylinders, and in open channels in some detail. If the Reynolds number is relatively low, the flow is laminar (see Sec. 3.3.3); if it is relatively high, then the flow is turbulent. For pipe flows, the flow is assumed to be laminar if R < 2000; for flow between wide parallel plates, it is laminar if Re < 1500; for flow between rotating concentric cylinders, it is laminar and flows in a circular motion below Re < 1700; and in the open channels of interest, it is assumed to be turbulent. The characteristic lengths and velocities will be defined later.

7.2 ENTRANCE FLOW

The comments and Reynolds numbers mentioned above refer to *developed flows*, flows in which the velocity profiles do not change in the stream-wise direction. In the region near a geometry change, such as an elbow or a valve or near an entrance, the velocity profile changes in the flow direction. Let us consider the changes in the entrance region for a laminar flow in a pipe or between parallel plates. The *entrance length* L_E is sketched in Fig. 7.1. The velocity profile very near the entrance is essentially uniform, the *viscous wall layer* grows until it permeates the entire cross section over the *inviscid core length* L_i; the profile continues to develop into a developed flow at the end of the *profile development region*.

   For a laminar flow in a pipe with a uniform velocity profile at the entrance,

$$\frac{L_E}{D} = 0.065\mathrm{Re} \qquad \mathrm{Re} = \frac{VD}{v} \qquad (7.2)$$

110

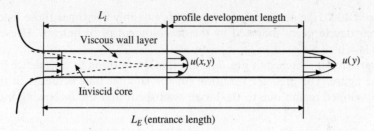

Figure 7.1   The laminar-flow entrance region in a pipe or between parallel plates.

where V is the average velocity and D is the diameter. The inviscid core is about half of the entrance length. It should be mentioned that laminar flows in pipes have been observed at Reynolds numbers as high as 40 000 in extremely controlled flows in smooth pipes in a building free of vibrations; for a conventional pipe with a rough wall, we use 2000 as the limit for a laminar flow.

For flow between wide parallel plates with a uniform profile at the entrance,

$$\frac{L_E}{h} = 0.04\text{Re} \qquad\qquad \text{Re} = \frac{Vh}{v} \qquad\qquad (7.3)$$

where h is the distance between the plates and V is the average velocity. A laminar flow cannot exist for $\text{Re} > 7700$; a value of 1500 is used as the limit for a conventional flow.

The entrance region for a developed turbulent flow is displayed in Fig. 7.2. The velocity profile is developed at the length L_d, but the characteristics of the turbulence in the flow require the additional length. For large Reynolds numbers exceeding 10^5 in a pipe, we use

$$\frac{L_i}{D} \cong 10 \qquad\qquad \frac{L_d}{D} \cong 40 \qquad\qquad \frac{L_E}{D} \cong 120 \qquad\qquad (7.4)$$

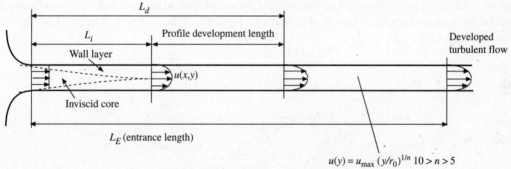

Figure 7.2   The turbulent-flow entrance region in a pipe.

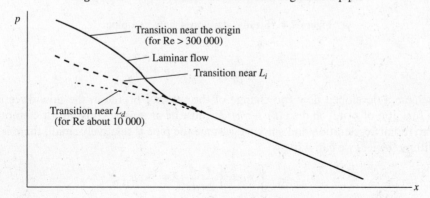

Figure 7.3   Pressure variation in a pipe for both laminar and turbulent flows.

For a flow with Re = 4000, the development lengths are possibly five times those listed in Eq. (7.4) due to the initial laminar development followed by the development of turbulence. (Research has not been reported for flows in which Re $< 10^5$).

The pressure variation is sketched in Fig. 7.3. The initial transition to turbulence from the wall of the pipe is noted in the figure. The pressure variation for the laminar flow is higher in the entrance region than in the fully developed region due to the larger wall shear and the increasing momentum flux.

7.3 LAMINAR FLOW IN A PIPE

Steady, developed laminar flow in a pipe will be derived applying Newton's second law to the element of Fig. 7.4 in Sec. 7.3.1 or using the appropriate Navier–Stokes equation of Chap. 5 in Sec. 7.3.2. Either derivation can be used since we arrive at the same equation using both approaches.

7.3.1 The Elemental Approach

The element of fluid shown in Fig. 7.4 can be considered a control volume into and from which the fluid flows or it can be considered a mass of fluid at a particular moment. Considering it to be an instantaneous mass of fluid that is not accelerating in this steady, developed flow, Newton's second law takes the form

$$\sum F_x = 0 \quad \text{or} \quad p\pi r^2 - (p + dp)\pi r^2 - \tau 2\pi r\, dx + \gamma\pi r^2\, dx \sin\theta = 0 \tag{7.5}$$

where τ is the shear on the wall of the element and γ is the specific weight of the fluid. The above equation simplifies to

$$\tau = -\frac{r}{2}\frac{d}{dx}(p + \gamma h) \tag{7.6}$$

using $dh = -\sin\theta\, dx$ with h measured in the vertical direction. Note that this equation can be applied to either a laminar or a turbulent flow. For a laminar flow, the shear stress τ is related to the velocity gradient[*] by Eq. (1.9):

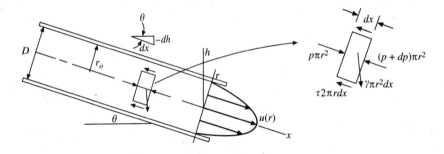

Figure 7.4   Steady, developed flow in a pipe.

$$-\mu\frac{du}{dr} = -\frac{r}{2}\frac{d}{dx}(p + \gamma h) \tag{7.7}$$

Because we assume a developed flow (no change of the velocity profile in the flow direction), the left-hand side is a function of r only and so $d(p + \gamma h)/dx$ must be at most a constant (it cannot depend on r since there is no radial acceleration and since we assume the pipe is relatively small, there is no variation of pressure with r); hence, we can write

$$\int du = \int \frac{r}{2\mu}\frac{d}{dx}(p + \gamma h)dr \tag{7.8}$$

---

[*] The minus sign is required since the stress is a positive quantity and du/dr is negative near the lower wall.

This is integrated to provide the velocity profile

$$u(r) = \frac{r^2}{4\mu} \frac{d}{dx}(p + \gamma h) + C \tag{7.9}$$

where the constant of integration C can be evaluated using $u(r_0) = 0$ so that

$$u(r) = \frac{(r^2 - r_0^2)}{4\mu} \frac{d}{dx}(p + \gamma h) \tag{7.10}$$

For a horizontal pipe for which $dh/dx = 0$, the velocity profile becomes

$$u(r) = \frac{1}{4\mu} \frac{dp}{dx}(r^2 - r_0^2) \tag{7.11}$$

The above velocity profile is a parabolic profile; the flow is sometimes referred to as a *Poiseuille flow*.

The same result can be obtained by solving the appropriate Navier–Stokes equation; if that is not of interest, go directly to Sec. 7.3.3.

7.3.2 Applying the Navier–Stokes Equations

The z-component differential momentum equation using cylindrical coordinates from Table 5.1 is applied to a steady, developed flow in a circular pipe. For the present situation, we wish to refer to the coordinate in the flow direction as x and the velocity component in the x-direction as $u(x)$; so, let us replace the z with x and the v_z with u. Then, the differential equation takes the form

$$\rho\left(\cancel{v_r} \frac{\partial u}{\partial r} + \frac{\cancel{v_\theta}}{r} \frac{\partial u}{\partial \theta} + u \cancel{\frac{\partial u}{\partial x}} + \cancel{\frac{\partial u}{\partial t}}\right) = -\frac{\partial p}{\partial x} + \rho g_x + \mu\left(\frac{\partial^2 u}{\partial r^2} + \frac{1}{r} \frac{\partial u}{\partial r} + \frac{1}{r^2} \cancel{\frac{\partial^2 u}{\partial \theta^2}} + \cancel{\frac{\partial^2 u}{\partial x^2}}\right) \tag{7.12}$$

no radial   no   developed   steady   symmetric   developed

velocity   swirl   flow   flow   flow   flow

Observe that the left-hand side is zero, i.e., the fluid particles are not accelerating. Using $\rho g_x = \gamma \sin\theta = -\gamma dh/dx$, the above equation simplifies to

$$\frac{1}{\mu} \frac{\partial}{\partial x}(p + \gamma h) = \frac{1}{r} \frac{\partial}{\partial r}\left(r \frac{\partial u}{\partial r}\right) \tag{7.13}$$

where the first two terms in the parentheses on the right-hand side of Eq. (7.12) have been combined, i.e.,

$$\frac{\partial^2 u}{\partial r^2} + \frac{1}{r} \frac{\partial u}{\partial r} = \frac{1}{r} \frac{\partial}{\partial r}\left(r \frac{\partial u}{\partial r}\right)$$

Now, we see that the left-hand side of Eq. (7.13) is at most a function of x and the right-hand side is a function of r. This means that each side is at most a constant, say λ, since x and r can be varied independently of each other. So, we replace the partial derivatives with ordinary derivatives and write Eq. (7.13) as

$$\lambda = \frac{1}{r} \frac{d}{dr}\left(r \frac{du}{dr}\right) \quad \text{or} \quad d\left(r \frac{du}{dr}\right) = \lambda r\, dr \tag{7.14}$$

This is integrated to provide

$$r \frac{du}{dr} = \lambda \frac{r^2}{2} + A \tag{7.15}$$

Multiply by dr/r and integrate again. We have

$$u(r) = \lambda \frac{r^2}{4} + A \ln r + B \tag{7.16}$$

Refer to Fig. 7.4: the two boundary conditions are u is finite at $r = 0$ and $u = 0$ at $r = r_0$. Thus, $A = 0$ and $B = -\lambda r_0^2/4$. Since λ is the left-hand side of Eq. (7.13), we can write Eq. (7.16) as

$$u(r) = \frac{1}{4\mu} \frac{d}{dx}(p + \gamma h)(r^2 - r_0^2) \tag{7.17}$$

This is the parabolic velocity distribution of a developed laminar flow in a pipe, sometimes called a *Poiseuille flow*. For a horizontal pipe, $dh/dx = 0$ and

$$u(r) = \frac{1}{4\mu} \frac{dp}{dx}(r^2 - r_0^2) \tag{7.18}$$

7.3.3 Quantities of Interest

The first quantity of interest in the flow in a pipe is the average velocity V. If we express the constant pressure gradient as $dp/dx = -\Delta p/L$, where Δp is the pressure drop (a positive number) over the length of pipe L, there results

$$V = \frac{1}{A} \int u(r) 2\pi r\, dr$$

$$= -\frac{2\pi}{\pi r_0^2} \frac{\Delta p}{4\mu L} \int_0^{r_0} (r^2 - r_0^2) r\, dr = \frac{r_0^2 \Delta p}{8\mu L} \tag{7.19}$$

The maximum velocity occurs at $r = 0$ and is

$$u_{\mathrm{max}} = \frac{r_0^2 \Delta p}{4\mu L} = 2V \tag{7.20}$$

The pressure drop, rewriting Eq. (7.19), is

$$\Delta p = \frac{8\mu L V}{r_0^2} \tag{7.21}$$

The shear stress at the wall can be found by considering a control volume of length L in the pipe. For a horizontal pipe, the pressure force balances the shear force so that the control volume yields

$$\pi r_0^2 \Delta p = 2\pi r_0 L \tau_0 \qquad \therefore \ \tau_0 = \frac{r_0 \Delta p}{2L} \tag{7.22}$$

Sometimes a dimensionless wall shear, called the *friction factor f*, is used. It is defined to be

$$f = \frac{\tau_0}{\frac{1}{8}\rho V^2} \tag{7.23}$$

We also refer to a *head loss h_L* defined as $\Delta p/\gamma$. By combining the above equations, it can be expressed as

$$h_L = \frac{\Delta p}{\gamma} = f \frac{L}{D} \frac{V^2}{2g} \tag{7.24}$$

This is sometimes referred to as the *Darcy–Weisbach equation*; it is valid for both a laminar and a turbulent flow in a pipe. In terms of the Reynolds number, the friction factor for a laminar flow is (combine Eqs. (7.21) and (7.24))

$$f = \frac{64}{\mathrm{Re}} \tag{7.25}$$

where $\mathrm{Re} = VD/\nu$. If this is substituted into Eq. (7.24), we see that the head loss is directly proportional to the average velocity in a laminar flow, a fact that is also applied to a laminar flow in a conduit of any cross section.

EXAMPLE 7.1 The pressure drop over a 30-m length of 1-cm-diameter horizontal pipe transporting water at 20°C is measured to be 2 kPa. A laminar flow is assumed. Determine (*a*) the maximum velocity in the pipe, (*b*) the Reynolds number, (*c*) the wall shear stress, and (*d*) the friction factor.

Solution: (*a*) The maximum velocity is found to be

$$u_{max} = \frac{r_0^2 \, \Delta p}{4\mu L} = \frac{0.005^2 \times 2000}{4 \times 10^{-3} \times 30} = 0.4167 \, \text{m/s}$$

Note: The pressure must be in pascals in order for the units to check. It is wise to make sure the units check when equations are used for the first time. The above units are checked as follows:

$$\frac{\text{m}^2 \times \text{N/m}^2}{(\text{N·s/m}^2) \times \text{m}} = \text{m/s}$$

(*b*) The Reynolds number, a dimensionless quantity, is

$$\text{Re} = \frac{VD}{\nu} = \frac{(0.4167/2)0.01}{10^{-6}} = 4167$$

This exceeds 2000 but a laminar flow can exist at higher Reynolds numbers if a smooth pipe is used and care is taken to provide a flow free of disturbances. But, note how low the velocity is in this relatively small pipe. Laminar flows are rare in most engineering applications unless the fluid is extremely viscous or the dimensions are quite small.

(*c*) The wall shear stress due to the viscous effects is found to be

$$\tau_0 = \frac{r_0 \, \Delta p}{2L} = \frac{0.005 \times 2000}{2 \times 30} = 0.1667 \, \text{Pa}$$

If we had used the pressure in kPa, the stress would have had units of kPa.

(*d*) Finally, the friction factor, a dimensionless quantity, is

$$f = \frac{\tau_0}{\frac{1}{2}\rho V^2} = \frac{0.1667}{0.5 \times 1000 \times [0.4167/2]^2} = 0.0077$$

7.4 LAMINAR FLOW BETWEEN PARALLEL PLATES

Steady, developed laminar flow between parallel plates (one plate is moving with velocity U) will be derived in Sec. 7.4.1 applying Newton's second law to the element of Fig. 7.5 or using the appropriate Navier–Stokes equation of Chap. 5 in Sec. 7.4.2. Either derivation can be used since we arrive at the same equation using both approaches.

7.4.1 The Elemental Approach

The element of fluid shown in Fig. 7.5 can be considered a control volume into and from which the fluid flows or it can be considered a mass of fluid at a particular moment. Considering it to be an instantaneous mass of fluid that is not accelerating in this steady, developed flow, Newton's second law takes the form

$$\sum F_x = 0 \quad \text{or} \quad p \, dy - (p + dp)dy + \tau \, dx - (\tau + d\tau)dx + \gamma \, dx \, dy \sin \theta = 0 \qquad (7.26)$$

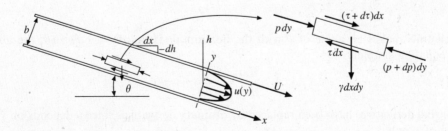

Figure 7.5 Steady, developed flow between parallel plates.

where τ is the shear on the wall of the element and γ is the specific weight of the fluid. We have assumed a unit length into the paper (in the z-direction). To simplify, divide by $dx\,dy$ and use $dh = -\sin\theta\,dx$ with h measured in the vertical direction:

$$\frac{d\tau}{dy} = \frac{d}{dx}(p + \gamma h) \tag{7.27}$$

For this laminar flow, the shear stress is related to the velocity gradient by $\tau = \mu\,du/dy$ so that Eq. (7.27) becomes

$$\mu\frac{d^2 u}{dy^2} = \frac{d}{dx}(p + \gamma h) \tag{7.28}$$

The left-hand side is a function of y only for this developed flow (we assume a wide channel with an aspect ratio in excess of 8) and the right-hand side is a function of x only. So, we can integrate twice on y to obtain

$$u(y) = \frac{1}{2\mu}\frac{d(p + \gamma h)}{dx}y^2 + Ay + B \tag{7.29}$$

Using the boundary conditions $u(0) = 0$ and $u(b) = U$, the constants of integration are evaluated and a parabolic profile results:

$$u(y) = \frac{1}{2\mu}\frac{d(p + \gamma h)}{dx}(y^2 - by) + \frac{U}{b}y \tag{7.30}$$

If the plates are horizontal and $U = 0$, the velocity profile simplifies to

$$u(y) = \frac{\Delta p}{2\mu L}(by - y^2) \tag{7.31}$$

where we have let $d(p + \gamma h)/dx = -\Delta p/L$ for the horizontal plates where Δp is the pressure drop, a positive quantity.

If the flow is due only to the top plate moving, with zero pressure gradient, it is a *Couette flow* so that $u(y) = Uy/b$. If both plates are stationary and the flow is due only to a pressure gradient, it is a Poiseuille flow.

The same result can be obtained by solving the appropriate Navier–Stokes equation; if that is not of interest, go directly to Sec. 7.4.3.

7.4.2 Applying the Navier–Stokes Equations

The x-component differential momentum equation in rectangular coordinates (see Eq. (5.18)) is selected for this steady, developed flow with streamlines parallel to the walls in a wide channel (at least an 8:1 aspect ratio):

$$\rho\left(\underbrace{\cancel{\frac{\partial u}{\partial t}}}_{\text{steady}} + \underbrace{u\cancel{\frac{\partial u}{\partial x}}}_{\text{developed}} + \underbrace{v\frac{\partial u}{\partial y} + \cancel{w\frac{\partial u}{\partial z}}}_{\substack{\text{streamlines}\\\text{parallel to wall}}}\right) = -\frac{\partial p}{\partial x} + \gamma\sin\theta + \mu\left(\underbrace{\cancel{\frac{\partial^2 u}{\partial x^2}}}_{\text{developed}} + \frac{\partial^2 u}{\partial y^2} + \underbrace{\cancel{\frac{\partial^2 u}{\partial z^2}}}_{\text{wide channel}}\right) \tag{7.32}$$

where the channel makes an angle of θ with the horizontal. Using $dh = -dx\sin\theta$, the above partial differential equation simplifies to

$$\frac{d^2 u}{dy^2} = \frac{1}{\mu}\frac{d}{dx}(p + \gamma h) \tag{7.33}$$

where the partial derivatives have been replaced by ordinary derivatives since u depends on y only and p is a function of x only.

Because the left-hand side is a function of y and the right-hand side is a function of x, both of which can be varied independent of each other, the two sides can be at most a constant, say λ, so that

$$\frac{d^2u}{dy^2} = \lambda \tag{7.34}$$

Integrating twice provides

$$u(y) = \frac{1}{2}\lambda y^2 + Ay + B \tag{7.35}$$

Refer to Fig. 7.5: the boundary conditions are $u(0) = 0$ and $u(b) = U$ provided

$$A = \frac{U}{b} - \lambda\frac{b}{2} \qquad B = 0 \tag{7.36}$$

The velocity profile is thus

$$u(y) = \frac{d(p + \gamma h)/dx}{2\mu}(y^2 - by) + \frac{U}{b}y \tag{7.37}$$

where λ has been used as the right-hand side of Eq. (7.33).

In a horizontal channel, we can write $d(p + \gamma h)/dx = -\Delta p/L$. If $U = 0$, the velocity profile is

$$u(y) = \frac{\Delta p}{2\mu L}(by - y^2) \tag{7.38}$$

This is the Poiseuille flow. If the pressure gradient is zero and the motion of the top plate causes the flow, it is a *Couette flow* with $u(y) = Uy/b$.

7.4.3 Quantities of Interest

Let us consider several quantities of interest for the case of two fixed plates with $U = 0$. The first quantity of interest in the flow is the average velocity V. The average velocity is, assuming unit width of the plates,

$$\begin{aligned}
V &= \frac{1}{b \times 1}\int u(y)dy \\
&= \frac{\Delta p}{2b\mu L}\int_0^b (by - y^2)dr = \frac{\Delta p}{2b\mu L}\left[b\frac{b^2}{2} - \frac{b^3}{3}\right] = \frac{b^2 \Delta p}{12\mu L}
\end{aligned} \tag{7.39}$$

The maximum velocity occurs at $y = b/2$ and is

$$u_{\max} = \frac{\Delta p}{2\mu L}\left(\frac{b^2}{2} - \frac{b^2}{4}\right) = \frac{b^2 \Delta p}{8\mu L} = \frac{2}{3}V \tag{7.40}$$

The pressure drop, rewriting Eq. (7.39), is for this horizontal* channel,

$$\Delta p = \frac{12\mu LV}{b^2} \tag{7.41}$$

The shear stress at either wall can be found by considering a free body of length L in the channel. For a horizontal channel, the pressure force balances the shear force:

$$(b \times 1)\Delta p = 2(L \times 1)\tau_0 \qquad \therefore \tau_0 = \frac{b\,\Delta p}{2L} \tag{7.42}$$

In terms of the friction factor f, defined by

$$f = \frac{\tau_0}{\frac{1}{8}\rho V^2} \tag{7.43}$$

---

* For a sloped channel simply replace p with $(p + \gamma h)$.

the head loss for the horizontal channel is

$$h_L = \frac{\Delta p}{\gamma} = f \frac{L}{2b} \frac{V^2}{2g} \qquad (7.44)$$

Several of the above equations can be combined to find

$$f = \frac{48}{\text{Re}} \qquad (7.45)$$

where $\text{Re} = bV/\nu$. If this is substituted into Eq. (7.44), we see that the head loss is directly proportional to the average velocity in a laminar flow.

The above equations were derived for a channel with aspect ratio > 8. For lower aspect-ratio channels, the sides would require additional terms since the shear acting on the side walls would influence the central part of the flow.

If interest is in a horizontal channel flow where the top plate is moving and there is no pressure gradient, then the velocity profile would be the linear profile

$$u(y) = \frac{U}{b} y \qquad (7.46)$$

EXAMPLE 7.2 The thin layer of rain at $20°\text{C}$ flows down a parking lot at a relatively constant depth of 4 mm. The area is 40 m wide with a slope of 8 cm over 60 m of length. Estimate (a) the flow rate, (b) shear at the surface, (c) the Reynolds number, and the velocity at the surface.

Solution: (a) The velocity profile can be assumed to be one-half of the profile shown in Fig. 7.5, assuming a laminar flow. The average velocity would remain as given by Eq. (7.39), i.e.,

$$V = \frac{b^2 \gamma h}{12 \mu L}$$

where Δp has been replaced with γh. The flow rate is

$$Q = AV = bw \frac{b^2 \gamma h}{12 \mu L} = 0.004 \times 40 \frac{0.004^2 \times 9810 \times 0.08}{12 \times 10^{-3} \times 60} = 2.80 \times 10^{-3} \, \text{m}^3/\text{s}$$

(b) The shear stress acts only at the solid wall, so Eq. (7.42) would provide

$$\tau_0 = \frac{b \gamma h}{L} = \frac{0.004 \times 9810 \times 0.08}{60} = 0.0523 \, \text{Pa}$$

(c) The Reynolds number is

$$\text{Re} = \frac{bV}{\nu} = \frac{0.004}{10^{-6}} \times \frac{0.004^2 \times 9810 \times 0.08}{12 \times 10^{-3} \times 60} = 697$$

The Reynolds number is below 1500, so the assumption of laminar flow is acceptable.

7.5 LAMINAR FLOW BETWEEN ROTATING CYLINDERS

Steady flow between concentric cylinders, as sketched in Fig. 7.6, is another relatively simple example of a laminar flow that we can solve analytically. Such a flow exists below a Reynolds number[*] of 1700. Above 1700, the flow might be a different laminar flow or a turbulent flow. This flow has application in lubrication in which the outer shaft is stationary. We will again solve this problem using a fluid element in Sec. 7.5.1 and using the appropriate Navier–Stokes equation in Sec. 7.5.2; either method may be used.

7.5.1 The Elemental Approach

The two rotating concentric cylinders are displayed in Fig. 7.6. We will assume vertical cylinders, so body forces will act normal to the circular flow in the θ-direction with the only nonzero velocity component v_θ.

---

[*] The Reynolds number is defined as $\text{Re} = \omega_1 r_1 \delta/\nu$, where $\delta = r_2 - r_1$.

The element of fluid selected, shown in Fig. 7.6, has no angular acceleration in this steady-flow condition. Consequently, the summation of torques acting on the element is zero:

$$\tau \times 2\pi r L \times r - (\tau + d\tau) \times 2\pi(r + dr)L \times (r + dr) = 0 \qquad (7.47)$$

where $\tau(r)$ is the shear stress and L is the length of the cylinders, which must be large when compared with the gap width $\delta = r_2 - r_1$. Equation (7.47) simplifies to

$$\tau 2r\,dr + r^2\,d\tau + 2r\,d\tau\,dr + d\tau(dr)^2 = 0 \qquad (7.48)$$

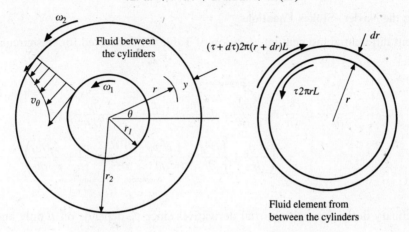

Figure 7.6   Flow between concentric cylinders.

The last two terms of Eq. (7.47) are higher-order terms that are negligible when compared with the first two terms, so that the simplified equation is

$$r\frac{d\tau}{dr} + 2\tau = 0 \qquad (7.49)$$

Now we must recognize that the τ of Eq. (7.49) is[*] $-\tau_{r\theta}$ of Table 5.1 with entry under "Stresses." For this simplified application, the shear stress is related to the velocity gradient by

$$\tau_{r\theta} = \mu r \frac{\partial(v_\theta/r)}{\partial r} \qquad (7.50)$$

This allows Eq. (7.49) to be written, writing the partial derivatives as ordinary derivatives since v_θ depends on r only, as

$$r\mu \frac{d}{dr} r \frac{d(v_\theta/r)}{dr} + 2\mu r \frac{d(v_\theta/r)}{dr} = 0 \qquad (7.51)$$

Multiply by dr, divide by μr, and integrate:

$$r\frac{d(v_\theta/r)}{dr} + 2\frac{v_\theta}{r} = A \qquad (7.52)$$

or, since $r\,d(v_\theta/r)/dr = dv_\theta/dr - v_\theta/r$, this can be written as

$$\frac{dv_\theta}{dr} + \frac{v_\theta}{r} = A \qquad \text{or} \qquad \frac{1}{r}\frac{d(rv_\theta)}{dr} = A \qquad (7.53)$$

Now integrate again and obtain

$$v_\theta(r) = \frac{A}{2}r + \frac{B}{r} \qquad (7.54)$$

---

[*] The minus sign results from the shear stress in Fig. 7.6 being on a negative face in the positive direction, the sign convention for a stress component.

Using the boundary conditions $v_\theta = r_1\omega_1$ at $r = r_1$ and $v_\theta = r_2\omega_2$ at $r = r_2$, the constants are found to be

$$A = 2\frac{\omega_2 r_2^2 - \omega_1 r_1^2}{r_2^2 - r_1^2} \qquad B = \frac{r_1^2 r_2^2(\omega_1 - \omega_2)}{r_2^2 - r_1^2} \qquad (7.55)$$

The same result can be obtained by solving the appropriate Navier–Stokes equation; if that is not of interest, go directly to Sec. 7.5.3.

7.5.2 Applying the Navier–Stokes Equations

The θ-component differential momentum equation of Table 5.1 is selected for this circular motion with $v_r = 0$ and $v_z = 0$:

$$\overset{\text{steady}}{\cancel{\frac{\partial v_\theta}{\partial t}}} + \cancel{v_r}\frac{\partial v_\theta}{\partial r} + \overset{\text{symmetric}}{\frac{v_\theta}{r}\cancel{\frac{\partial v_\theta}{\partial \theta}}} + \cancel{v_z}\cancel{\frac{\partial v_\theta}{\partial z}} + \cancel{\frac{v_\theta v_r}{r}} = -\frac{1}{\rho r}\cancel{\frac{\partial p}{\partial \theta}} + g_\theta$$

$$+ v\left(\frac{\partial^2 v_\theta}{\partial r^2} + \frac{1}{r}\frac{\partial v_\theta}{\partial r} + \frac{1}{r^2}\underset{\text{symmetric}}{\cancel{\frac{\partial^2 v_\theta}{\partial \theta^2}}} + \underset{\substack{\text{away from}\\\text{end walls}}}{\cancel{\frac{\partial^2 v_\theta}{\partial z^2}}} - \frac{v_\theta}{r^2} + \frac{2}{r^2}\cancel{\frac{\partial v_r}{\partial \theta}}\right) \qquad (7.56)$$

Replace the ordinary derivatives with partial derivatives since v_θ depends on θ only and the equation becomes

$$0 = \frac{d^2 v_\theta}{dr^2} + \frac{1}{r}\frac{dv_\theta}{dr} - \frac{v_\theta}{r^2} \qquad (7.57)$$

which can be written in the form

$$\frac{d}{dr}\frac{dv_\theta}{dr} = -\frac{d(v_\theta/r)}{dr} \qquad (7.58)$$

Multiply by dr and integrate:

$$\frac{dv_\theta}{dr} = -\frac{v_\theta}{r} + A \quad \text{or} \quad \frac{1}{r}\frac{d(rv_\theta)}{dr} = A \qquad (7.59)$$

Integrate once again:

$$v_\theta(r) = \frac{A}{2}r + \frac{B}{r} \qquad (7.60)$$

The boundary conditions $v_\theta(r_1) = r\omega_1$ and $v_\theta(r_2) = r\omega_2$ allow

$$A = 2\frac{\omega_2 r_2^2 - \omega_1 r_1^2}{r_2^2 - r_1^2} \qquad B = \frac{r_1^2 r_2^2(\omega_1 - \omega_2)}{r_2^2 - r_1^2} \qquad (7.61)$$

7.5.3 Quantities of Interest

Many applications of rotating cylinders involve the outer cylinder being fixed, that is, $\omega_2 = 0$. The velocity distribution, found in the preceding two sections, with A and B simplified, becomes

$$v_\theta(r) = \frac{\omega_1 r_1^2}{r_2^2 - r_1^2}\left(\frac{r_2^2}{r} - r\right) \qquad (7.62)$$

The shear stress τ_1 ($-\tau_{r\theta}$ from Table 5.1) acts on the inner cylinder. It is

$$\tau_1 = -\left[\mu r\frac{d(v_\theta/r)}{dr}\right]_{r=r_1} = \frac{2\mu r_2^2 \omega_1}{r_2^2 - r_1^2} \qquad (7.63)$$

The torque T needed to rotate the inner cylinder is

$$T = \tau_1 A r_1 = \frac{2\mu r_2^2 \omega_1}{r_2^2 - r_1^2} 2\pi r_1 L \times r_1 = \frac{4\pi\mu r_1^2 r_2^2 L \omega_1}{r_2^2 - r_1^2} \qquad (7.64)$$

The power $\dot{W}$ required to rotate the inner cylinder with rotational speed ω_1 is then

$$\dot{W} = T\omega_1 = \frac{4\pi\mu r_1^2 r_2^2 L \omega_1^2}{r_2^2 - r_1^2} \qquad (7.65)$$

This power, required because of the viscous effects in between the two cylinders, heats up the fluid in bearings and often demands cooling to control the temperature.

For a small gap δ between the cylinders, as occurs in lubrication problems, it is acceptable to approximate the velocity distribution as a linear profile, a *Couette flow*. Using the variable y of Fig. 7.6 the velocity distribution is

$$v_\theta(r) = \frac{r_1 \omega_1}{\delta} y \qquad (7.66)$$

where y is measured from the outer cylinder in towards the center.

EXAMPLE 7.3   The viscosity is to be determined by rotating a long 6-cm-diameter, 30-cm-long cylinder inside a 6.2-cm-diameter cylinder. The torque is measured to be 0.22 N·m and the rotational speed is measured to be 3000 rpm. Use Eqs. (7.62) and (7.66) to estimate the viscosity. Assume that $S = 0.86$.

Solution: The torque is found from Eq. (7.64) based on the velocity distribution of Eq. (7.62):

$$T = \frac{4\pi\mu r_1^2 r_2^2 L \omega_1}{r_2^2 - r_1^2} = \frac{4\pi\mu \times 0.03^2 \times 0.031^2 \times 0.3 \times (3000 \times 2\pi/60)}{0.031^2 - 0.03^2} = 0.22$$

$$\therefore \mu = 0.0131 \, (\text{N·s/m}^2)$$

Using Eq. (7.66), the torque is found to be

$$T = \tau_1 A r_1 = \mu \frac{r_1 \omega_1}{\delta} 2\pi r_1 L \times r_1$$

$$0.22 = \mu \frac{0.03(3000 \times 2\pi/60)}{0.031 - 0.03} 2\pi \times 0.03^2 \times 0.3 \quad \therefore \mu = 0.0138 \, (\text{N·s/m}^2)$$

The error assuming the linear profile is 5.3 percent.
The Reynolds number is, using $\nu = \mu/\rho$,

$$\text{Re} = \frac{\omega_1 r_1 \delta}{\nu} = \frac{(3000 \times 2\pi/60) \times 0.03 \times 0.001}{0.0131/(1000 \times 0.86)} = 619$$

The laminar flow assumption is acceptable since $\text{Re} < 1700$.

7.6   TURBULENT FLOW IN A PIPE

The Reynolds numbers for most flows of interest in conduits exceed those at which laminar flows cease to exist. If a flow starts from rest, it rather quickly undergoes transition to a turbulent flow. The objective of this section is to express the velocity distribution in a turbulent flow in a pipe and to determine quantities associated with such a flow.

A *turbulent flow* is a flow in which all three velocity components are nonzero and exhibit random behavior. In addition, there must be a correlation between the randomness of at least two of the velocity components; if there is no correlation, it is simply a fluctuating flow. For example, a turbulent boundary layer usually exists near the surface of an airfoil but the flow outside the boundary layer is not referred to as "turbulent" even though there are fluctuations in the flow; it is the free stream.

Let us present one way of describing a turbulent flow. The three velocity components at some point are written as

$$u = \bar{u} + u' \qquad v = \bar{v} + v' \qquad w = \bar{w} + w' \qquad (7.67)$$

where $\bar{u}$ denotes a time-average part of the x-component velocity and u' denotes the fluctuating random part. The *time average* of u is

$$\bar{u} = \frac{1}{T} \int_0^T u(t)\,dt \qquad (7.68)$$

where T is sufficiently large when compared with the fluctuation time. For a developed turbulent pipe flow, the three velocity components would appear as in Fig. 7.7. The only time-average component would be $\bar{u}$ in the flow direction. Yet there must exist a correlation between at least two of the random velocity fluctuations, e.g., $\overline{u'v'} \neq 0$; such velocity correlations result in turbulent shear.

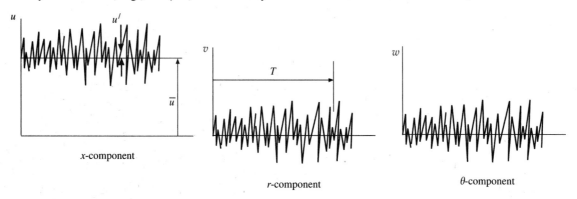

Figure 7.7   The three velocity components in a turbulent flow at a point where the flow is in the x-direction so that $\bar{v} = \bar{w} = 0$ and $\bar{u} \neq 0$.

We can derive an equation that relates $\overline{u'v'}$ and the time-average velocity component $\bar{u}$ in the flow direction of a turbulent flow, but we cannot solve the equation even for the simplest case of steady[*] flow in a pipe. So, we will present experimental data for the velocity profile and define some quantities of interest for a turbulent flow in a pipe.

First, let us describe what we mean by a "smooth" wall. Sketched in Fig. 7.8 is a "smooth" wall and a "rough" wall. The *viscous wall layer* is a thin layer near the pipe wall in which the viscous effects are significant. If this viscous layer covers the wall roughness elements, the wall is "smooth," as in Fig. 7.8(*a*); if the roughness elements protrude out from the viscous layer, the wall is "rough," as in Fig. 7.8(*b*).

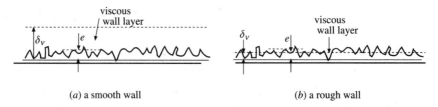

(*a*) a smooth wall   (*b*) a rough wall

Figure 7.8   A smooth wall and a rough wall.

There are two methods commonly used to describe the turbulent velocity profile in a pipe. These are presented in the following sections.

---

[*] Steady turbulent flow means the time-average quantities are independent of time.

7.6.1 The Semi-Log Profile

The time-average velocity profile in a pipe is presented for a smooth pipe as a semi-log plot in Fig. 7.9 with empirical relationships near the wall and centerline that allow $\bar{u}(0) = 0$ and $d\bar{u}/dy = 0$ at $y = r_0$. In the wall region, the characteristic velocity is the *shear velocity*[*] $u_\tau = \sqrt{\tau_0/\rho}$ and the characteristic length is the *viscous length* v/u_τ; the profiles are

$$\frac{\bar{u}}{u_\tau} = \frac{u_\tau y}{v} \qquad 0 \le \frac{u_\tau y}{v} \le 5 \qquad \text{(the viscous wall layer)} \tag{7.69}$$

$$\frac{\bar{u}}{u_\tau} = 2.44 \ln \frac{u_\tau y}{v} + 4.9 \qquad 30 < \frac{u_\tau y}{v}, \frac{y}{r_0} < 0.15 \qquad \text{(the turbulent region)} \tag{7.70}$$

The interval $5 < u_\tau y/v < 30$ is a *buffer zone* in which the experimental data do not fit either of the curves. The outer edge of the wall region may be as low as $u_\tau y/v = 3000$ for a low-Reynolds-number flow.

The viscous wall layer plays no role for a rough pipe. The characteristic length is the average roughness height e and the wall region is represented by

$$\frac{\bar{u}}{u_\tau} = 2.44 \ln \frac{y}{e} + 8.5 \qquad \frac{y}{r_0} < 0.15 \qquad \text{(the wall region, rough pipe)} \tag{7.71}$$

The outer region is independent of the wall effects and thus is normalized for both smooth and rough walls using the radius as the characteristic length and is given by

$$\frac{u_{\max} - \bar{u}}{u_\tau} = -2.44 \ln \frac{y}{r_0} + 0.8 \qquad \frac{y}{r_0} \le 0.15 \qquad \text{(the outer region)} \tag{7.72}$$

An additional empirical relationship $h(y/r_0)$ is needed to complete the profile for $y > 0.15 r_0$. Most relationships that satisfy $d\bar{u}/dy = 0$ at $y = r_0$ will do.

The wall region of Fig. 7.9(a) and the outer region of Fig. 7.9(b) overlap as displayed in Fig. 7.9(a). For smooth and rough pipes respectively

$$\frac{u_{\max}}{u_\tau} = 2.44 \ln \frac{u_\tau r_0}{v} + 5.7 \qquad \text{(smooth pipes)} \tag{7.73}$$

$$\frac{u_{\max}}{u_\tau} = 2.44 \ln \frac{r_0}{e} + 9.3 \qquad \text{(rough pipes)} \tag{7.74}$$

We do not often desire the velocity at a particular location, but if we do, before $u_{\max}$ can be found u_τ must be known. To find u_τ we must know τ_0. To find τ_0 we can use (see Eq. (7.6))

$$\tau_0 = \frac{r_0 \Delta p}{2L} \quad \text{or} \quad \tau_0 = \frac{1}{8} \rho V^2 f \tag{7.75}$$

The friction factor f can be estimated using the power-law profile that follows if the pressure drop is not known.

7.6.2 The Power-Law Profile

Another approach, although not quite as accurate, involves using the *power-law profile* given by

$$\frac{\bar{u}}{u_{\max}} = \left(\frac{y}{r_0}\right)^{1/n} \tag{7.76}$$

---

[*] The shear velocity is a fictitious velocity that allows experimental data to be presented in dimensionless form that is valid for all turbulent pipe flows. The viscous length is also a fictitious length.

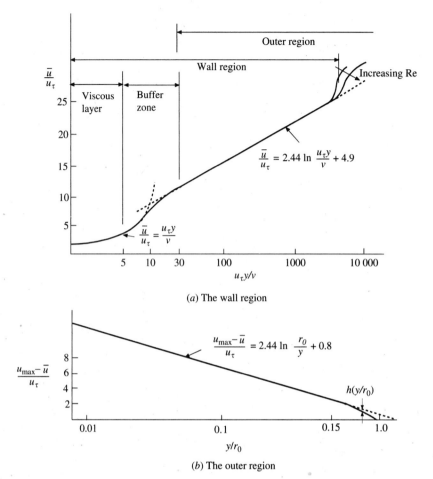

(a) The wall region

(b) The outer region

Figure 7.9   Experimental data for a smooth wall in a developed pipe flow.

where n is between 5 and 10, usually an integer. This can be integrated to yield the average velocity

$$V = \frac{1}{\pi r_0^2} \int_0^{r_0} \bar{u}(r) 2\pi r \, dr = \frac{2n^2}{(n+1)(2n+1)} u_{\max} \qquad (7.77)$$

The value of n in Eq. (7.76) is related empirically to f by

$$n = f^{-1/2} \qquad (7.78)$$

For smooth pipes, n is related to the Reynolds number as shown in Table 7.1.

Table 7.1   Exponent n for Smooth Pipes

$Re = VD/v$	4×10^3	10^5	10^6	$> 2 \times 10^6$
n	6	7	9	10

The power-law profile cannot be used to estimate the wall shear since it has an infinite slope at the wall for all values of n. It also does not have a zero slope at the pipe centerline, so it is not valid near the centerline. It is used to estimate the energy flux and momentum flux of pipe flows.

Finally, it should be noted that the kinetic-energy correction factor is 1.03 for $n = 7$; hence, it is often taken as unity for turbulent flows.

EXAMPLE 7.4   Water at $20°C$ flows in a 4-cm-diameter pipe with a flow rate of $0.002\ \mathrm{m^3/s}$. Estimate (a) the wall shear stress, (b) the maximum velocity, (c) the pressure drop over 20 m, (d) the viscous layer thickness, and (e) determine if the wall is smooth or rough assuming the roughness elements to have a height of 0.0015 mm. Use the power-law profile.

Solution: First, the average velocity and Reynolds number are

$$V = \frac{Q}{A} = \frac{0.002}{\pi 0.02^2} = 1.464\ \mathrm{m/s}, \qquad \mathrm{Re} = \frac{VD}{\nu} = \frac{1.464 \times 0.04}{10^{-6}} = 5.85 \times 10^4$$

(a) To find the wall shear stress, first let us find the friction factor. From Table 7.1, the value $n = 6.8$ is selected and from Eq. (7.78)

$$f = \frac{1}{n^2} = \frac{1}{6.8^2} = 0.0216$$

The wall shear stress is, see Eq. (7.75),

$$\tau_0 = \frac{1}{2}\rho V^2 f = \frac{1}{2} \times 1000 \times 1.464^2 \times 0.0216 = 23.2\ \mathrm{Pa}$$

(b) The maximum velocity is found using Eq. (7.77):

$$u_{max} = \frac{(n+1)(2n+1)}{2n^2} V = \frac{7.8 \times 14.6}{2 \times 6.8^2} 1.464 = 1.80\ \mathrm{m/s}$$

(c) The pressure drop is

$$\Delta p = \frac{2L\tau_0}{r_0} = \frac{2 \times 20 \times 23.2}{0.02} = 46\,400\ \mathrm{Pa} \quad \text{or} \quad 46.4\ \mathrm{kPa}$$

(d) The friction velocity is

$$u_\tau = \sqrt{\frac{\tau_0}{\rho}} = \sqrt{\frac{23.2}{1000}} = 0.152\ \mathrm{m/s}$$

and the viscous layer thickness is

$$\delta_v = \frac{5\nu}{u_\tau} = \frac{5 \times 10^{-6}}{0.152} = 3.29 \times 10^{-5}\ \mathrm{m} \quad \text{or} \quad 0.0329\ \mathrm{mm}$$

(e) The height of the roughness elements is given as 0.0015 mm (drawn tubing), which is less than the viscous layer thickness. Hence, the wall is smooth. Note: If the height of the wall elements was 0.046 mm (wrought iron), the wall would be rough.

7.6.3   Losses in Pipe Flow

The head loss is of considerable interest in pipe flows. It was presented in Eqs. (7.24) and (4.23) and is

$$h_L = f\frac{L}{D}\frac{V^2}{2g} \quad \text{or} \quad h_L = \frac{\Delta p}{\gamma} + z_2 - z_1 \tag{7.79}$$

So, once the friction factor is known, the head loss and pressure drop can be determined. The friction factor depends on a number of properties of the fluid and the pipe:

$$f = f(\rho, \mu, V, D, e) \tag{7.80}$$

where the roughness height e accounts for the turbulence generated by the roughness elements. A dimensional analysis allows Eq. (7.80) to be written as

$$f = f\left(\frac{e}{D}, \frac{VD\rho}{\mu}\right) \tag{7.81}$$

where e/D is termed the *relative roughness*.

Experimental data has been collected and presented in the form of the *Moody diagram*, displayed in Fig. 7.10 for developed flow in a conventional pipe. The roughness heights are also included in the diagram. There are several features of this diagram that should be emphasized. They follow:

- A laminar flow exists up to $Re \cong 2000$ after which there is a *critical zone* in which the flow is undergoing transition to a turbulent flow. This may involve transitory flow that alternates between laminar and turbulent flows.
- The friction factor in the *transition zone*, which begins at about Re = 4000 and decreases with increasing Reynolds numbers, becomes constant at the end of the zone as signified by the dashed line in Fig. 7.10.
- The friction factor in the *completely turbulent zone* is constant and depends on the relative roughness e/D. Viscous effects, and thus the Reynolds number, do not affect the friction factor.
- The height e of the roughness elements in the Moody diagram is for new pipes. Pipes become fouled with age changing both e and the diameter D resulting in an increased friction factor. Designs of piping systems should include such aging effects.

An alternate to using the Moody diagram is to use formulas developed by Swamee and Jain for pipe flow; the particular formula selected depends on the information given. The formulas to determine quantities in long reaches of developed pipe flow (these formulas are not used in short lengths or in pipes with numerous fittings and geometry changes) are as follows:

$$h_L = 1.07 \frac{Q^2 L}{gD^5} \left\{ \ln \left[\frac{e}{3.7D} + 4.62 \left(\frac{vD}{Q} \right)^{0.9} \right] \right\}^{-2} \qquad \begin{array}{l} 10^{-6} < \dfrac{e}{D} < 10^{-2} \\[4pt] 3000 < \mathrm{Re} < 3 \times 10^8 \end{array} \qquad (7.82)$$

$$Q = -0.965 \sqrt{\frac{gD^5 h_L}{L}} \ln \left[\frac{e}{3.7D} + \left(\frac{3.17 v^2 L}{gD^3 h_L} \right)^{0.5} \right] \qquad 2000 < \mathrm{Re} \qquad (7.83)$$

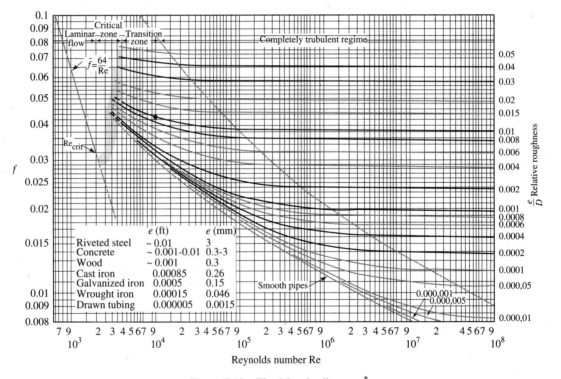

Figure 7.10 The Moody diagram.[*]

---

[*] Note: If $e/D = 0.01$ and $\mathrm{Re} = 10^4$, the dot locates $f = 0.043$.
Source: From L.F. Moody, *Trans. ASME,* v. 66, 1944.

$$D = 0.66\left[e^{1.25}\left(\frac{LQ^2}{gh_L}\right)^{4.75} + vQ^{9.4}\left(\frac{L}{gh_L}\right)^{5.2}\right]^{0.04} \qquad \begin{array}{l} 10^{-6} < \frac{e}{D} < 10^{-2} \\ 5000 < \text{Re} < 3 \times 10^8 \end{array} \qquad (7.84)$$

Either SI or English units can be used in the above equations. Note also that the Moody diagram and the above equations are accurate to within about 5 percent, sufficiently accurate for most engineering applications.

EXAMPLE 7.5 A pressure drop of 500 kPa is measured over 200 m of a horizontal length of 8-cm-diameter cast iron pipe transporting water at $20°C$. Estimate the flow rate using (a) the Moody diagram and (b) an alternate equation.

Solution: (a) The relative roughness (find e in Fig. 7.10) is

$$\frac{e}{D} = \frac{0.26}{80} = 0.00325$$

Assuming a completely turbulent flow, the friction factor from Fig. 7.10 is $f = 0.026$. The head loss is

$$h_L = \frac{\Delta p}{\gamma} = \frac{500\,000}{9800} = 51\text{ m}$$

The average velocity, from Eq. (7.79), is

$$V = \sqrt{\frac{2gDh_L}{fL}} = \sqrt{\frac{2 \times 9.8 \times 0.08 \times 51}{0.026 \times 200}} = 3.92\text{ m/s}$$

We must check the Reynolds number to make sure the flow is completely turbulent, and it is

$$\text{Re} = \frac{VD}{v} = \frac{3.92 \times 0.08}{10^{-6}} = 3.14 \times 10^5$$

This is just acceptable and requires no iteration to improve the friction factor. So, the flow rate is

$$Q = AV = \pi \times 0.04^2 \times 3.92 = 0.0197\text{ m}^3/\text{s}$$

(b) Use the alternate equation that relates Q to the other quantities, i.e., Eq. (7.83). We use the head loss from part (a):

$$Q = -0.965\sqrt{\frac{9.8 \times 0.08^5 \times 51}{200}}\ln\left[\frac{0.26}{3.7 \times 80} + \left(\frac{3.17 \times 10^{-12} \times 200}{9.8 \times 0.08^3 \times 51}\right)^{0.5}\right] = 0.0193\text{ m}^3/\text{s}$$

This equation was easier to use and gave an acceptable result.

7.6.4 Losses in Noncircular Conduits

To determine the head loss in a relatively "open" noncircular conduit, we use the *hydraulic radius R*, defined as

$$R = \frac{A}{P} \qquad (7.85)$$

where A is the cross-sectional area and P is the *wetted perimeter*, the perimeter of the conduit that is in contact with the fluid. The Reynolds number, relative roughness, and head loss are respectively

$$\text{Re} = \frac{4VR}{v} \qquad \frac{\text{relative}}{\text{roughness}} = \frac{e}{4R} \qquad h_L = f\frac{L}{4R}\frac{V^2}{2g} \qquad (7.86)$$

A rectangular area should have an aspect ratio < 4. This method should not be used with shapes like an annulus.

7.6.5 Minor Losses

The preceding losses were for the developed flow in long conduits. Most piping systems, however, include sudden changes such as elbows, valves, inlets, etc., that add additional losses to the system.

Table 7.2 Minor Loss Coefficients K for Selected Devices[*]

Type of fitting	Screwed			Flanged		
Diameter	2.5 cm	5 cm	10 cm	5 cm	10 cm	20 cm
Globe value (fully open)	8.2	6.9	5.7	8.5	6.0	5.8
(half open)	20	17	14	21	15	14
(one-quarter open)	57	48	40	60	42	41
Angle valve (fully open)	4.7	2.0	1.0	2.4	2.0	2.0
Swing check valve (fully open)	2.9	2.1	2.0	2.0	2.0	2.0
Gate valve (fully open)	0.24	0.16	0.11	0.35	0.16	0.07
Return bend	1.5	0.95	0.64	0.35	0.30	0.25
Tee (branch)	1.8	1.4	1.1	0.80	0.64	0.58
Tee (line)	0.9	0.9	0.9	0.19	0.14	0.10
Standard elbow	1.5	0.95	0.64	0.39	0.30	0.26
Long sweep elbow	0.72	0.41	0.23	0.30	0.19	0.15
45° elbow	0.32	0.30	0.29			

Square-edged entrance		0.5
Reentrant entrance		0.8
Well-rounded entrance		0.03
Pipe exit		1.0
	Area ratio	
Sudden contraction[†]	2:1	0.25
	5:1	0.41
	10:1	0.46
	Area ratio A/A_0	
Orifice plate	1.5:1	0.85
	2:1	3.4
	4:1	29
	≥6:1	$2.78\left(\dfrac{A}{A_0} - 0.6\right)^2$
Sudden enlargement[‡]		$\left(1 - \dfrac{A_1}{A_2}\right)^2$
90° miter bend (without vanes)		1.1
(with vanes)		0.2
General contraction	(30° included anlge)	0.02
	(70° included angle)	0.07

[*] Values for other geometries can be found in *Technical Paper 410*. The Crane Company, 1957.

[†] Based on exit velocity V_2.

[‡] Based on entrance velocity V_1.

These losses are called *minor losses* that may, in fact, add up to exceed the head loss found in the preceding sections. These minor losses are expressed in terms of a *loss coefficient K*, defined for most devices by

$$h_L = K\frac{V^2}{2g} \tag{7.87}$$

A number of loss coefficients are included in Table 7.2. Note that relatively low loss coefficients are associated with gradual contractions, whereas relatively large coefficients with enlargements. This is due to the separated flows in enlargements. Separated and secondary flows also occur in elbows resulting in relatively large loss coefficients. Vanes that eliminate such separated or secondary flows can substantially reduce the losses, as noted in the table.

We often equate the losses in a device to an *equivalent length* of pipe, i.e.,

$$h_L = K\frac{V^2}{2g} = f\frac{L_e}{D}\frac{V^2}{2g} \tag{7.88}$$

This provides the relationship

$$L_e = K\frac{D}{f} \tag{7.89}$$

A last comment relating to minor losses is in order: if the pipe is quite long, >1000 diameters, the minor losses are usually neglected. For lengths as short as 100 diameters, the minor losses usually exceed the frictional losses. For intermediate lengths, the minor losses should be included.

EXAMPLE 7.6　A 1.5-cm-diameter, 20-m-long plastic pipe transports water from a pressurized 400-kPa tank out a free open end located 3 m above the water surface in the tank. There are three elbows in the water line and a square-edged inlet from the tank. Estimate the flow rate.

Solution: The energy equation is applied between the tank and the faucet exit:

$$0 = \frac{V_2^2 - V_1^2}{2g} + \frac{p_2 - p_1}{\gamma} + z_2 - z_1 + h_L$$

where

$$h_L = \left(f\frac{L}{D} + 3K_{\text{elbow}} + K_{\text{entrance}}\right)\frac{V^2}{2g}$$

Assume that the pipe has $e/D = 0$ and that $\text{Re} \cong 2 \times 10^5$ so that the Moody diagram yields $f = 0.016$. The energy equation yields

$$0 = \frac{V_2^2}{2 \times 9.8} - \frac{400\,000}{9800} + 3 + \left(0.016 \times \frac{20}{0.015} + 3 \times 1.6 + 0.5\right)\frac{V^2}{2 \times 9.8} \qquad \therefore \ V = 5.18\,\text{m/s}$$

The Reynolds number is then $\text{Re} = 5.18 \times 0.15/10^{-6} = 7.8 \times 10^4$. Try $f = 0.018$. Then

$$0 = \frac{V_2^2}{2 \times 9.8} - \frac{400\,000}{9800} + 3 + \left(0.018 \times \frac{20}{0.015} + 3 \times 1.6 + 0.5\right)\frac{V^2}{2 \times 9.8} \qquad \therefore \ V = 4.95\,\text{m/s}$$

Thus $\text{Re} = 4.95 \times 0.15/10^{-6} = 7.4 \times 10^4$. This is close enough so use $V = 5.0$ m/s. The flow rate is

$$Q = AV = \pi \times 0.0075^2 \times 5 = 8.8 \times 10^{-4}\,\text{m}^3/\text{s}$$

7.6.6　Hydraulic and Energy Grade Lines

The energy equation is most often written so that each term has dimensions of length, i.e.,

$$-\frac{\dot{W}_S}{\dot{m}g} = \frac{V_2^2 - V_1^2}{2g} + \frac{p_2 - p_1}{\gamma} + z_2 - z_1 + h_L \tag{7.90}$$

In piping systems, it is often conventional to refer to the *hydraulic grade line* (HGL) and the *energy grade line* (EGL). The HGL, the dashed line in Fig. 7.11, is the locus of points located a distance p/γ above the centerline of a pipe. The EGL, the solid line in Fig. 7.11, is the locus of points located a distance $V^2/2$ above the HGL. The following observations relate to the HGL and the EGL.

- The EGL approaches the HGL as the velocity goes to zero. They are identical on the surface of a reservoir.
- Both the EGL and the HGL slope downward in the direction of the flow due to the losses in the pipe. The greater the losses, the greater the slope.
- A sudden drop occurs in the EGL and the HGL equal to the loss due to a sudden geometry change, such as an entrance, an enlargement, or a valve.
- A jump occurs in the EGL and the HGL due to a pump and a drop due to a turbine.
- If the HGL is below the pipe, there is a vacuum in the pipe, a condition that is most often avoided in the design of piping systems because of possible contamination.

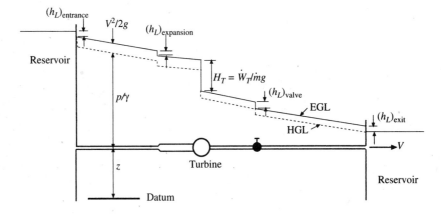

Figure 7.11 The hydraulic grade line (HGL) and the energy grade line (EGL) for a piping system.

7.7 OPEN CHANNEL FLOW

Consider the developed turbulent flow in an open channel, sketched in Fig. 7.12. The water flows at a depth of y and the channel is on a slope S, which is assumed to be small so that $\sin \theta = S$. The cross section could be trapezoidal, as shown, or it could be circular, rectangular, or triangular. Let us apply the energy equation between the two sections:

$$0 = \frac{V_2^2 - V_1^2}{2g} + \frac{p_2 - p_1}{\gamma} + z_2 - z_1 + h_L \qquad (7.91)$$

The head loss is the elevation change, i.e.,

$$h_L = z_1 - z_2$$
$$= L \sin \theta = LS \qquad (7.92)$$

where L is the distance between the two selected sections. Using the head loss expressed by Eq. (7.86), we have

$$h_L = f \frac{L}{4R} \frac{V^2}{2g} = LS \quad \text{or} \quad V^2 = \frac{8g}{f} RS \qquad (7.93)$$

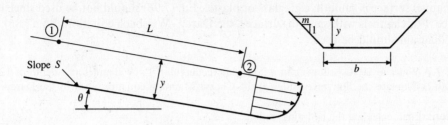

Figure 7.12   Flow in an open channel.

The Reynolds number of the flow in an open channel is invariably large and the channel is rough so that the friction factor is a constant independent of the velocity (see the Moody diagram of Fig. 7.10) for a particular channel. Consequently, the velocity is related to the slope and hydraulic radius by

$$V = C\sqrt{RS} \qquad (7.94)$$

where C is a dimensional constant called the *Chezy coefficient*; it has been related experimentally to the channel roughness and the hydraulic radius by

$$C = \frac{1}{n}R^{1/6} \qquad (7.95)$$

The dimensionless constant n is a measure of the wall roughness and is called the *Manning n*. Values for a variety of wall materials are listed in Table 7.3.

The flow rate in an open channel follows from $Q = AV$ and is

$$Q = \frac{1}{n}AR^{2/3}S^{1/2} \qquad (7.96)$$

This is referred to as the *Chezy–Manning equation*. It can be applied using English units by replacing the "1" in the numerator with "1.49."

Table 7.3   Values* of the Manning n

Wall material	Manning n
Brick	0.016
Cast or wrought iron	0.015
Concrete pipe	0.015
Corrugated metal	0.025
Earth	0.022
Earth with stones and weeds	2.035
Finished concrete	0.012
Mountain streams	0.05
Planed wood	0.012
Sewer pipe	0.013
Riveted steel	0.017
Rubble	0.03
Unfinished concrete	0.014
Rough wood	0.013

---

* The values in this table result in flow rates too large for $R > 3$ m. The Manning n should be increased by 10 to 15 percent for the larger channels.

If the channel surface is smooth, e.g., glass or plastic, Eq. (7.96) should not be used since it assumes a rough surface. For channels with smooth surfaces, the Darcy–Weisbach equation, Eq. (7.86), along with the Moody diagram should be used.

EXAMPLE 7.7 Water at $20°C$ is flowing in a 2-m-wide rectangular, brick channel at a depth of 120 cm. The slope is 0.0012. Estimate the flow rate using (a) the Chezy–Manning equation and (b) the Darcy–Weisbach equation.

Solution: First, calculate the hydraulic radius

$$R = \frac{A}{P} = \frac{by}{b+2y} = \frac{2 \times 1.2}{2 + 2 \times 1.2} = 0.545 \,\text{m}$$

(a) The Chezy–Manning equation provides

$$Q = \frac{1}{n} A R^{2/3} S^{1/2}$$

$$= \frac{1}{0.016} \times (2 \times 1.2) \times 0.545^{2/3} \times 0.0012^{1/2} = 3.47 \,\text{m}^3/\text{s}$$

(b) To use the Darcy–Weisbach equation, we must find the friction factor f. The Moody diagram requires a value for e. Use a relatively large value such as that for rougher concrete, i.e., $e = 1$ mm. Since the hydraulic radius $R = D/4$ for a circle, we use

$$\frac{e}{D} = \frac{e}{4R} = \frac{0.001}{4 \times 0.545} = 0.00046$$

The Moody diagram yields $f \cong 0.0165$. The Darcy–Weisbach equation takes the form of Eq. (7.93):

$$V = \sqrt{\frac{8g}{f} RS} = \sqrt{\frac{8 \times 9.8}{0.0165} \times 0.545 \times 0.0012} = 1.76 \,\text{m/s}$$

The flow rate is then

$$Q = AV = 2 \times 1.2 \times 1.76 = 4.23 \,\text{m}^3/\text{s}$$

Check the Reynolds number

$$\text{Re} = \frac{4VR}{v} = \frac{4 \times 1.76 \times 0.545}{10^{-6}} = 3.8 \times 10^6$$

This is sufficiently large so that f is acceptable. Note that the Q of part (a) is about 18 percent lower than that of part (b), and that of part (b) is considered more accurate.

Solved Problems

7.1 A 4-mm-diameter horizontal, 40-m-long pipe is attached to a reservoir containing $20°C$ water. The surface of the water in the reservoir is 4 m above the pipe outlet. Assume a laminar flow and estimate the average velocity in the pipe. Also, calculate the length of the entrance region.

Using Eq. (7.21), the average velocity in the pipe is

$$V = \frac{r_0^2 \Delta p}{8\mu L} = \frac{0.002^2 \times (9800 \times 4)}{8 \times 10^{-3} \times 40} = 0.49 \,\text{m/s}$$

where the pressure at the pipe inlet is $p = \gamma h = 9800 \times 4 \,\text{N/m}^2$, neglecting the velocity head $V^2/2g$ at the entrance. Check the Reynolds number; it is

$$\text{Re} = \frac{Vd}{v} = \frac{0.49 \times 0.004}{10^{-6}} = 1960$$

This is acceptable for a laminar flow to exist. We have assumed the velocity head at the entrance to be small; it is

$$\frac{V^2}{2g} = \frac{0.49^2}{2 \times 9.81} = 0.102 \, \text{m}$$

This is quite small compared with the pressure head of 4 m. So, the calculations are acceptable provided the entrance region is not very long.

We have neglected the effects of the entrance region's non-parabolic velocity profile (see Fig. 7.1). The entrance region's length is

$$L_E = 0.065 \times \text{Re} \times D = 0.065 \times 1960 \times 0.004 = 0.51 \, \text{m}$$

so the effect of the entrance region is negligible.

7.2   A developed, steady laminar flow exists between horizontal concentric pipes. The flow is in the direction of the axis of the pipes. Derive the differential equations and solve for the velocity profile.

The element selected, upon which the forces would be placed, would be a hollow cylindrical shell (a sketch may be helpful for visualization purposes), that would appear as a ring from an end view, with length dx. The ring would have an inner radius r and an outer radius $r + dr$. The net pressure force acting on the two ends would be

$$p2\pi\left(r + \frac{dr}{2}\right)dr - (p + dp)2\pi\left(r + \frac{dr}{2}\right)dr = -2\pi r \, dr \, dp$$

The shear stress forces on the inner and the outer cylinder sum as follows (the shear stress is assumed to oppose the flow):

$$-\tau 2\pi r \, dx + (\tau + d\tau)2\pi(r + dr)dx = 2\pi\tau \, dr \, dx + 2\pi r \, d\tau \, dx$$

For a steady flow, the pressure and shear stress forces must balance. This provides

$$-2\pi r \, dr \, dp = 2\pi\tau \, dr \, dx + 2\pi r \, d\tau \, dx \qquad \therefore \frac{dp}{dx} = -\frac{\tau}{r} - \frac{d\tau}{dr}$$

Substitute the constitutive equation $\tau = -\mu \, du/dr$ (see footnote associated with Eq. (7.7) assuming the element is near the outer pipe) and obtain

$$\frac{dp}{dx} = \mu\left(\frac{1}{r}\frac{du}{dr} + \frac{d^2u}{dr^2}\right) = \frac{\mu}{r}\frac{d}{dr}\left(r\frac{du}{dr}\right)$$

This can now be integrated to yield

$$\frac{r^2}{2\mu}\frac{dp}{dx} = r\frac{du}{dr} + A \quad \text{or} \quad \frac{r}{2\mu}\frac{dp}{dx} = \frac{du}{dr} + \frac{A}{r}$$

Integrate once more to find the velocity profile to be

$$u(r) = \frac{r^2}{4\mu}\frac{dp}{dx} - A\ln r + B$$

The constants A and B can be evaluated by using $u(r_1) = 0$ and $u(r_2) = 0$.

7.3   What pressure gradient would provide a zero shear stress on the stationary lower plate in Fig. 7.5 assuming horizontal plates with the top plate moving to the right with velocity U.

The shear stress is $\tau = -\mu \, du/dy$ so that the boundary conditions are $du/dy \, (0) = 0$, $u(0) = 0$, and $u(b) = U$. These are applied to Eq. (7.29) to provide the following:

$$\frac{du}{dy}(0) = \frac{1}{\mu}\frac{dp}{dx}0 + A = 0 \qquad \therefore A = 0$$

$$u(0) = \frac{1}{2\mu}\frac{dp}{dx}0 + B = 0 \qquad \therefore B = 0 \quad \text{and} \quad u(y) = \frac{1}{2\mu}\frac{dp}{dx}y^2$$

Now, $u(b) = U$, resulting in

$$u = \frac{1}{2\mu}\frac{dp}{dx}b^2 \quad \text{or} \quad \frac{dp}{dx} = \frac{2\mu u}{b^2}$$

This is a positive pressure gradient, so the pressure increases in the direction of U.

7.4   Show that the velocity distribution given by Eq. (7.62) approximates a straight line when the gap between the two cylinders is small relative to the radii of the cylinders.

Since the gap is small relative to the two radii, we can let $R \cong r_1 \cong r_2$. Also, let $\delta = r_2 - r_1$ and $y = r_2 - r$ (refer to Fig. 7.6) in the velocity distribution of Eq. (7.62). The velocity distribution takes the form

$$v_\theta(r) = \frac{\omega_1 r_1^2}{r_2^2 - r_1^2}\left(\frac{r_2^2}{r} - r\right) = \frac{\omega_1 r_1^2}{(r_2 - r_1)(r_2 + r_1)}\frac{(r_2 - r)(r_2 + r)}{r}$$

$$\cong \frac{\omega_1 R^2}{2R\delta} \times \frac{y(2R - y)}{R - y} \cong \frac{\omega_1 R}{\delta}y$$

where we have used the approximation

$$\frac{2R - y}{R - y} \cong 2$$

since y is small compared with R and $2R$. The above velocity distribution is a straight-line distribution with slope $\omega_1 R/\delta$.

7.5   Water at $15°C$ is transported in a 6-cm-diameter wrought iron pipe at a flow rate of $0.004 \text{ m}^3/\text{s}$. Estimate the pressure drop over 300 m of horizontal pipe using (a) the Moody diagram and (b) an alternate equation.

The average velocity and Reynolds number are

$$V = \frac{Q}{A} = \frac{0.004}{\pi \times 0.03^2} = 1.415\,\text{m/s} \qquad \text{Re} = \frac{VD}{\nu} = \frac{1.415 \times 0.06}{1.14 \times 10^{-6}} = 7.44 \times 10^4$$

(a) The value of e is found on the Moody diagram so that

$$\frac{e}{D} = \frac{0.046}{60} = 0.00077$$

The friction factor is found from the Moody diagram to be

$$f = 0.0225$$

The pressure drop is then

$$\Delta p = \gamma h_L = \rho f\frac{L}{D}\frac{V^2}{2} = 1000 \times 0.0225\frac{300}{0.06}\frac{1.415^2}{2} = 113\,000\,\text{Pa or } 113\,\text{kPa}$$

(b) Using Eq. (7.82), the pressure drop is

$$\Delta p = \gamma h_L = 1.07 \times 1000\frac{0.004^2 \times 300}{0.06^5}\left\{\ln\left[\frac{0.00077}{3.7} + 4.62\left(\frac{1.14 \times 10^{-6} \times 0.06}{0.004}\right)^{0.9}\right]\right\}^{-2}$$

$$= 111\,000\,\text{Pa or } 111\,\text{kPa}$$

These two results are within 2 percent and are essentially the same.

7.6   A pressure drop of 200 kPa is measured over a 400-m length of 8-cm-diameter horizontal cast iron pipe that transports $20°C$ water. Determine the flow rate using (a) the Moody diagram and (b) an alternate equation.

The relative roughness is

$$\frac{e}{D} = \frac{0.26}{80} = 0.00325$$

and the head loss is

$$h_L = \frac{\Delta p}{\gamma} = \frac{200\,000}{9800} = 20.41\,\text{m}$$

(a) Assuming a completely turbulent flow, Moody's diagram yields

$$f = 0.026$$

The average velocity in the pipe is found, using Eq. (7.79), to be

$$V = \sqrt{\frac{2h_L D g}{fL}} = \sqrt{\frac{2 \times 20.41 \times 0.08 \times 9.81}{0.026 \times 400}} = 1.76\,\text{m/s}$$

resulting in a Reynolds number of

$$\text{Re} = \frac{VD}{\nu} = \frac{1.76 \times 0.08}{10^{-6}} = 1.4 \times 10^5$$

At this Reynolds number and $e/D = 0.0325$, Moody's diagram provides $f \cong 0.026$, so the friction factor does not have to be adjusted. The flow rate is then expected to be

$$Q = AV = \pi \times 0.04^2 \times 1.76 = 0.0088\,\text{m}^3/\text{s}$$

(b) Since the head loss was calculated using the pressure drop, Eq. (7.83) can be used to find the flow rate:

$$Q = -0.965\sqrt{\frac{9.81 \times 0.08^5 \times 20.41}{400}}\ln\left[\frac{0.00325}{3.7} + \left(\frac{3.17 \times 10^{-12} \times 400}{9.81 \times 0.08^3 \times 20.41}\right)^{0.5}\right] = 0.00855\,\text{m}^3/\text{s}$$

These two results are within 3 percent and either is acceptable.

7.7   A farmer needs to provide a volume of 500 L every minute of 20°C water from a lake through a wrought iron pipe a distance of 800 m to a field 4 m below the surface of the lake. Determine the diameter of pipe that should be selected. Use (a) the Moody diagram and (b) an alternate equation.

(a) The average velocity is related to the unknown diameter D by

$$V = \frac{Q}{A} = \frac{0.5/60}{\pi D^2/4} = \frac{0.0106}{D^2}$$

The head loss is 4 m (the energy equation from the lake surface to the pipe exit provides this. We assume that $V^2/2g$ is negligible at the pipe exit), so

$$h_L = f\frac{L}{D}\frac{V^2}{2g} \qquad 4 = f\frac{800}{D} \times \frac{0.0106^2/D^4}{2g} \qquad \therefore\ D^5 = 0.00114f$$

The Reynolds number and relative roughness are

$$\text{Re} = \frac{VD}{\nu} = \frac{0.0106D}{D^2 \times 10^{-6}} = \frac{10\,600}{D} \qquad \frac{e}{D} = \frac{0.046}{D}$$

This requires a trial-and-error solution. We can select a value for f and check to see if the equations and the Moody diagram agree with that selection. Select $f = 0.02$. Then, the above equations yield

$$D = (0.00114 \times 0.02)^{0.2} = 0.118\,\text{m}, \quad \text{Re} = \frac{10\,600}{0.118} = 90\,000, \quad \frac{e}{D} = \frac{0.046}{118} = 0.00039$$

The above match very well on the Moody diagram. Usually, another selection for f and a recalculation of the diameter, Reynolds number, and relative roughness are required.

(b) Since the diameter is unknown, Eq. (7.84) is used which provides

$$D = 0.66\left[0.000046^{1.25}\left(\frac{800(0.5/60)^2}{9.81 \times 4}\right)^{4.75} + 10^{-6}\left(\frac{0.5}{60}\right)^{9.4}\left(\frac{800}{9.81 \times 4}\right)^{5.2}\right]^{0.04} = 0.12\,\text{m}$$

The two results are within 2 percent, so are essentially the same.

7.8 A smooth rectangular duct that measures 10×20 cm transports 0.4 m^3/s of air at standard conditions horizontal distance of 200 m. Estimate the pressure drop in the duct.

The hydraulic radius is

$$R = \frac{A}{P} = \frac{0.1 \times 0.2}{2(0.1 + 0.2)} = 0.0333 \, \text{m}$$

The average velocity and Reynolds number in the duct are

$$V = \frac{Q}{A} = \frac{0.4}{0.1 \times 0.2} = 20 \, \text{m/s} \qquad \text{Re} = \frac{4VR}{\nu} = \frac{4 \times 20 \times 0.0333}{1.5 \times 10^{-5}} = 1.8 \times 10^5$$

The Moody diagram provides $f = 0.016$. The pressure drop is then

$$\Delta p = \gamma h_L = \gamma f \frac{L}{4R} \frac{V^2}{2g} = 1.23 \times 9.81 \times 0.016 \frac{200}{4 \times 0.0333} \times \frac{20^2}{2 \times 9.81} = 5900 \, \text{Pa}$$

7.9 Sketch the hydraulic grade line for the piping system of Example 7.6 if the three elbows are spaced equally between the pressurized tank and the exit of the pipe.

The hydraulic grade line is a distance p/γ above the surface of the water in the tank at the beginning of the pipe. The hydraulic grade line is sketched in Fig. 7.13.

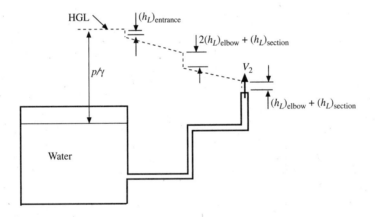

Figure 7.13

7.10 An 80-cm-diameter sewer pipe (finished concrete) is selected to transport water at a flow rate of 0.24 m^3/s at a slope of 0.0012. Estimate the depth at which the water will flow.

Assume the water flows with the pipe half full. The flow rate would be

$$Q = \frac{1}{n} A R^{2/3} S^{1/2} = \frac{1}{0.013} \frac{\pi \times 0.4^2}{2} \left(\frac{0.08\pi}{0.4\pi}\right)^{2/3} \times 0.0012^{1/2} = 0.229 \, \text{m}^3/\text{s}$$

Consequently, the pipe is over half full. A sketch of the area is shown in Fig. 7.14. For this pipe we have

$$0.24 = \frac{1}{0.013} A R^{2/3} 0.0012^{1/2} \qquad \therefore \ A R^{2/3} = 0.09$$

with

$$A = 0.8\pi\frac{180 - \alpha}{180} + (y - 0.4)0.4\sin\alpha \qquad R = \frac{A}{0.8\pi\dfrac{180 - \alpha}{180}}$$

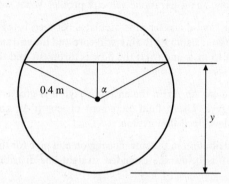

Figure 7.14

Trial and error is needed for a solution.

Try $y = 0.46$m : Then $A = 0.299$ $R = 0.217$ $AR^{2/3} = 0.108$

Try $y = 0.44$m : Then $A = 0.283$ $R = 0.211$ $AR^{2/3} = 0.100$

Hence, $y = 0.42$ m is an acceptable result.

Supplementary Problems

Laminar or Turbulent Flow

7.11 Calculate the maximum average velocity in a 2-cm-diameter pipe for a laminar flow using a critical Reynolds number of 2000 if the fluid is (a) water at 20°C, (b) water at 80°C, (c) SAE-30 oil at 80°C, and (d) atmospheric air at 20°C.

7.12 The Red Cedar River flows placidly through MSU's campus at a depth of 80 cm. A leaf is observed to travel about 1 m in 4 s. Decide if the flow is laminar or turbulent. Make any assumptions needed.

7.13 A drinking fountain has an opening of 4 mm in diameter. The water rises a distance of about 20 cm in the air. Is the flow laminar or turbulent as it leaves the opening? Make any assumptions needed.

7.14 SAE-30 oil at 80°C occupies the space between two cylinders, 2 and 2.2 cm in diameter. The outer cylinder is stationary and the inner cylinder rotates at 100 rpm. Is the oil in a laminar or turbulent state if $\mathrm{Re}_{\mathrm{crit}} = 1700$? Use $\mathrm{Re} = \omega r_1\delta/\nu$, where $\delta = r_2 - r_1$.

Entrance Flow

7.15 Water is flowing in a 2-cm-diameter pipe with a flow rate of 0.0002 m³/s. For an entrance that provides a uniform velocity profile, estimate inviscid core length and the entrance length if the water temperature is (a) 20, (b) 40, (c) 60, and (d) 80°C.

7.16   A parabolic velocity profile is desired at the end of a 10-m-long, 8-mm-diameter tube attached to a tank filled with 20°C water. An experiment is run during which 60 L is collected in 90 min. Is the laminar flow assumption reasonable? If so, would the tube be sufficiently long?

7.17   A parabolic profile is desired in 20°C air as it passes between two parallel plates that are 80 mm apart in a university laboratory. If the Reynolds number $Vh/v = 1500$, how long would the channel need to be to observe a fully developed flow, i.e., a parabolic velocity profile? What would be the average velocity?

7.18   The flow of 20°C water in a 2-cm-diameter pipe oscillates between being laminar and turbulent as it flows through the pipe from a reservoir. Estimate the inviscid core and the entrance lengths (a) if the flow is laminar and the average velocity is 0.15 m/s, and (b) if the flow is turbulent and the average velocity is 0.6 m/s (use the results of Eq. (7.4)).

7.19   Argue that the pressure gradient $\Delta p/\Delta x$ in the entrance region is greater than the pressure gradient in the developed flow region of a pipe. Use a fluid increment of length Δx and cross-sectional area πr_0^2 in the entrance region and in the developed-flow region.

7.20   Explain why the pressure distribution in the entrance region of a pipe for the relatively low-Reynolds-number turbulent flow (Re $\approx$ 10 000) is below the extended straight-line distribution of developed flow. Refer to Fig. 7.3.

Laminar Flow in a Pipe

7.21   Show that the right-hand side of Eq. (7.19) does indeed follow from the integration.

7.22   Show that $f = 64/\text{Re}$ for a laminar flow in a pipe.

7.23   Show that the head loss in a laminar flow in a pipe is directly proportional to the average velocity in the pipe.

7.24   The pressure drop over a 15-m length of 8-mm-diameter horizontal pipe transporting water at 40°C is measured to be 1200 Pa. A laminar flow is assumed. Determine (a) the maximum velocity in the pipe, (b) the Reynolds number, (c) the wall shear stress, and (d) the friction factor.

7.25   A liquid flows through a 2-cm-diameter pipe at a rate of 20 L every minute. Assume a laminar flow and estimate the pressure drop over 20 m of length in the horizontal pipe for (a) water at 40°C, (b) SAE-10 oil at 20°C, and (c) glycerin at 40°C. Decide if a laminar flow is a reasonable assumption.

7.26   Water at 20°C flows through a 12-mm-diameter pipe on a downward slope so that Re = 2000. What angle would result in a zero pressure drop?

7.27   Water at 40°C flows in a vertical 8-mm-diameter pipe at 2 L/min. Assuming a laminar flow, calculate the pressure drop over a length of 20 m if the flow is (a) upwards and (b) downwards.

7.28   Atmospheric air at 25°C flows in a 2-cm-diameter horizontal pipe at Re = 1600. Calculate the wall shear stress, the friction factor, the head loss, and the pressure drop over 20 m of pipe.

7.29   A liquid flows in a 4-cm-diameter pipe. At what radius does the velocity equal the average velocity assuming a laminar flow? At what radius is the shear stress equal to one-half the wall shear stress?

7.30   Find an expression for the angle θ that a pipeline would require such that the pressure is constant assuming a laminar flow. Then, find the angle of a 10-mm-diameter pipe transporting 20°C water at Re = 2000 so that a constant pressure occurs.

7.31   Solve for the constants A and B in Solved Problem 7.2 using cylinder radii of $r_1 = 4$ cm and $r_2 = 5$ cm assuming that 20°C water has a pressure drop of 40 Pa over a 10-m length. Also find the flow rate. Assume laminar flow.

7.32   SAE-10 oil at 20°C flows between two concentric cylinders parallel to the axes of the horizontal cylinders having radii of 2 and 4 cm. The pressure drop is 60 Pa over a length of 20 m. Assume laminar flow. What is the shear stress on the inner cylinder?

Laminar Flow Between Parallel Plates

7.33   What pressure gradient would provide a zero shear stress on the stationary lower plate in Fig. 7.5 for horizontal plates with the top plate moving to the right with velocity U. Assume a laminar flow.

7.34   What pressure gradient is needed so that the flow rate is zero for laminar flow between horizontal parallel plates if the lower plate is stationary and the top plate moves with velocity U. See Fig. 7.5.

7.35   Fluid flows in a horizontal channel that measures 1×40 cm. If Re = 1500, calculate the flow rate and the pressure drop over a length of 10 m if the fluid is (a) water at 20°C, (b) air at 25°C, and (c) SAE-10 oil at 40°C. Assume laminar flow.

7.36   Water at 20°C flows down an 80-m-wide parking lot at a constant depth of 5 mm. The slope of the parking lot is 0.0002. Estimate the flow rate and the maximum shear stress. Is a laminar flow assumption reasonable?

7.37   Water at 20°C flows between two parallel horizontal plates separated by a distance of 8 mm. The lower plate is stationary and the upper plate moves at 4 m/s to the right (see Fig. 7.5). Assuming a laminar flow, what pressure gradient is needed such that:

   (a) The shear stress at the upper plate is zero
   (b) The shear stress at the lower plate is zero
   (c) The flow rate is zero
   (d) The velocity at $y = 4$ mm is 4 m/s

7.38   Atmospheric air at 40°C flows between two parallel horizontal plates separated by a distance of 6 mm. The lower plate is stationary and the pressure gradient is -3 Pa/m. Assuming a laminar flow, what velocity of the upper plate (see Fig. 7.5) is needed such that:

   (a) The shear stress at the upper plate is zero
   (b) The shear stress at the lower plate is zero
   (c) The flow rate is zero
   (d) The velocity at $y = 4$ mm is 2 m/s

7.39   SAE-30 oil at 40°C fills the gap between the stationary plate and the 20-cm-diameter rotating plate shown in Fig. 7.15. Estimate the torque needed assuming a linear velocity profile if $\Omega = 100$ rad/s.

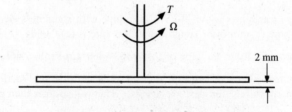

Figure 7.15

7.40   SAE-10 oil at 20°C fills the gap between the moving 120-cm-long cylinder and the fixed outer surface. Assuming a zero pressure gradient, estimate the force needed to move the cylinder at 10 m/s. Assume a laminar flow.

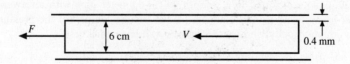

Figure 7.16

Laminar Flow Between Rotating Cylinders

7.41   Assuming a Couette flow between a stationary and a rotating cylinder, determine the expression for the power needed to rotate the inner rotating cylinder. Refer to Fig. 7.6.

7.42   SAE-10 oil at 20°C fills the gap between the rotating cylinder and the fixed outer cylinder shown in Fig. 7.17. Estimate the torque needed to rotate the 20-cm-long cylinder at 40 rad/s (*a*) using the profile of Eq. (7.62) and (*b*) assuming a Couette flow.

Figure 7.17

7.43   A 3-cm-diameter cylinder rotates inside a fixed 4-cm-diameter cylinder with 40°C SAE-30 oil filling the space between the 30-cm-long concentric cylinders. Write the velocity profile and calculate the torque and the power required to rotate the inner cylinder at 2000 rpm assuming a laminar flow.

7.44   Determine the expressions for the torque and the power required to rotate the outer cylinder if the inner cylinder of Fig. 7.6 is fixed. Assume a laminar flow.

Turbulent Flow in a Pipe

7.45   Time average the differential continuity equation for an incompressible flow and prove that two continuity equations result:

$$\frac{\partial u'}{\partial x} + \frac{\partial v'}{\partial y} + \frac{\partial w'}{\partial z} = 0 \quad \text{and} \quad \frac{\partial \bar{u}}{\partial x} + \frac{\partial \bar{v}}{\partial y} + \frac{\partial \bar{w}}{\partial z} = 0$$

7.46   A 12-cm-diameter pipe transports water at 25°C in a pipe with roughness elements averaging 0.26 mm in height. Decide if the pipe is smooth or rough if the flow rate is (*a*) 0.0004, (*b*) 0.004, and (*c*) 0.04 m³/s.

7.47   Estimate the maximum velocity in the pipe of (*a*) Prob. 7.46*a*, (*b*) Prob. 7.46*b*, and (*c*) Prob. 7.46*c*.

7.48   Draw a cylindrical control volume of length L and radius r in a horizontal section of pipe and show that the shear stress varies linearly with r, that is, $\tau = r\,\Delta p/(2L)$. The wall shear is then given by $\tau_0 = r_0\,\Delta p/(2L)$ (see Eq. (7.75)).

7.49   Estimate the velocity gradient at the wall, the pressure drop, and the head loss over 20 m of length for the water flow of (*a*) Prob. 7.46*a*, (*b*) Prob. 7.46*b*, and (*c*) Prob. 7.46*c*. Note: Since turbulence must be zero at the wall, the wall shear stress is given by $\mu\,\partial \bar{u}/\partial y|_{y=0}$.

7.50   Water at 20°C flows in a 10-cm-diameter smooth horizontal pipe at the rate of 0.004 m³/s. Estimate the maximum velocity in the pipe and the head loss over 40 m of length. Use the power-law velocity distribution.

7.51   SAE-30 oil at 20°C is transported in a smooth 40-cm-diameter pipe with an average velocity of 10 m/s. Using the power-law velocity profile, estimate (*a*) the friction factor, (*b*) the pressure drop over 100 m of pipe, (*c*) the maximum velocity, and (*d*) the viscous wall layer thickness.

7.52   Rework Prob. 7.51 using the semi-log velocity profile.

7.53   If the pipe of Prob. 7.51 is a cast iron pipe, rework the problem using the semi-log velocity profile.

Losses in Pipe Flow

7.54 Water at 20°C flows at 0.02 m³/s in an 8-cm-diameter galvanized iron pipe. Calculate the head loss over 40 m of horizontal pipe using (*a*) the Moody diagram and (*b*) the alternate equation.

7.55 Rework Prob. 7.54 using (*a*) SAE-10 oil at 80°C, (*b*) glycerin at 70°C, and (*c*) SAE-30 oil at 40°C.

7.56 Water at 30°C flows down a 30° incline in a smooth 6-cm-diameter pipe at a flow rate of 0.006 m³/s. Find the pressure drop and the head loss over an 80-m length of pipe.

7.57 If the pressure drop in a 100-m section of horizontal 10-cm-diameter galvanized iron pipe is 200 kPa, estimate the flow rate if the liquid flowing is (*a*) water at 20°C, (*b*) SAE-10 oil at 80°C, (*c*) glycerin at 70°C, and (*d*) SAE-30 oil at 20°C. Because the Moody diagram requires a trial-and-error solution, one of the alternate equations is recommended.

7.58 Air at 40°C and 200 kPa enters a 300-m section of 10-cm-diameter galvanized iron pipe. If a pressure drop of 200 Pa is measured over the section, estimate the mass flux and the flow rate. Because the Moody diagram requires a trial-and-error solution, one of the alternate equations is recommended. Assume the air to be incompressible.

7.59 A pressure drop of 100 kPa is desired in 80 m of smooth pipe transporting 20°C water at a flow rate of 0.0016 m³/s. What diameter pipe should be used? Because the Moody diagram requires a trial-and-error solution, one of the alternate equations is recommended.

7.60 Rework Prob. 7.59 using (*a*) SAE-10 oil at 80°C, (*b*) glycerin at 70°C, and (*c*) SAE-30 oil at 20°C.

7.61 A farmer wishes to siphon 20°C water from a lake, the surface of which is 4 m above the plastic tube exit. If the total distance is 400 m and 300 L of water is desired per minute, what size tubing should be selected? Because the Moody diagram requires a trial-and-error solution, one of the alternate equations is recommended.

7.62 Air at 35°C and 120 kPa enters a 20×50 cm sheet metal conduit at a rate of 6 m³/s. What pressure drop is to be expected over a length of 120 m?

7.63 A pressure drop of 6000 Pa is measured over a 20 m length as water at 30°C flows through the 2×6 cm smooth conduit. Estimate the flow rate.

Minor Losses

7.64 The loss coefficient of the standard elbow listed in Table 7.2 appears quite large compared with several of the other loss coefficients. Explain why the elbow has such a relatively large loss coefficient by inferring a secondary flow after the bend. Refer to Eq. (3.31).

7.65 Water at 20°C flows from a reservoir out a 100-m-long, 4-cm-diameter galvanized iron pipe to the atmosphere. The outlet is 20 m below the surface of the reservoir. What is the exit velocity (*a*) assuming no losses in the pipe and (*b*) including the losses? There is a square-edged entrance. Sketch the EGL and the HGL for both (*a*) and (*b*).

7.66 Add a nozzle with a 2-cm-diameter outlet to the pipe of Prob. 7.65. Calculate the exit velocity.

7.67 The horizontal pipe of Prob. 7.65 is fitted with three standard screwed elbows equally spaced. Calculate the flow including all losses. Sketch the HGL.

7.68 A 4-cm-diameter cast iron pipe connects two reservoirs with the surface of one reservoir 10 m below the surface of the other. There are two standard screwed elbows and one wide-open angle valve in the 50-m-long pipe. Assuming a square-edged entrance, estimate the flow rate between the reservoirs. Assume a temperature of 20°C.

7.69 An 88% efficient pump is used to transport 30°C water from a lower reservoir through an 8-cm-diameter galvanized iron pipe to a higher reservoir whose surface is 40 m above the surface of the lower one. The pipe has a total length of 200 m. Estimate the power required for a flow rate of 0.04 m³/s. What is the maximum distance from the lower reservoir that the pump can be located if the horizontal pipe is 10 m below the surface of the lower reservoir?

7.70 A 90% efficient turbine operates between two reservoirs connected by a 200-m length of 40-cm-diameter cast iron pipe that transports 0.8 m³/s of 20°C water. Estimate the power output of the turbine if the elevation difference between the surfaces of the reservoirs is 40 m.

7.71 The *pump characteristic curves*, shown in Fig. 7.18, relate the efficiency and the pump head (see Eq. (4.25)) for the pump of this problem to the flow rate. If the pump is used to move 20°C water from a lower reservoir at elevation 20 m to a higher reservoir at elevation 60 m through 200 m of 16-cm-diameter cast iron pipe, estimate the flow rate and the power required.

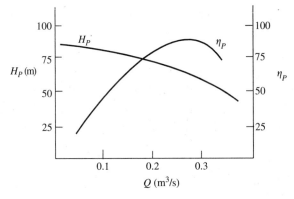

Figure 7.18

Open Channel Flow

7.72 Water flows at a depth of 80 cm in an open channel on a slope of 0.0012. Find the average shear stress acting on the channel walls if the channel cross section is (*a*) a 140-cm-wide rectangle and (*b*) a 3.2-m-diameter circle. (Draw a control volume and sum forces.)

7.73 Water flows in a 2-m-wide rectangular finished concrete channel with a slope of 0.001 at a depth of 80 cm. Estimate the flow rate using (*a*) the Chezy–Manning equation and (*b*) the Moody diagram.

7.74 Water is not to exceed a depth of 120 cm in a 2-m-wide finished rectangular concrete channel on a slope of 0.001. What would the flow rate be at that depth? (*a*) Use the Chezy–Manning equation and (*b*) the Moody diagram.

7.75 Estimate the flow rate in the channel shown in Fig. 7.19 if the slope is 0.0014. The sides are on a slope of 45°. (*a*) Use the Chezy–Manning equation and (*b*) the Darcy–Weisbach equation. (*c*) Also, calculate the average shear stress on the walls.

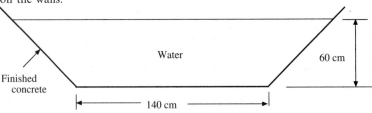

Figure 7.19

7.76   Water flows in a 2-m-diameter sewer (finished concrete) with $S = 0.0016$. Estimate the flow rate if the depth is (a) 50, (b) 100, (c) 150, and (d) 199 cm.

7.77   Water flows in a 120-cm-diameter sewer (finished concrete) with $S = 0.001$ at a flow rate of 0.4 m^3/s. What is the expected depth of flow?

Answers to Supplementary Problems

7.11   (a) 0.1007 m/s   (b) 0.0367 m/s   (c) 1.8 m/s   (d) 1.51 m/s

7.12   Highly turbulent

7.13   Turbulent

7.14   Laminar

7.15   (a) 16.4 m, 8.2 m   (b) 25.1 m, 12.5 m   (c) 34.7 m, 17.4 m   (d) 45.1 m, 22.6 m

7.16   Yes, yes

7.17   4.8 m, 0.283 m/s

7.18   (a) 1.94 m, 3.87 m   (b) 0.2 m, 2.4 m

7.19   See problem statement

7.20   See problem statement

7.21   See problem statement

7.22   See problem statement

7.23   See problem statement

7.24   (a) 0.448 m/s   (b) 2950   (c) 0.16 Pa   (d) 0.0217

7.25   (a) 1114 Pa, not laminar   (b) 153 kPa, laminar   (c) 594 kPa, laminar

7.26   0.219° downward

7.27   (a) 200.6 kPa   (b) −191.8 kPa

7.28   0.0091 Pa, 0.04, 3.13 m, 36.4 Pa

7.29   1.414 cm, 1 cm

7.30 $\sin^{-1}(8\mu V/\gamma r_0^2)$, 0.376°

7.31   4.1, 14.8, 0.0117 m^3/s

7.32   0.035 N/m^2

7.33 $2U/b^2$

7.34 $6\mu U/b^2$

7.35   (a) 0.0006 m^3/s, 180 Pa   (b) 0.0093 m^3/s, 51.5 Pa   (c) 0.0024 m^3/s, 264 kPa

7.36   0.000653 m^3/s, 0.00098 N/m^2

7.37   (a) −125 Pa/m   (b) 125 Pa/m   (c) 375 Pa/m   (d) −0.25 Pa/m

7.38   (a) 2.83 m/s   (b) −2.83 m/s   (c) −0.942 m/s   (d) 1.37 m/s

7.39   12.6 N·m

7.40   565 N

7.41 $2\pi\mu r_1^3\omega^2 L/\delta$

7.42 (a) 0.346 N·m   (b) 0.339 N·m

7.43 $539(0.0016/r - r)$, 0.325 N·m, 136 W

7.44 $T = 4\pi\mu r_1^2 r_2^2 L\omega_2/(r_2^2 - r_1^2)$

7.45 See problem statement

7.46 (a) smooth   (b) rough   (c) rough

7.47 (a) 0.047 m/s   (b) 0.474 m/s   (c) 4.6 m/s

7.48 See problem statement

7.49 (a) 7.3 s^{-1}, 4.4 Pa, 0.00045 m   (b) 490 s^{-1}, 290 Pa, 0.03 m   (c) 45,200 s^{-1}, 27 kPa, 2.7 m

7.50 0.63 m/s, 0.125 m

7.51 (a) 0.024   (b) 275 kPa   (c) 12.4 m/s   (d) 1.82 mm

7.52 (a) 0.0255   (b) 292 kPa   (c) 11.9 m/s   (d) 1.77 mm

7.53 (a) 0.0275   (b) 315 kPa   (c) 12.5 m/s   (d) 1.71 mm

7.54 (a) 9.7 m   (b) 9.55 m

7.55 (a) 11.3 m, 11.1 m   (b) 15.1 m, 15.2 m   (c) 16.1 m, 16.5 m

7.56 -343 kPa, 4.98 m

7.57 (a) 0.033 m^3/s   (b) 0.032 m^3/s   (c) 0.022 m^3/s   (d) 0.022 m^3/s

7.58 0.0093 m^3/s, 0.031 kg/s

7.59 5.6 cm

7.60 (a) 3.5 cm   (b) 3.9 cm   (c) 3.9 cm

7.61 8.4 cm

7.62 18.8 kPa

7.63 0.00063 m^3/s

7.64 See problem statement

7.65 (a) 19.8 m/s   (b) 2.01 m/s

7.66 0.995 m/s

7.67 1.56 m/s

7.68 1.125 m/s

7.69 138 hp, 6.7 m

7.70 173 hp

7.71 0.3 m^3/s, 290 hp

7.72 (a) 4.39 Pa   (b) 4.92 Pa

7.73 (a) 2.45 m^3/s   (b) 2.57 m^3/s

7.74 (a) 0.422 m^3/s   (b) 0.435 m^3/s

7.75 (a) 1.99 m^3/s   (b) 2.09 m^3/s   (c) 5.32 Pa

7.76 (a) 0.747 m^3/s   (b) 3.30 m^3/s   (c) 6.27 m^3/s   (d) 6.59 m^3/s

7.77 0.45 m

Chapter 8

External Flows

8.1 INTRODUCTION

The subject of external flows involves both low- and high-Reynolds number flows. Low-Reynolds number flows are not of interest in most engineering applications and will not be considered in this book; flow around spray droplets, river sediment, filaments, and red blood cells would be examples that are left to the specialists. High-Reynolds number flows are of interest to many engineers and include flow around airfoils, vehicles, buildings, bridge cables, stadiums, turbine blades, and signs, to name a few.

It is quite difficult to solve for the flow field external to a body, even the simplest of bodies like a long cylinder or a sphere. We can, however, develop equations that allow us to estimate the growth of the thin viscous layer, the boundary layer, which grows on a flat plate or the rounded nose of a vehicle. Also, coefficients have been determined experimentally that allow the drag and the lift to be objects of interest. We will begin this chapter by presenting such coefficients. But first, some definitions are needed.

The flow around a blunt body involves a *separated region*, a region in which the flow separates from the body and forms a recirculating region downstream, as sketched in Fig. 8.1. A *wake*, a region influenced by viscosity, is also formed; it is a diffusive region that continues to grow (some distance downstream the velocity is less than the free-stream velocity V). A laminar boundary layer exists near the front of the body followed by a turbulent boundary layer as shown in Fig. 8.1. An inviscid flow, often referred to as the *free stream*, exists on the front of the body and outside the boundary layer, separated region, and wake. The flow around a streamlined body has all the same components as that of Fig. 8.1 except that it does not have a significant separated region and the wake is much smaller.

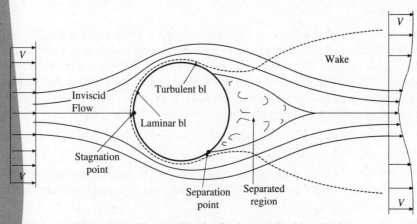

Figure 8.1 The details of a flow around a blunt body.

The free-stream inviscid flow is usually irrotational although it could be a rotational flow with vorticity, e.g., the flow of air near the ground around a tree trunk or water near the ground around a post in a river; the water digs a depression in the sand in front of the post and the air digs a similar depression in snow in front of the tree, a rather interesting observation. The vorticity in the approaching air or water accounts for the observed phenomenon.

It should be noted that the boundary of the separated region is shown at an average location. It is, however, highly unsteady and is able to slowly exchange mass with the free stream even though the time-average streamlines remain outside the separated region. Also, the separated region is always located inside the wake.

Interest in the flow around a blunt object is focused on the *drag*, the force the flow exerts on the body in the direction of the flow[*]. *Lift* is the force exerted normal to the flow direction and is of interest on airfoils and streamlined bodies. The drag F_D and lift F_L are specified in terms of the *drag coefficient C_D* and *lift coefficient C_L*, respectively, by

$$F_D = \frac{1}{2}\rho A V^2 C_D \quad \text{and} \quad F_L = \frac{1}{2}\rho A V^2 C_L \tag{8.1}$$

where, for a blunt body, the area A is the area projected on a plane normal to the flow direction, and for an airfoil the area A is the *chord* (the distance from the nose to the trailing edge) times the length.

The force due to the lower pressure in the separated region dominates the drag force on a blunt body, the subject of Sec. 8.2. The viscous stress that acts on and parallel to each boundary element is negligible and thus little, if any, attention is paid to the boundary layer on the surface of a blunt body. The opposite is true for an airfoil, the subject of Sec. 8.3; the drag force is due primarily to the viscous stresses that act on the boundary elements. Consequently, there is considerable interest in the boundary layer that develops on a streamlined body. It is this interest that has motivated much study of boundary layers. The basics of boundary-layer theory will be presented in Sec. 8.5. But first, the inviscid flow outside the boundary layer (Fig. 8.1) must be known. Therefore, inviscid flow theory will be presented in Sec. 8.4. The boundary layer is so thin that it can be ignored when solving for the inviscid flow. The inviscid flow solution provides the lift, which is not significantly influenced by the viscous boundary layer, and the pressure distribution on the body's surface as well as the velocity on that surface (since the inviscid solution ignores the effects of viscosity, the fluid does not stick to the boundary but slips by the boundary). The pressure and the velocity at the surface are needed in the boundary-layer solution.

8.2 FLOW AROUND BLUNT BODIES

8.2.1 Drag Coefficients

The primary flow parameter that influences the drag around a blunt body is the Reynolds number. If there is no free surface, the drag coefficients for both smooth and rough long cylinders and spheres are presented in Fig. 8.2; the values for streamlined cylinders and spheres are also included.

Separation always occurs in the flow of a fluid around a blunt body if the Reynolds number is sufficiently high. However, at low Reynolds numbers (it is called a *Stokes flow* if Re < 5), there is no separation and the drag coefficient, for a sphere, is given by

$$C_D = \frac{24}{\text{Re}} \qquad \text{Re} < 1 \tag{8.2}$$

Separation occurs for Re ≥ 10 beginning over a small area on the rear of the sphere until the separated region reaches a maximum at Re $\cong 1000$. The drag coefficient is then relatively constant until a sudden

---

[*] Actually, the body moves through the stationary fluid. To create a steady flow, the fluid is moved past the stationary body, as in a laboratory; the pressures and force remains the same. To obtain the actual velocity, the flow velocity is subtracted from the velocity at each point.

drop occurs in the vicinity of Re $= 2 \times 10^5$. This sudden drop is due to the transition of the boundary layer just before separation undergoing transition from a laminar flow to a turbulent flow. A turbulent boundary layer contains substantially more momentum and is able to move the separation region further to the rear; see the comparison in Fig. 8.3. The sudden decrease in drag could be as much as 80 percent. The surface of an object can be roughened to cause the boundary layer to undergo transition prematurely; the dimples on a golf ball accomplish this and increase the flight by up to 100 percent when compared with the flight of a smooth ball.

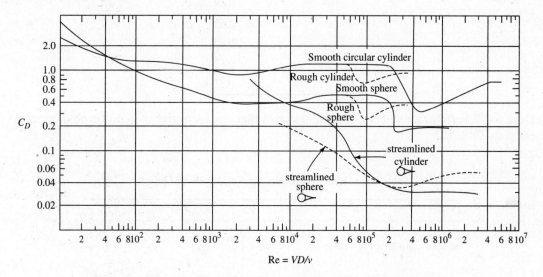

Figure 8.2   Drag coefficients for flow around spheres and long cylinders.

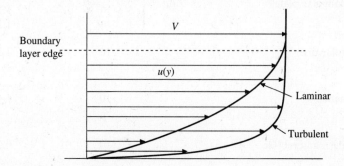

Figure 8.3   Laminar and turbulent velocity profiles for the same boundary-layer thickness.

After the sudden drop, the drag coefficient again increases with increased Reynolds number. Experimental data do not provide the drag coefficients for either the sphere or the cylinder for high Reynolds numbers. The values of 0.2 for smooth spheres and 0.4 for long smooth cylinders for Reynolds numbers exceeding 10^6 are often used.

Streamlining can substantially reduce the drag coefficients of blunt bodies. The drag coefficients for streamlined cylinders and spheres are shown in Fig. 8.2. The included angle at the trailing edge should not exceed about 20° if the separated region is to be minimized. The drag due to the shear stress acting on the enlarged surface will certainly increase for a streamlined body, but the drag due to the low pressure will be reduced much more so that the total drag will be less. Also, streamlining eliminates the vibrations that often occur when vortices are shed from a blunt body.

For cylinders of finite length with free ends, the drag coefficient must be reduced using the data of Table 8.1. If a finite-length cylinder has one end fixed to a solid surface, the length of the cylinder is doubled. Note that the L/D of a cylinder with free ends has to be quite large before the end effects are not significant.

Table 8.1   Drag Coefficients for Finite-Length Circular Cylinders* with Free Ends†

L/D	$C_D/C_{D\infty}$
∞	1
40	0.82
20	0.76
10	0.68
5	0.62
3	0.62
2	0.57
1	0.53

Drag coefficients for a number of common shapes that are insensitive to high Reynolds numbers are presented in Table 8.2.

Table 8.2   Drag Coefficients for Various Blunt Objects

Object	Re	C_D
Square cylinder of width w	$L/w = \begin{cases} \infty > 10^4 \\ 1 > 10^4 \end{cases}$	2.0 1.1
Rectangular plates	$L/w = \begin{cases} \infty > 10^3 \\ 20 > 10^3 \\ 5 > 10^3 \\ 1 > 10^3 \end{cases}$	2.0 1.5 1.2 1.1
Circular disc	$>10^3$	1.1
Parachute	$>10^7$	1.4
Modern automobile	$>10^5$	0.29
Van	$>10^5$	0.42
Bicycle	$\begin{cases} \text{upright rider} \\ \text{bent over rider} \\ \text{drafting rider} \end{cases}$	1.1 0.9 0.5
Semitruck	$\begin{cases} \text{standard} \\ \text{with streamlined deflector} \\ \text{with deflector and gap seal} \end{cases}$	0.96 0.76 0.70

---

* $C_{D\infty}$ is the drag coefficient from Fig. 8.2.
† If one end is fixed to a solid surface, double the length of the cylinder.

EXAMPLE 8.1 A 5-cm-diameter, 6-m-high pole fixed in concrete supports a flat, circular 4-m-diameter sign. Estimate the maximum moment that must be resisted by the concrete for a wind speed of 30 m/s.

Solution: To obtain the maximum moment, the wind is assumed normal to the sign. From Table 8.2, the drag coefficient for a disk is 1.1. The moment due to the drag force, which acts at the center of the sign, is

$$M_1 = F_{D1} \times L_1 = \frac{1}{2}\rho A_1 V^2 C_{D1} \times L_1 = \frac{1}{2} \times 1.22 \times \pi \times 2^2 \times 30^2 \times 1.1 \times 8 = 60\,700\,\text{N·m}$$

where the density at an elevation above sea level of 0 is used since the elevation is not given. The moment due to the pole is

$$M_2 = F_{D2} \times L_2 = \frac{1}{2}\rho A_2 V^2 C_{D2} \times L_2 = \frac{1}{2} \times 1.22 \times 0.05 \times 6 \times 30^2 \times 0.7 \times 3 = 346\,\text{N·m}$$

using a Reynolds number of Re $= 30 \times 0.05/1.5 \times 10^{-5} = 10^{-5}$ and assuming high-intensity fluctuations in the air flow, i.e., a rough cylinder. The factor from Table 8.1 was not used since neither end was free.

The moment that must be resisted by the concrete base is

$$M = M_1 + M_2 = 60\,700 + 346 = 61\,000\,\text{N·m}$$

8.2.2 Vortex Shedding

Long cylindrical bodies exposed to a fluid flow can exhibit the phenomenon of vortex shedding at relatively low Reynolds numbers. Vortices are shed from electrical wires, bridges, towers, and underwater communication wires, and can actually cause significant damage. We will consider the vortices shed from a long circular cylinder. The shedding occurs alternately from each side of the cylinder, as sketched in Fig. 8.4. The shedding frequency f, Hz, is given by the Strouhal number,

$$\text{St} = \frac{fD}{V} \tag{8.3}$$

If this shedding frequency is the same or a multiple of a structure's frequency, then there is the possibility that damage may occur due to resonance.

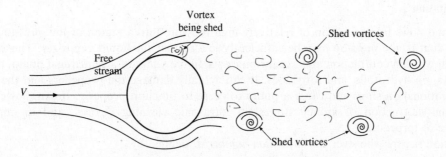

Figure 8.4 Vortices shed from a cylinder.

The Strouhal number cannot be calculated from equations; it is determined experimentally and shown in Fig. 8.5. Note that vortex shedding initiates at Re ≈ 40 and for Re ≥ 300 the Strouhal number is essentially independent of Reynolds number and is equal to about 0.21. The vortex shedding phenomenon disappears for Re $> 10^4$.

EXAMPLE 8.2 A 6-cm-diameter cylinder is used to measure the velocity of a slow-moving air stream. Two pressure taps are used to determine that the vortices are shed with a frequency of 4 Hz. Determine the velocity of the air stream.

Solution: Assume the Strouhal number to be in the range $300 < \text{Re} < 10\,000$. Then

$$\frac{fD}{V} = 0.21 \qquad \text{so that} \qquad V = \frac{4 \times 0.06}{0.21} = 1.14 \, \text{m/s}$$

It is quite difficult to measure the velocity of an air stream this low. The measurement of the shed vortices is one method of doing so.

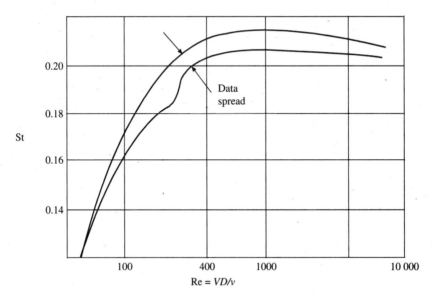

St

Re = VD/ν

Figure 8.5   Strouhal number for vortex shedding from a cylinder.

8.2.3   Cavitation

When a liquid flows from a region of relatively high pressure into a region of low pressure, *cavitation* may occur, that is, the pressure may be sufficiently low so that the liquid vaporizes. This can occur in pipe flows in which a contraction and expansion exists, in the vanes of a centrifugal pump, near the tips of propellers, on hydrofoils, and torpedoes. It can actually damage the propellers and the steel shafts (due to vibrations) on ships and cause a pump to cease to function properly. It can, however, also be useful in the destruction of kidney stones, in ultrasonic cleaning devices, and in improving the performance of torpedoes.

Cavitation occurs whenever the *cavitation number* σ, defined by

$$\sigma = \frac{p_\infty - p_v}{\frac{1}{2}\rho V^2} \tag{8.4}$$

is less than the critical cavitation number σ_{crit}, which depends on the geometry and the Reynolds number. In Eq. (8.4) p_∞ is the absolute pressure in the free stream and p_v the vapor pressure of the liquid.

The drag coefficient of a body that experiences cavitation is given by

$$C_D(\sigma) = C_D(0)(1 + \sigma) \tag{8.5}$$

where $C_D(0)$ is given in Table 8.3 for several bodies for Re $\cong 10^5$.

The *hydrofoil*, an airfoil-type shape that is used to lift a vessel above the water surface, invariably cannot operate without cavitation. The area and Reynolds number are based on the chord length. The drag and lift coefficients along with the critical cavitation numbers are presented in Table 8.4.

Table 8.3   Drag Coefficients for Zero Cavitation Numbers at Re $\cong 10^5$

Geometry	Angle	$C_D(0)$
Sphere		0.30
Disk (circular)		0.8
Circular cylinder		0.50
Flat plate (rectangular)		0.88
Two-dimensional Wedge	120	0.74
	90	0.64
	60	0.49
	30	0.28
Cone (axisymmetric)	120	0.64
	90	0.52
	60	0.38
	30	0.20

Table 8.4   Drag and Lift Coefficients and Critical Cavitation Numbers for Hydrofoils for $10^5 < Re < 10^6$

Angle (°)	Lift coefficient	Drag coefficient	Critical cavitation number
−2	0.2	0.014	0.5
0	0.4	0.014	0.6
2	0.6	0.015	0.7
4	0.8	0.018	0.8
6	0.95	0.022	1.2
8	1.10	0.03	1.8
10	1.22	0.04	2.5

EXAMPLE 8.3   A 2-m-long hydrofoil with chord length 40 cm operates 30 cm below the water's surface with an angle of attack of 6°. For a speed of 16 m/s determine the drag and lift and decide if cavitation exists on the hydrofoil.

   Solution: The pressure p_∞ must be absolute. It is

$$p_\infty = \gamma h + p_{\text{atm}} = 9800 \times 0.3 + 100\,000 = 102\,900 \text{ Pa abs}$$

Assuming that the water temperature is about 15°C, the vapor pressure is 1600 Pa (Table C.1) and the cavitation number is

$$\sigma = \frac{p_\infty - p_v}{\frac{1}{2}\rho V^2} = \frac{102\,900 - 1705}{0.5 \times 1000 \times 16^2} = 0.79$$

This is less than the critical cavitation number of 1.2 given in Table 8.4 and hence cavitation is present. Note that we could have used $p_v = 0$, as is often done, with sufficient accuracy.

   The drag and lift are

$$F_D = \frac{1}{2}\rho V^2 A C_D = \frac{1}{2} \times 1000 \times 16^2 \times 2 \times 0.4 \times 0.022 = 2250 \text{ N}$$

$$F_L = \frac{1}{2}\rho V^2 A C_L = \frac{1}{2} \times 1000 \times 16^2 \times 2 \times 0.4 \times 0.95 = 97\,300 \text{ N}$$

8.2.4 Added Mass

When a body is accelerated in a fluid, some of the surrounding fluid is also accelerated. This requires a larger force than that required to accelerate only the body. To account for the increased mass that must be accelerated, an *added mass* m_a is simply added to the body to calculate the force. For motion in the horizontal plane, the force needed to accelerate a body is given by

$$F - F_D = (m + m_a)\frac{dV}{dt} \tag{8.6}$$

where F_D is the drag force. If the body is accelerating from rest, the drag force would be 0.

The added mass is related to the mass of fluid m_f displaced by the body. The relationship

$$m_a = k m_f \tag{8.7}$$

provides the added mass if the factor k is known. For a sphere, $k = 0.5$; for an ellipsoid with major axis twice the minor axis and moving in the direction of the major axis, $k = 0.2$; and for a long cylinder moving normal to its axis, $k = 1.0$. These values are for inviscid flows so they are used when starting from rest or at very low speeds.

For dense bodies accelerating in air the added mass can be ignored, but for bodies accelerating in a liquid, the added mass must be included.

EXAMPLE 8.4 A sphere with a specific gravity of 3 is released from rest in a body of water. Determine its initial acceleration.

Solution: Apply Newton's second law including the buoyant force:

$$W - F_B = (m + m_a)\frac{dV}{dt}$$

$$(S\gamma_{\text{water}} - \gamma_{\text{water}})V\!\!\!/_{\text{sphere}} = (S\rho_{\text{water}} + 0.5\rho_{\text{water}})V\!\!\!/_{\text{sphere}}\frac{dV}{dt}$$

$$\therefore \frac{dV}{dt} = \frac{(3-1)g}{3+0.5} = 5.6\,\text{m/s}^2$$

where we have used $\gamma = \rho g$.

8.3 FLOW AROUND AIRFOILS

Airfoils are streamlined so that separation does not occur. Airfoils designed to operate at subsonic speeds are rounded at the leading edge whereas those designed for supersonic speeds may have sharp leading edges. The drag on an airfoil is due primarily to the shear stress that acts on the surface; there is some drag due to the pressure distribution. The boundary layer, in which all the shear stresses are confined, that develops on an airfoil is very thin (see the sketch in Fig. 8.6) and can be ignored when solving for the inviscid flow surrounding the airfoil. The pressure distribution that is determined from the inviscid flow solution is influenced very little by the presence of the boundary layer. Consequently, the lift is estimated on an airfoil by ignoring the boundary layer and integrating the pressure distribution of the inviscid flow. The inviscid flow solution also provides the velocity at the outer edge of the thin boundary layer, a boundary condition needed when solving the boundary-layer equations; the solution of the boundary-layer equations will be presented in Sec. 8.5.

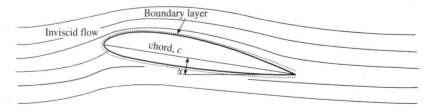

Figure 8.6 Flow around an airfoil at an angle of attack α.

The lift and drag on airfoils will not be calculated from the flow conditions but from graphical values of the lift and drag coefficients. These are displayed in Fig. 8.7 for a conventional airfoil with $\text{Re} = 9 \times 10^6$. The lift and drag coefficients are defined as

$$C_L = \frac{F_L}{\frac{1}{2}\rho c L V^2} \qquad\qquad C_D = \frac{F_D}{\frac{1}{2}\rho c L V^2} \tag{8.8}$$

Conventional airfoils are not symmetric and are designed to have positive lift at zero angle of attack, as shown in Fig. 8.7. The lift is directly proportional to the angle of attack until just before stall is encountered. The drag coefficient is also directly proportional to the angle of attack up to about 5°. The cruise condition is at an angle of attack of about 5° where the drag is a minimum at $C_L = 0.3$ as noted. Mainly the wings supply the lift on an aircraft but an effective length is the tip-to-tip distance, the *wingspan*, since the fuselage also supplies some lift.

The drag coefficient is essentially constant up to a Mach number of about 0.75. It then increases by over a factor of 10 until a Mach number of 1 is reached at which point it begins to slowly decrease. So, cruise Mach numbers between 0.75 and 1.5 are avoided to stay away from the high drag coefficients. Swept-back airfoils are used since it is the normal component of velocity that is used when calculating the Mach number, which allows a higher plane velocity before the larger drag coefficients are encountered.

Slotted flaps are also used to provide larger lift coefficients during takeoff and landing. Air flows from the high-pressure region on the bottom of the airfoil through a slot to energize the slow-moving air in the boundary layer on the top side of the airfoil thereby reducing the tendency to separate and stall. The lift coefficient can reach 2.5 with a single-slotted flap and 3.2 with two slots.

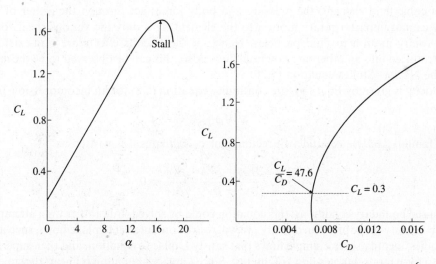

Figure 8.7   Lift and drag coefficients for a conventional airfoil at $\text{Re} = 9 \times 10^6$.

EXAMPLE 8.5   Determine the takeoff speed for an aircraft that weighs 15 000 N including its cargo if its wingspan is 15 m with a 2-m chord. Assume an angle of 8° at takeoff.

   Solution: Assume a conventional airfoil and use the lift coefficient of Fig. 8.7 of about 0.95. The velocity is found from the equation for the lift coefficient

$$C_L = \frac{F_L}{\frac{1}{2}\rho c L V^2} \qquad 0.95 = \frac{15\,000}{\frac{1}{2} \times 1.2 \times 2 \times 15 \times V^2}. \qquad \therefore \ V = 30\,\text{m/s}$$

The answer is rounded off to two significant digits since the lift coefficient of 0.95 is read from the figure.

8.4 POTENTIAL FLOW

8.4.1 Basics

When a body is moving in an otherwise stationary fluid, there is no vorticity present in the undisturbed fluid. To create a steady flow, a uniform flow with the body's velocity is superimposed on the flow field so that the vorticity-free flow moves by the stationary body, as in a wind tunnel. The only way vorticity is introduced into the flow is through the effects of viscosity. For high-Reynolds number flows, the viscous effects are concentrated in the boundary layer and the wake (the wake includes the separated region). For a streamlined body and over the front part of a blunt body, the flow outside the boundary layer is free of viscous effects and thus vorticity so it is an inviscid flow. The solution of the inviscid flow problem provides the velocity field and the pressure field in the vicinity of the body. The pressure is not significantly influenced by the boundary layer so it will provide the lift when integrated over the body's surface. The velocity at the boundary of the body* from the inviscid flow solution will be the velocity at the outer edge of the thin boundary layer, needed in the boundary-layer solution (to be presented in Sec. 8.5). So, before the boundary layer can be analyzed on a body, the inviscid flow must be known.

A *potential flow* (or irrotational flow) is one in which the velocity field can be expressed as the gradient of a scalar function, that is,

$$\mathbf{V} = \nabla \phi \tag{8.9}$$

where ϕ is the *velocity potential*. For a potential flow, the vorticity is 0

$$\boldsymbol{\omega} = \nabla \times \mathbf{V} = 0 \tag{8.10}$$

This can be shown to be true by expanding in rectangular coordinates and using Eq. (8.9).

To understand why an irrotational flow cannot generate vorticity consider the three types of forces that act on a cubic fluid element: the pressure and body forces act through the center of element and consequently cannot impart a rotary motion to the element. It is only the viscous shear forces that are able to give rotary motion to fluid particles. Hence, if the viscous effects are non-existent, vorticity cannot be introduced into an otherwise potential flow. Also, this can be observed to be the case by taking the curl of the Navier–Stokes equation (5.20).

If the velocity is given by Eq. (8.9), the continuity equation (5.8) for an incompressible flow provides

$$\nabla \cdot \nabla \phi = \nabla^2 \phi = 0 \tag{8.11}$$

which is the famous *Laplace equation*. In rectangular coordinates it is written as

$$\frac{\partial^2 \phi}{\partial x^2} + \frac{\partial^2 \phi}{\partial y^2} + \frac{\partial^2 \phi}{\partial z^2} = 0 \tag{8.12}$$

With the required boundary conditions, this equation could be solved. But, rather than attempting to solve the resulting boundary-value problem directly, we will restrict our interest to plane flows, such as airfoils and cylindrical bodies, identify several simple flows that satisfy Laplace's equation, and then superimpose those simple flows to form more complex flows of interest. Since Laplace's equation is linear, the superposed flows will also satisfy Laplace's equation.

First, however, let us define another scalar function that will be quite useful in our study. For the plane flows of interest, the *stream function* ψ, is defined by

$$u = \frac{\partial \psi}{\partial y} \quad \text{and} \quad v = -\frac{\partial \psi}{\partial x} \tag{8.13}$$

so that the continuity equation (5.8) with $\partial w/\partial z = 0$ (for a plane flow) is satisfied for all plane flows. The vorticity, Eqs. (8.10) and (3.14), then provides

---

* If there are insignificant viscous effects, the fluid does not stick to a boundary but is allowed to slip.

$$\omega_z = \frac{\partial v}{\partial x} - \frac{\partial u}{\partial y} = -\frac{\partial^2 \psi}{\partial x^2} - \frac{\partial^2 \psi}{\partial y^2} = 0 \tag{8.14}$$

so that

$$\frac{\partial^2 \psi}{\partial x^2} + \frac{\partial^2 \psi}{\partial y^2} = 0 \tag{8.15}$$

The stream function also satisfies the Laplace equation. So, from the above equations we have

$$u = \frac{\partial \phi}{\partial x} = \frac{\partial \psi}{\partial y} \quad \text{and} \quad v = \frac{\partial \phi}{\partial y} = -\frac{\partial \psi}{\partial x} \tag{8.16}$$

The equations between ϕ and ψ in Eq. (8.16) form the *Cauchy–Riemann equations* and ϕ and ψ are referred to as *harmonic functions*. The function $\phi + i\psi$ is the *complex velocity potential*. The powerful mathematical theory of complex variables is thus applicable to this subset of fluid flows: steady, incompressible plane flows.

　　Three items of interest contained in the above equations are:

- The stream function is constant along a streamline.
- The streamlines and lines of constant potential lines intersect at right angles.
- The difference of the stream functions between two streamlines is the flow rate q per unit depth between the two streamlines, i.e., $q = \psi_2 - \psi_1$.

These items will be shown to be true in the examples and solved problems.

EXAMPLE 8.6 Show that ψ is constant along a streamline.

　　Solution: A streamline is a line to which the velocity vector is tangent. This is expressed in vector form as $\mathbf{V} \times d\mathbf{r} = 0$, which, for a plane flow (no z variation), using $d\mathbf{r} = dx\,\mathbf{i} + dy\,\mathbf{j}$ takes the form $u\,dy - v\,dx = 0$. Using Eq. (8.13), this becomes

$$\frac{\partial \psi}{\partial y} dy + \frac{\partial \psi}{\partial x} dx = 0$$

This is the definition of $d\psi$ from calculus, thus $d\psi = 0$ along a streamline, or, in other words, ψ is constant along a streamline.

8.4.2　Several Simple Flows

Several of the simple flows to be presented are much easier understood using polar (cylindrical) coordinates. The Laplace equation, the continuity equation, and the expressions for the velocity components for a plane flow (Table 5.1) are

$$\nabla^2 \psi = \frac{1}{r}\frac{\partial}{\partial r}\left(r\frac{\partial \psi}{\partial r}\right) + \frac{1}{r^2}\frac{\partial^2 \psi}{\partial \theta^2} = 0 \tag{8.17}$$

$$\frac{1}{r}\frac{\partial}{\partial r}(rv_r) + \frac{1}{r}\frac{\partial v_\theta}{\partial \theta} = 0 \tag{8.18}$$

$$v_r = \frac{\partial \phi}{\partial r} = \frac{1}{r}\frac{\partial \psi}{\partial \theta} \quad \text{and} \quad v_\theta = \frac{1}{r}\frac{\partial \phi}{\partial \theta} = -\frac{\partial \psi}{\partial r} \tag{8.19}$$

where the expressions relating the velocity components to the stream function are selected so that the continuity equation is always satisfied. We now define four simple flows that satisfy the Laplace equation.

$$\textit{Uniform flow:} \quad \psi = U_\infty y \quad \phi = U_\infty x \tag{8.20}$$

$$\textit{Line source}\ : \quad \psi = \frac{q}{2\pi}\theta \quad \phi = \frac{q}{2\pi}\ln r \tag{8.21}$$

$$\text{Vortex}: \qquad \psi = \frac{\Gamma}{2\pi}\ln r \qquad \phi = \frac{\Gamma}{2\pi}\theta \qquad\qquad (8.22)$$

$$\text{Doublet}: \qquad \psi = -\frac{\mu\sin\theta}{r} \qquad \phi = -\frac{\mu\cos\theta}{r} \qquad\qquad (8.23)$$

These simple plane flows are sketched in Fig. 8.8. If a y-component is desired for the uniform flow, an appropriate term is added. The *source strength q* in the line source is the flow rate per unit depth; adding a minus sign creates a sink. The *vortex strength* Γ is the *circulation* about the origin, defined as

$$\Gamma = \oint_L \mathbf{V}\cdot d\mathbf{s} \qquad\qquad (8.24)$$

where L is a closed curve, usually a circle, about the origin with clockwise being positive. The heavy arrow in the negative x-direction represents the *doublet strength* μ in Fig. 8.8(d). (A doublet can be thought of as a source and a sink of equal strengths separated by a very small distance.)

The velocity components are used quite often for the simple flows presented. They follow for both polar and rectangular coordinates:

$$\text{Uniform flow}: \qquad \begin{aligned} u &= U_\infty & v &= 0 \\ v_r &= U_\infty\cos\theta & v_\theta &= -U_\infty\sin\theta \end{aligned} \qquad\qquad (8.25)$$

$$\text{Line source}: \qquad \begin{aligned} v_r &= \frac{q}{2\pi r} & v_\theta &= 0 \\ u &= \frac{q}{2\pi}\frac{x}{x^2+y^2} & v &= \frac{q}{2\pi}\frac{y}{x^2+y^2} \end{aligned} \qquad\qquad (8.26)$$

$$\text{Vortex}: \qquad \begin{aligned} v_r &= 0 & v_\theta &= -\frac{\Gamma}{2\pi r} \\ u &= -\frac{\Gamma}{2\pi}\frac{y}{x^2+y^2} & v &= \frac{\Gamma}{2\pi}\frac{x}{x^2+y^2} \end{aligned} \qquad\qquad (8.27)$$

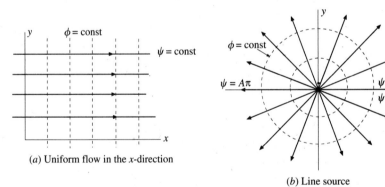

(a) Uniform flow in the x-direction

(b) Line source

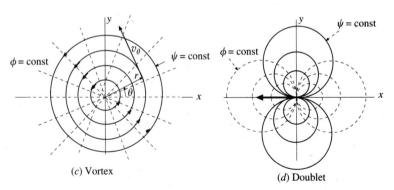

(c) Vortex

(d) Doublet

Figure 8.8   Four simple plane potential flows.

$$Doublet: \quad \begin{array}{ll} v_r = -\dfrac{\mu \cos\theta}{r^2} & v_\theta = -\dfrac{\mu \sin\theta}{r^2} \\[4mm] u = -\mu\dfrac{x^2 - y^2}{(x^2 + y^2)^2} & v = -\mu\dfrac{2xy}{(x^2 + y^2)^2} \end{array} \qquad (8.28)$$

These four simple flows can be superimposed to create more complicated flows of interest. This will be done in the following section.

EXAMPLE 8.7 If the stream function of a flow is given as $\psi = A\theta$, determine the potential function ϕ.

Solution: We use Eq. (8.19) to relate the stream function to the potential function assuming polar coordinates because of the presence of θ:

$$\frac{\partial \phi}{\partial r} = \frac{1}{r}\frac{\partial \psi}{\partial \theta} = \frac{A}{r}. \qquad \therefore \ \phi(r,\theta) = A\ln r + f(\theta)$$

Now, use the second equation of Eq. 8.19:

$$\frac{1}{r}\frac{\partial \phi}{\partial \theta} = \frac{1}{r}\frac{df}{d\theta} = -\frac{\partial \psi}{\partial r} = 0 \qquad \text{implying that} \qquad \frac{df}{d\theta} = 0 \qquad \text{so that} \qquad f = \text{Const}$$

Since we are only interested in the derivatives of the potential functions to provide the velocity and pressure fields, we simply let the constant be 0 and thus

$$\phi(r,\theta) = A\ln r$$

So, we see that the potential function can be found if the stream function is known. Also, the stream function can be found if the potential function is known.

8.4.3 Superimposed flows

Combining the simple flows introduced in Sec. 8.4.2 can create the most complicated plane flows. Divide a surface, such as an airfoil, into a large number of segments and place sources or sinks or doublets at the center of each segment; in addition, add a uniform flow and a vortex. Then, adjust the various strengths so that the normal velocity component at each segment is 0 and the rear stagnation point is located at the trailing edge. Obviously, a computer program must be used to create flow around an airfoil. We will not attempt it in this book but will demonstrate how flow around a circular cylinder can be created.

Superimpose the stream functions of a uniform flow and a doublet

$$\psi(r,\theta) = U_\infty y - \frac{\mu \sin\theta}{r} \qquad (8.29)$$

The velocity component v_r is (let $y = r\sin\theta$)

$$v_r = \frac{1}{r}\frac{\partial \psi}{\partial \theta} = U_\infty \cos\theta - \frac{\mu}{r^2}\cos\theta \qquad (8.30)$$

A circular cylinder exists if there is a circle on which there is no radial velocity component, i.e., $v_r = 0$ at $r = r_c$. Set $v_r = 0$ in Eq. (8.30) and find

$$U_\infty \cos\theta - \frac{\mu}{r_c^2}\cos\theta = 0 \qquad \text{so that} \qquad r_c = \sqrt{\frac{\mu}{U_\infty}} \qquad (8.31)$$

At this radius $v_r = 0$ for all θ and thus $r = r_c$ is a streamline and the result is flow around a cylinder. The stagnation points occur where the velocity is 0; if $r = r_c$ this means where $v_\theta = 0$, that is,

$$v_\theta = -\frac{\partial \psi}{\partial r}\bigg|_{r=r_c} = -U_\infty \sin\theta - \frac{\mu \sin\theta}{r_c^2} = 0. \qquad \therefore \ -2U_\infty \sin\theta = 0 \qquad (8.32)$$

Thus, two stagnation points occur at $\theta = 0°$ and $180°$. The streamline pattern would appear as in the sketch of Fig. 8.9. The circular streamline represents the cylinder, which is typically a solid, and hence our interest is in the flow outside the circle. For a real flow, there would be a separated region on the rear of the cylinder but the flow over the front part (perhaps over the whole front half, depending on the Reynolds number) could be approximated by the potential flow shown in the sketch. The velocity that exists outside the thin boundary layer that would be present on a real cylinder would be approximated as the velocity on the cylinder of the potential flow, i.e., it would be given by

$$v_\theta = -2U_\infty \sin\theta \tag{8.33}$$

The pressure that would exist on the cylinder's surface would be found by applying Bernoulli's equation between the stagnation point where the pressure is p_0 and $V = 0$ and some general point at r_c and θ:

$$p_c = p_0 - \rho\frac{v_\theta^2}{2} \tag{8.34}$$

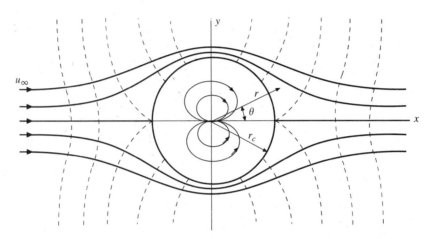

Figure 8.9   Potential flow around a circular cylinder. (The dashed lines are lines of constant ϕ.)

This pressure would approximate the actual pressure for high-Reynolds number flows up to separation. For low-Reynolds number flows, say below Re $\approx$ 50, viscous effects are not confined to a thin boundary layer so potential flow does not approximate the real flow.

   To create flow around a rotating cylinder, as in Fig. 8.10, add a vortex to the stream function of Eq. (8.29) (use the cylinder's radius of Eq. (8.31)):

$$\psi(r,\theta) = U_\infty y - r_c^2 U_\infty \frac{\sin\theta}{r} + \frac{\Gamma}{2\pi}\ln r \tag{8.35}$$

recognizing that the cylinder's radius remains unchanged since a vortex does not effect v_r. The stagnation points change, however, and are located by letting $v_\theta = 0$ on $r = r_c$:

$$v_\theta = -\frac{\partial\psi}{\partial r}\bigg|_{r=r_c} = -U_\infty \sin\theta - r_c^2 U_\infty \frac{\sin\theta}{r_c^2} - \frac{\Gamma}{2\pi r_c} = 0 \tag{8.36}$$

This locates the stagnation points at

$$\theta = \sin^{-1}\frac{-\Gamma}{4\pi r_c U_\infty} \tag{8.37}$$

If $\Gamma > 4\pi r_c U_\infty$, Eq. (8.37) is not valid (this would give $|\sin\theta| > 1$) so the stagnation point exists off the cylinder as sketched in Fig. 8.10(b). The angle $\theta = 270°$ and the radius are found by setting the velocity components equal to 0. Problems will illustrate.

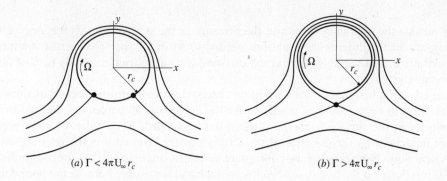

Figure 8.10   Flow around a rotating cylinder.

The pressure on the surface of the rotating cylinder using Bernoulli's equation is found to be

$$p_c = p_0 - \rho\frac{v_c^2}{2} = p_0 - \rho\frac{U_\infty^2}{2}\left(2\sin\theta + \frac{\Gamma}{2\pi r_c U_\infty}\right)^2 \tag{8.38}$$

If this is integrated around the surface of the cylinder, the component in the flow direction, the drag, would be 0 and the component normal to the flow direction, the lift, would be

$$F_L = \int_0^{2\pi} p_c\sin\theta\, r_c\, d\theta = \rho U_\infty\Gamma \tag{8.39}$$

It turns out that this expression for the lift is applicable for all cylinders including the airfoil. It is known as the *Kutta–Joukowski theorem*; it is exact for potential flows and is an approximation for real flows.

EXAMPLE 8.8   A 20-cm-diameter cylinder rotates clockwise at 200 rpm in an atmospheric air stream flowing at 10 m/s. Locate any stagnation points and find the minimum pressure.

Solution: First, let us find the circulation. It is $\Gamma = \oint_L \mathbf{V}\cdot d\mathbf{s}$, the velocity $r_c\Omega$ multiplied by $2\pi r_c$, recognizing that $\mathbf{V}$ is in the direction of $d\mathbf{s}$ on the cylinder's surface:

$$\Gamma = 2\pi r_c^2\Omega = 2\pi\times0.1^2\times(200\times2\pi/60) = 1.318\,\text{m}^2/\text{s}$$

This is less than $4\pi r_c U_\infty = 12.57\,\text{m}^2/\text{s}$ so the two stagnation points are on the cylinder at

$$\theta = \sin^{-1}\frac{-\Gamma}{4\pi r_c U_\infty} = \sin^{-1}\frac{-1.318}{4\pi\times0.1\times10} = -6° \quad\text{and}\quad 186°$$

The minimum pressure exists at the very top of the cylinder (Fig. 8.10 and Eq. (8.38)), so let us apply Bernoulli's equation between the free stream and the point on the top where $\theta = 90°$:

$$p_c = p_\infty + \rho\frac{U_\infty^2}{2} - \rho\frac{U_\infty^2}{2}\left(2\sin\theta + \frac{\Gamma}{2\pi r_c U_\infty}\right)^2$$

$$= 0 + 1.2\times\frac{10^2}{2}\left[1 - \left(2\sin90° + \frac{1.318}{2\pi\times0.1\times10}\right)^2\right] = -233\,\text{Pa}$$

using $\rho = 1.2\,\text{kg/m}^3$ for atmospheric air. (If the temperature is not given, assume standard conditions.)

8.5   BOUNDARY-LAYER FLOW

8.5.1   General Information

Undoubtedly, the identification of a boundary layer resulted from interest in the airfoil. The observation that for a high-Reynolds number flow all the viscous effects can be confined to a thin layer of fluid near the surface gave rise to boundary-layer theory. Outside the boundary layer the fluid acts as an inviscid fluid since viscous effects are negligible. So, the potential flow theory of the previous section provides the velocity just outside the boundary layer and the pressure at the surface. In this section, we will provide both the integral and the differential equations needed to solve for the velocity distribution. But, since those equations are difficult to solve for curved surfaces, we will restrict our study to flow on a flat plate with zero pressure gradient.

The outer edge of a boundary layer cannot be observed so we arbitrarily assign its thickness $\delta(x)$, as shown in Fig. 8.11, to be the locus of points where the velocity is 99 percent of the *free-stream velocity* $U(x)$ (the velocity at the surface from the inviscid flow solution). Recall also that the pressure at the surface is not influenced by the presence of the thin boundary layer so it is the pressure on the surface from the inviscid flow. Note that the xy-coordinate system is oriented so that the x-coordinate is along the surface; this is done for the boundary-layer equations and is possible because the boundary layer is so thin that curvature terms do not appear in the describing equations.

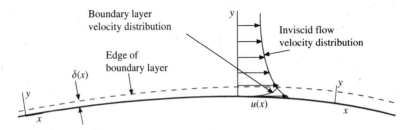

Figure 8.11   A boundary layer.

A boundary layer is laminar near the leading edge or near a stagnation point. It undergoes transition at x_T to a turbulent flow if there is sufficient length, as shown in Fig. 8.12. This transition occurs when the *critical Reynolds number* $U_\infty x_T/\nu = 5 \times 10^5$ on smooth rigid flat plates in a zero pressure-gradient flow with low free-stream fluctuation intensity[*] and $U_\infty x_T/\nu = 3 \times 10^5$ for flow on rough flat plates or with high free-stream fluctuation intensity (intensity of at least 0.1). The transition region from laminar to turbulent flow is relatively short and is typically ignored so a turbulent flow is assumed to exist at the location of the first burst.

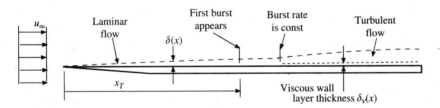

Figure 8.12   A boundary layer undergoing transition.

The turbulent boundary layer thickens more rapidly than a laminar boundary layer and contains significantly more momentum (if it has the same thickness), as observed from a sketch of the velocity profiles in Fig. 8.13. It also has a much greater slope at the wall resulting in a much larger wall shear stress. The instantaneous turbulent boundary layer varies randomly with time and position and can be 20 percent thicker or 60 percent thinner at any position at an instant in time or at any time at a given position. So, we usually sketch a time-average boundary-layer thickness. The *viscous wall layer* with thickness δ_ν in which the viscous effects are thought to be concentrated in a turbulent boundary layer, is quite thin when compared with the boundary-layer thickness, as sketched.

It should be kept in mind that a turbulent boundary layer is very thin for most applications. On a flat plate with $U_\infty = 5\,\mathrm{m/s}$ the boundary layer would be about 7 cm thick after 4 m. If this were drawn to

---

[*] Fluctuation intensity is $\sqrt{\overline{u'^2}}/U_\infty$ [see Eq. (7.67)].

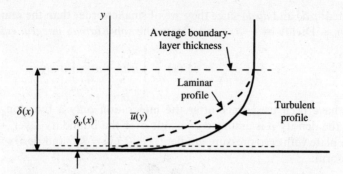

Figure 8.13   Laminar and turbulent boundary-layer profiles.

scale, the fact that the boundary layer is very thin would be quite apparent. Because the boundary layer is so thin and the velocity varies from 0 at the wall to $U(x)$ at the edge of the boundary layer, it is possible to approximate the velocity profile in the boundary layer by assuming a parabolic or cubic profile for a laminar layer and a power-law profile for a turbulent layer. With the velocity profile assumed, the integral equations, which follow, give the quantities of interest.

8.5.2   The Integral Equations

An infinitesimal control volume of thickness dx is shown in Fig. 8.14 with mass fluxes in (b) and momentum fluxes in (d). The continuity equation provides the mass flux $\dot{m}_{top}$ that crosses into the control volume through the top; it is

$$\dot{m}_{top} = \dot{m}_{out} - \dot{m}_{in} = \frac{\partial}{\partial x}\left(\int_0^\delta \rho u\, dy\right) dx \tag{8.40}$$

The x-component momentum equation (Newton's second law) is written as

$$\sum F_x = \dot{mom}_{out} - \dot{mom}_{in} - \dot{mom}_{top} \tag{8.41}$$

which becomes

$$-\tau_0\, dx - \delta\, dp = \frac{\partial}{\partial x}\left(\int_0^\delta \rho u^2\, dy\right) dx - U(x)\frac{\partial}{\partial x}\left(\int_0^\delta \rho u\, dy\right) dx \tag{8.42}$$

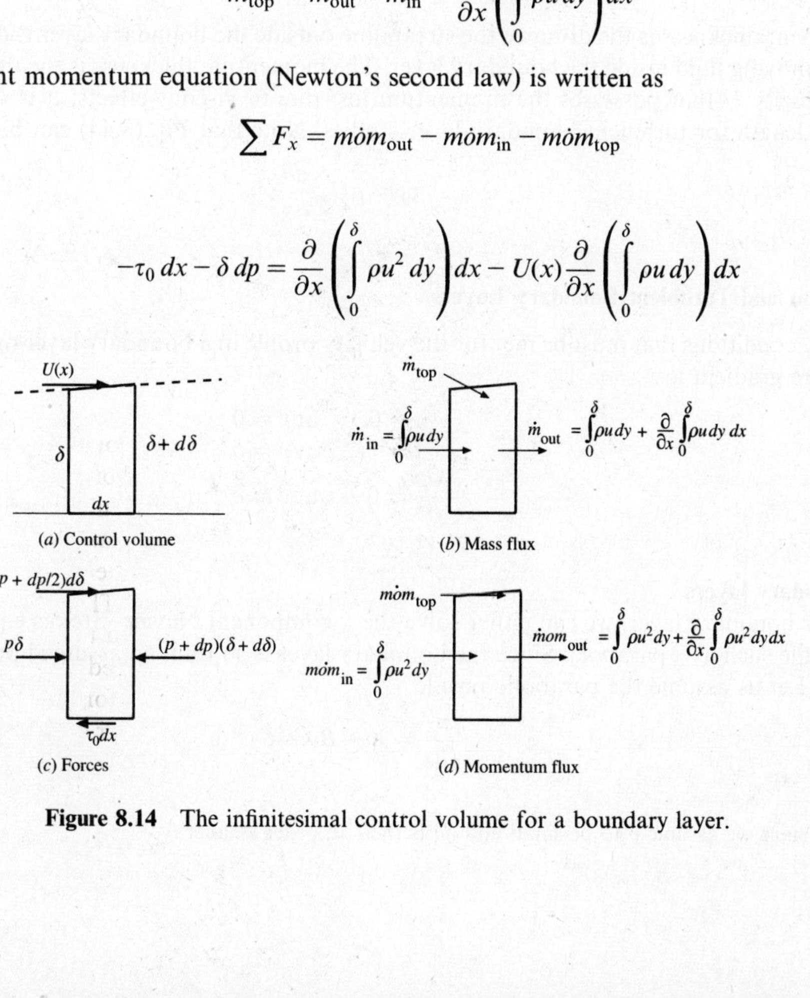

(a) Control volume

(b) Mass flux

$\dot{m}_{in} = \int_0^\delta \rho u\, dy$ $\dot{m}_{out} = \int_0^\delta \rho u\, dy + \frac{\partial}{\partial x}\int_0^\delta \rho u\, dy\, dx$

(c) Forces

(d) Momentum flux

$\dot{mom}_{in} = \int_0^\delta \rho u^2\, dy$ $\dot{mom}_{out} = \int_0^\delta \rho u^2\, dy + \frac{\partial}{\partial x}\int_0^\delta \rho u^2\, dy\, dx$

Figure 8.14   The infinitesimal control volume for a boundary layer.

where we have neglected* $p\,d\delta$ and $dp\,d\delta$ since they are of smaller order than the remaining terms; we also used $\dot{m}\dot{o}m_{top} = U(x)\dot{m}_{top}$. Divide by $(-dx)$ and obtain the *von Karman integral equation*:

$$\tau_0 + \delta\frac{dp}{dx} = \rho U(x)\frac{d}{dx}\int_0^\delta u\,dy - \rho\frac{d}{dx}\int_0^\delta u^2\,dy \qquad (8.43)$$

Ordinary derivatives have been used since after the integration only a function of x remains (δ is a function of x). Also, the density ρ is assumed constant over the boundary layer.

For flow on a flat plate with zero pressure gradient, i.e., $U(x) = U_\infty$ and $\partial p/\partial x = 0$, Eq. (8.43) can be put in the simplified form

$$\tau_0 = \rho\frac{d}{dx}\int_0^\delta u(U_\infty - u)dy \qquad (8.44)$$

If a velocity profile $u(x, y)$ is assumed for a particular flow, Eq. (8.44) along with $\tau_0 = \mu\partial u/\partial y|_{y=0}$ allows both $\delta(x)$ and $\tau_0(x)$ to be determined.

Two additional lengths are used in the study of boundary layers. They are the *displacement thickness* δ_d and the *momentum thickness* θ defined by

$$\delta_d = \frac{1}{U}\int_0^\delta (U - u)dy \qquad (8.45)$$

$$\theta = \frac{1}{U^2}\int_0^\delta u(U - u)dy \qquad (8.47)$$

The displacement thickness is the distance the streamline outside the boundary layer is displaced because of the slower moving fluid inside the boundary layer. The momentum thickness is the thickness of a fluid layer with velocity U that possesses the momentum lost due to viscous effects; it is often used as the characteristic length for turbulent boundary-layer studies. Note that Eq. (8.44) can be written as

$$\tau_0 = \rho U_\infty^2 \frac{d\theta}{dx} \qquad (8.48)$$

8.5.3 Laminar and Turbulent Boundary Layers

The boundary conditions that must be met for the velocity profile in a boundary layer on a flat plate with a zero pressure gradient are

$$\begin{aligned} u &= 0 & \text{at } y = 0\\ u &= U_\infty & \text{at } y = \delta\\ \frac{\partial u}{\partial y} &= 0 & \text{at } y = \delta \end{aligned} \qquad (8.49)$$

Laminar boundary layers

For a laminar boundary layer, we can either solve the x-component Navier–Stokes equation or we can assume a profile such as a parabola. Since the boundary layer is so thin, an assumed profile gives rather good results. Let us assume the parabolic profile

$$\frac{u}{U_\infty} = A + By + Cy^2 \qquad (8.50)$$

---

* $p\,d\delta$ is small since we assume δ to be small and $d\delta$ is then an order smaller.

The above three boundary conditions require

$$0 = A$$
$$1 = A + B\delta + C\delta^2 \tag{8.51}$$
$$0 = B + 2C\delta$$

the solution of which is

$$A = 0 \qquad B = \frac{2}{\delta} \qquad C = -\frac{1}{\delta^2} \tag{8.52}$$

resulting in the laminar-flow velocity profile

$$\frac{u}{U_\infty} = 2\frac{y}{\delta} - \frac{y^2}{\delta^2} \tag{8.53}$$

Substitute this profile into the integral equation (8.44) and integrate:

$$\tau_0 = \frac{d}{dx}\int_0^\delta \rho U_\infty^2 \left(\frac{2y}{\delta} - \frac{y^2}{\delta^2}\right)\left(1 - \frac{2y}{\delta} + \frac{y^2}{\delta^2}\right)dy = \frac{2}{15}\rho U_\infty^2 \frac{d\delta}{dx} \tag{8.54}$$

The wall shear stress is also given by

$$\tau_0 = \mu \frac{\partial u}{\partial y}\bigg|_{y=0} = \mu U_\infty \frac{2}{\delta} \tag{8.55}$$

Equate the two expressions for τ_0 above to obtain

$$\delta\, d\delta = \frac{15\nu}{U_\infty} dx \tag{8.56}$$

Integrate the above with $\delta = 0$ at $x = 0$ and find the expression for $\delta(x)$

$$\delta(x) = 5.48\sqrt{\frac{\nu x}{U_\infty}} \tag{8.57}$$

This is about 10 percent higher than the more accurate solution of $5\sqrt{\nu x/U_\infty}$ found by solving the Navier–Stokes equation in the next Sec. 8.5.4.

The wall shear stress is found by substituting Eq. (8.57) into Eq. (8.55) and is

$$\tau_0(x) = 0.365\rho U_\infty^2 \sqrt{\frac{\nu}{xU_\infty}} \tag{8.58}$$

The *local skin friction coefficient* c_f is often of interest and is

$$c_f(x) = \frac{\tau_0}{\frac{1}{2}\rho U_\infty^2} = 0.730\sqrt{\frac{\nu}{xU_\infty}} \tag{8.59}$$

The *skin friction coefficient* C_f is a dimensionless drag force and is

$$C_f = \frac{F_D}{\frac{1}{2}\rho U_\infty^2 L} = \frac{\int_0^L \tau_0\, dx}{\frac{1}{2}\rho U_\infty^2 L} = 1.46\sqrt{\frac{\nu}{U_\infty L}} \tag{8.60}$$

The more accurate coefficients for τ_0, c_f, and C_f are 0.332, 0.664, and 1.33, respectively, so the assumption of a parabolic velocity profile for laminar boundary-layer flow has an error of about 10 percent.

Turbulent boundary layers

For a turbulent boundary layer we often assume a power-law velocity profile[*] as we did for flow in a pipe. It is

$$\frac{\bar{u}}{U_\infty} = \left(\frac{y}{\delta}\right)^{1/n} \qquad n = \begin{cases} 7 & \text{Re}_x < 10^7 \\ 8 & 10^7 < \text{Re}_x < 10^8 \\ 9 & 10^8 < \text{Re}_x < 10^9 \end{cases} \qquad (8.61)$$

where $\text{Re}_x = U_\infty x/\nu$. Substitute this velocity profile with $n = 7$ into Eq. (8.44) and integrate to obtain

$$\tau_0 = \frac{7}{72}\rho U_\infty^2 \frac{d\delta}{dx} \qquad (8.62)$$

The power-law velocity profile yields $\tau_0 = \mu\, \partial\bar{u}/\partial y = \infty$ at $y = 0$ so it cannot be used at the wall. A second expression for τ_0 is needed; we select the *Blasius formula*, given by

$$c_f = 0.046\left(\frac{\nu}{U_\infty \delta}\right)^{1/4} \qquad \text{giving} \qquad \tau_0 = 0.023\rho U_\infty^2\left(\frac{\nu}{U_\infty \delta}\right)^{1/4} \qquad (8.63)$$

Combine Eqs. (8.62) and (8.63) and find

$$\delta^{1/4}\, d\delta = 0.237\left(\frac{\nu}{U_\infty}\right)^{1/4} dx \qquad (8.64)$$

Assume a turbulent flow from the leading edge (the laminar portion is often quite short) and integrate from 0 to x:

$$\delta = 0.38x\left(\frac{\nu}{U_\infty x}\right)^{1/5} \qquad \text{Re}_x < 10^7 \qquad (8.65)$$

Substitute this into the Blasius formula and find the local skin friction coefficient to be

$$c_f = 0.059\left(\frac{\nu}{U_\infty x}\right)^{1/5} \qquad \text{Re}_x < 10^7 \qquad (8.66)$$

The skin friction coefficient becomes

$$C_f = 0.073\left(\frac{\nu}{U_\infty L}\right)^{1/5} \qquad \text{Re}_x < 10^7 \qquad (8.67)$$

The above formulae can actually be used up to $\text{Re} = 10^8$ without substantial error.

If there is a significant laminar part of the boundary layer, it should be included. If transition occurs at $\text{Re}_{\text{crit}} = 5 \times 10^5$, then the skin friction coefficient should be modified as

$$C_f = 0.073\left(\frac{\nu}{U_\infty L}\right)^{1/5} - 1700\frac{\nu}{U_\infty L} \qquad \text{Re}_x < 10^7 \qquad (8.68)$$

For a rough plate, $\text{Re}_{\text{crit}} = 3 \times 10^5$ and the constant of 1700 should be replaced with 1060.

The displacement and momentum thicknesses can be evaluated using the power-law velocity profile to be

$$\begin{aligned} \delta_d &= 0.048x\left(\frac{\nu}{U_\infty x}\right)^{1/5} \\ \theta &= 0.037x\left(\frac{\nu}{U_\infty x}\right)^{1/5} \end{aligned} \qquad \text{Re} < 10^7 \qquad (8.69)$$

---

[*] There are other more detailed and complicated methods for considering the turbulent boundary layer. They are all empirical since there are no analytical solutions of the turbulent boundary layer.

There are additional quantities often used in the study of turbulent boundary layers. We will introduce two such quantities here. One is the *shear velocity* u_τ defined to be

$$u_\tau = \sqrt{\frac{\tau_0}{\rho}} \tag{8.70}$$

It is a fictitious velocity and often appears in turbulent boundary-layer relationships. The other is the thickness δ_v of the highly fluctuating viscous wall layer, displayed in Figs. 8.12 and 8.13. It is in this very thin layer that the turbulent bursts are thought to originate. It has been related to the shear velocity through experimental observations by

$$\delta_v = \frac{5v}{u_\tau} \tag{8.71}$$

EXAMPLE 8.9 Atmospheric air at 20°C flows at 10 m/s over a smooth, rigid 2-m-wide, 4-m-long flat plate aligned with the flow. How long is the laminar portion of the boundary layer? Predict the drag force on the laminar portion on one side of the plate.

Solution: Assuming the air to free of high-intensity disturbances, use the critical Reynolds number to be 5×10^5, i.e.,

$$\frac{U_\infty x_T}{v} = 5 \times 10^5$$

so that

$$x_T = 5 \times 10^5 \times 1.51 \times 10^{-5}/10 = 0.755 \text{ m}$$

The drag force, using Eq. (8.60) and a coefficient of 1.33 rather than the 1.46 (the coefficient of 1.33 is more accurate as stated), is

$$F_D = \frac{1.33}{2} \rho U_\infty^2 Lw \sqrt{\frac{v}{U_\infty L}}$$

$$= 0.665 \times 1.2 \times 10^2 \times 0.755 \times 2 \times \sqrt{\frac{1.51 \times 10^{-5}}{10 \times 0.755}} = 0.017 \text{ N}$$

a rather small force.

EXAMPLE 8.10 Water at 20°C flows over a 2-m-long, 3-m-wide flat plate at 12 m/s. Estimate the shear velocity, the viscous wall layer thickness, and the boundary-layer thickness at the end of the plate (assume a turbulent layer from the leading edge). Also, predict the drag force on one side of the plate.

Solution: The Reynolds number is $\text{Re} = U_\infty x/v = 12 \times 2/10^{-6} = 2.4 \times 10^7$. So, with $n = 7$ Eq. (8.66) provides

$$\tau_0 = \frac{0.059}{2} \rho U_\infty^2 \left(\frac{v}{U_\infty x}\right)^{1/5} = 0.0295 \times 1000 \times 12^2 \times \left(\frac{10^{-6}}{12 \times 2}\right)^{0.2} = 142 \text{ Pa}$$

The shear velocity is then

$$u_\tau = \sqrt{\frac{\tau_0}{\rho}} = \sqrt{\frac{142}{1000}} = 0.377 \text{ m/s}$$

The viscous wall layer thickness is

$$\delta_v = \frac{5v}{u_\tau} = \frac{5 \times 10^{-6}}{0.377} = 1.33 \times 10^{-5} \text{ m}$$

The boundary-layer thickness is, assuming a turbulent layer from the leading edge

$$\delta = 0.38x \left(\frac{v}{U_\infty x}\right)^{1/5} = 0.38 \times 2 \times \left(\frac{10^{-6}}{12 \times 2}\right)^{0.2} = 0.0254 \text{ m}$$

The drag force on one side of the plate is

$$F_D = \frac{0.073}{2} \rho U_\infty^2 Lw \left(\frac{v}{U_\infty L} \right)^{1/5}$$

$$= 0.0365 \times 1000 \times 12^2 \times 2 \times 3 \times \left(\frac{10^{-6}}{12 \times 2} \right)^{0.2} = 1050 \, \text{N}$$

8.5.4 Laminar Boundary-Layer Differential Equation

The laminar flow solution given in Sec. 8.5.3 was an approximate solution. In this section, we will present a more accurate solution using the x-component Navier–Stokes equation. It is, for horizontal plane flow (no z-variation)

$$u \frac{\partial u}{\partial x} + v \frac{\partial u}{\partial y} = -\frac{1}{\rho} \frac{\partial p}{\partial x} + v \left(\frac{\partial^2 u}{\partial x^2} + \frac{\partial^2 u}{\partial y^2} \right) \tag{8.72}$$

We can simplify this equation and actually obtain a solution. First, recall that the boundary layer is very thin so that there is no pressure variation normal to the boundary layer, i.e., the pressure depends on x only and it is the pressure at the wall from the potential flow solution. Since the pressure is considered known, the unknowns in Eq. (8.72) are u and v. The continuity equation

$$\frac{\partial u}{\partial x} + \frac{\partial v}{\partial y} = 0 \tag{8.73}$$

also relates u and v. So, we have two equations and two unknowns. Consider Figs. 8.12 and 8.13; u changes from 0 to U_∞ over the very small distance δ resulting in very large gradients in the y-direction, whereas u changes quite slowly in the x-direction (holding y fixed). Consequently, we conclude that

$$\frac{\partial^2 u}{\partial y^2} >> \frac{\partial^2 u}{\partial x^2} \tag{8.74}$$

The differential equation (8.72) can then be written as

$$u \frac{\partial u}{\partial x} + v \frac{\partial u}{\partial y} = -\frac{1}{\rho} \frac{dp}{dx} + v \frac{\partial^2 u}{\partial y^2} \tag{8.75}$$

The two acceleration terms on the left are retained since v may be quite small but the gradient $\partial u / \partial y$ is quite large and hence the product is retained. Equation (8.75) is the *Prandtl boundary-layer equation*.

For flow on a flat plate with $dp/dx = 0$, and in terms of the stream function ψ (recall that $u = \partial \psi / \partial y$ and $v = -\partial \psi / \partial x$), Eq. (8.75) takes the form

$$\frac{\partial \psi}{\partial y} \frac{\partial^2 \psi}{\partial x \partial y} - \frac{\partial \psi}{\partial x} \frac{\partial^2 \psi}{\partial y^2} = v \frac{\partial^3 \psi}{\partial y^3} \tag{8.76}$$

If we let (trial-and-error and experience was used to find this transformation)

$$\xi = x \quad \text{and} \quad \eta = y \sqrt{\frac{U_\infty}{vx}} \tag{8.77}$$

Eq. (8.76) becomes[*]

$$-\frac{1}{2\xi} \left(\frac{\partial \psi}{\partial \eta} \right)^2 + \frac{\partial \psi}{\partial \eta} \frac{\partial^2 \psi}{\partial \xi \partial \eta} - \frac{\partial \psi}{\partial \xi} \frac{\partial^2 \psi}{\partial \eta^2} = v \frac{\partial^3 \psi}{\partial \eta^3} \sqrt{\frac{U_\infty}{v\xi}} \tag{8.78}$$

This equation appears more formidable than Eq. (8.76), but if we let

$$\psi(\xi, \eta) = \sqrt{U_\infty v \xi} F(\eta) \tag{8.79}$$

---

[*] Note that $\dfrac{\partial \psi}{\partial y} = \dfrac{\partial \psi}{\partial \eta} \dfrac{\partial \eta}{\partial y} + \dfrac{\partial \psi}{\partial \xi} \dfrac{\partial \xi}{\partial y} = \dfrac{\partial \psi}{\partial \eta} \sqrt{\dfrac{U_\infty}{vx}}$

and substitute this into Eq. (8.78), there results

$$F\frac{d^2F}{d\eta^2} + 2\frac{d^3F}{d\eta^3} = 0 \qquad (8.80)$$

This ordinary differential equation can be solved numerically with the appropriate boundary conditions. They are

$$F = F' = 0 \text{ at } \eta = 0 \qquad \text{and} \qquad F' = 1 \text{ at large } \eta \qquad (8.81)$$

which result from the velocity components

$$u = \frac{\partial\psi}{\partial y} = U_\infty F'(\eta)$$

$$v = -\frac{\partial\psi}{\partial x} = \frac{1}{2}\sqrt{\frac{\nu U_\infty}{x}}(\eta F' - F) \qquad (8.82)$$

The numerical solution to the boundary-value problem is presented in Table 8.5. The last two columns allow the calculation of v and τ_0, respectively. We defined the boundary-layer thickness to be that thickness where $u = 0.99U_\infty$ and we observe that this occurs at $\eta = 5$, so, from this numerical solution

$$\delta = 5\sqrt{\frac{\nu x}{U_\infty}} \qquad (8.83)$$

Also

$$\frac{\partial u}{\partial y} = \frac{\partial u}{\partial \eta}\frac{\partial \eta}{\partial y} = U_\infty F''\sqrt{\frac{U_\infty}{\nu x}} \qquad (8.84)$$

so that the wall shear stress for this boundary layer with $dp/dx = 0$ is

$$\tau_0 = \mu\frac{\partial u}{\partial y}\bigg|_{y=0} = 0.332\rho U_\infty\sqrt{\frac{\nu U_\infty}{x}} \qquad (8.85)$$

The friction coefficients are

$$c_f = 0.664\sqrt{\frac{\nu}{U_\infty x}} \qquad C_f = 1.33\sqrt{\frac{\nu}{U_\infty L}} \qquad (8.86)$$

and the displacement and momentum thicknesses are (these require numerical integration)

$$\delta_d = 1.72\sqrt{\frac{\nu x}{U_\infty}} \qquad \theta = 0.644\sqrt{\frac{\nu x}{U_\infty}} \qquad (8.87)$$

Table 8.5   The Laminar Boundary-Layer Solution with $dp/dx = 0$

$\eta = y\sqrt{U_\infty/\nu x}$	F	$F' = u/U_\infty$	$\frac{1}{2}(\eta F' - F)$	F''
0	0	0	0	0.3321
1	0.1656	0.3298	0.0821	0.3230
2	0.6500	0.6298	0.3005	0.2668
3	1.397	0.8461	0.5708	0.1614
4	2.306	0.9555	0.7581	0.0642
5	3.283	0.9916	0.8379	0.0159
6	4.280	0.9990	0.8572	0.0024
7	5.279	0.9999	0.8604	0.0002
8	6.279	1.000	0.8605	0.0000

EXAMPLE 8.11 Air at 30°C flows over a 2-m-wide, 4-m-long flat plate with a velocity of 2 m/s and $dp/dx = 0$. At the end of the plate, estimate (a) the wall shear stress, (b) the maximum value of v in the boundary layer, and (c) the flow rate through the boundary layer. Assume laminar flow over the entire length.

Solution: The Reynolds number is $Re = U_\infty L/\nu = 2 \times 4/1.6 \times 10^{-5} = 5 \times 10^5$ so laminar flow is reasonable.

(a) The wall shear stress (this requires F'' at the wall) at $x = 4$ m is

$$\tau_0 = 0.332 \rho U_\infty \sqrt{\frac{\nu U_\infty}{x}} = 0.332 \times 1.164 \times 2 \times \sqrt{\frac{1.6 \times 10^{-5} \times 2}{4}} = 0.00219 \, \text{Pa}$$

(b) The maximum value of v requires the use of $(\eta F' - F)$. Its maximum value occurs at the outer edge of the boundary layer and is 0.860. The maximum value of v is

$$v = \frac{1}{2}\sqrt{\frac{\nu U_\infty}{x}}(\eta F' - F) = \frac{1}{2} \times \sqrt{\frac{1.6 \times 10^{-5} \times 2}{4}} \times 0.860 = 0.0012 \, \text{m/s}$$

Note the small value of v compared to $U_\infty = 2$ m/s.

(c) To find the flow rate through the boundary layer, integrate the $u(y)$ at $x = 4$ m

$$Q = \int_0^\delta u \times 2 \, dy = \int_0^5 U_\infty \frac{dF}{d\eta} \times 2 \times \sqrt{\frac{\nu x}{U_\infty}} \, d\eta$$

$$= 2 \times 2 \times \sqrt{\frac{1.6 \times 10^{-5} \times 4}{2}} \int_0^{3.283} dF = 0.0743 \, \text{m}^3/\text{s}$$

Solved Problems

8.1   A 20-cm-diameter sphere with specific gravity $S = 1.06$ is dropped in 20°C water. Estimate the terminal velocity if it is (a) smooth and (b) rough.

At terminal velocity the sphere will not be accelerating so the forces, including the buoyant force [Eq. (2.24)], will sum as follows

$$W = F_D + F_B$$

$$\gamma_{\text{sphere}} \times \text{volume} = C_D \times \frac{1}{2}\rho A V^2 + \gamma_{\text{water}} \times \text{volume}$$

Using $\gamma_{\text{sphere}} = S_{\text{sphere}} \gamma_{\text{water}}$ there results

$$C_D \times \frac{\gamma_{\text{water}}}{2g} \times \pi R^2 \times V^2 = (S-1)\gamma_{\text{water}} \times \frac{4}{3}\pi R^3$$

Substituting in the known values gives

$$V = \left(\frac{8R(S-1)g}{3C_D}\right)^{1/2} = \left(\frac{8 \times 0.1 \times (1.06-1) \times 9.81}{3C_D}\right)^{1/2} = \frac{0.396}{\sqrt{C_D}}$$

(a) For a smooth sphere, Fig. 8.2 suggests we assume for $2 \times 10^4 < Re < 2 \times 10^5$ that $C_D = 0.6$. Then

$$V = \frac{0.396}{\sqrt{0.6}} = 0.511 \, \text{m/s} \quad \text{and} \quad Re = \frac{VD}{\nu} = \frac{0.511 \times 0.2}{10^{-6}} = 1.02 \times 10^5$$

Hence, the terminal velocity of 0.511 m/s is to be expected.

(b) For a rough sphere, Fig. 8.2 suggests that for $Re \cong 105$ we assume $C_D \cong 0.3$. Then

$$V = \frac{0.396}{\sqrt{0.3}} = 0.723 \, \text{m/s} \quad \text{and} \quad Re = \frac{VD}{\nu} = \frac{0.723 \times 0.2}{10^{-6}} = 1.4 \times 10^5$$

This is quite close to $C_D = 0.3$ so it is an approximate value for the velocity, almost 50 percent greater than the velocity of a smooth sphere. A golf ball is roughened for this very reason: a higher velocity over much of its trajectory results in a greater flight distance.

8.2 Calculate the power required to move a 10-m-long, 10-cm-diameter smooth circular cylinder that protrudes vertically from the deck of a ship at a speed of 30 kn (15.4 m/s). Then streamline the cylinder and recalculate the power.

First, find the Reynolds number. It is

$$\text{Re} = \frac{VD}{v} = \frac{15.4 \times 0.1}{1.6 \times 10^{-5}} = 9.6 \times 10^4$$

The drag coefficient found using Fig. 8.2 and Table 8.1 to be

$$C_D \cong 1.2 \times 0.85 = 1.02$$

The power is then

$$\dot{W} = F_D \times V = \frac{1}{2} C_D \rho V^3 A_{\text{projected}}$$
$$= \frac{1}{2} \times 1.02 \times 1.2 \times 15.4^3 \times \pi \times 0.1 \times 10 = 7020 \,\text{W}$$

For the streamlined cylinder, the drag coefficient reduces to

$$C_D \cong 0.06 \times 0.85 = 0.051$$

and the power to

$$\dot{W} = F_D \times V = \frac{1}{2} C_D \rho V^3 A_{\text{projected}}$$
$$= \frac{1}{2} \times 0.051 \times 1.2 \times 15.4^3 \times \pi \times 0.1 \times 10 = 350 \,\text{W}$$

The effect of streamlining is to significantly reduce the drag coefficient.

8.3 Estimate the power required for the conventional airfoil of Example 8.2 to fly at a speed of 150 knots.

Converted to m/s, the speed is $V = 150 \times 1.688/3.281 = 77.2$ m/s. The power is the drag force times the velocity. The drag coefficient from Fig. 8.7 is $C_D = 0.3/47.6 = 0.0063$. The power is then

$$\dot{W} = F_D \times V = \frac{1}{2} \rho c L V^2 C_D \times V$$
$$= \frac{1}{2} \times 1.2 \times 2 \times 15 \times 77.2^3 \times 0.0063 = 52\,000 \,\text{W} \quad \text{or} \quad 70 \,\text{hp}$$

8.4 Show that streamlines and potential lines of an inviscid flow intersect at right angles.

If the streamlines and potential lines intersect at right angles, calculus says that the slope of a streamline will be the negative reciprocal of the potential line. We know that a velocity vector **V** is tangent to a streamline so that the slope of the streamline would be given by (Fig. 8.15)

$$\frac{v}{u} = \frac{dy}{dx}$$

The slope of a potential line is found from

$$d\phi = \frac{\partial \phi}{\partial x} dx + \frac{\partial \phi}{\partial y} dy = 0$$

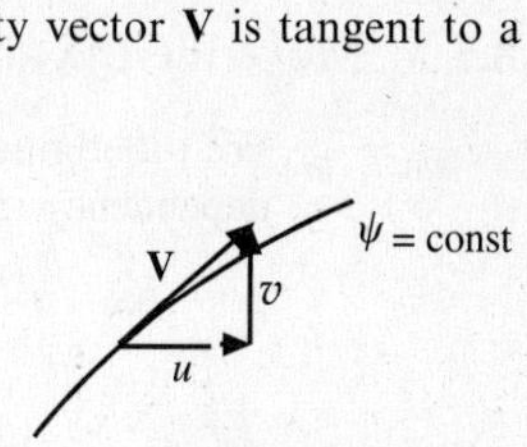

Figure 8.15

so that for the potential line

$$\frac{dy}{dx} = -\frac{\partial \phi/\partial x}{\partial \phi/\partial y} = -\frac{u}{v}$$

Therefore, we see that the slope of the potential line is negative the reciprocal of the slope of the streamline. Hence, the potential line intersects the streamline at a right angle.

8.5   A tornado is approximated as an irrotational vortex (except near its "eye" where it rotates as a rigid body). Estimate the force tending to lift the flat 5 m×10 m roof off a building (the pressure inside the building is assumed to be atmospheric, i.e., 0) if the pressure on the roof is approximated by the pressure at $r = 4$ m. The velocity at a distance of 60 m from the center of the building is observed to be 8 m/s.

The circulation is found from Eq. (8.27) to be

$$\Gamma = -2\pi r v_\theta = -2\pi \times 60 \times 8 = -3320 \,\text{m}^2/\text{s}$$

The velocity at $r = 4$ m is then

$$v_\theta = -\frac{\Gamma}{2\pi r} = \frac{-(-3320)}{2\pi \times 4} = 132 \,\text{m/s}$$

The pressure using Bernoulli's equation is found to be, assuming

$$V_\infty = 0 \text{ and } p_\infty = 0,$$

$$p + \frac{V^2}{2} = \cancel{p}_\infty + \frac{\cancel{V}_\infty^2}{2}\rho = 0. \qquad \therefore p = -\frac{132^2}{2} \times 1.2 = -10\,500 \,\text{Pa}$$

The lifting force is then

$$F = pA = 10\,500 \times 5 \times 10 = 520\,000 \,\text{N}$$

8.6   A 40-cm-diameter cylinder rotates clockwise at 800 rpm in an atmospheric air stream flowing at 8 m/s. Locate any stagnation points and find the minimum pressure.

The circulation $\Gamma = \oint_L \mathbf{V} \cdot d\mathbf{s}$ is the velocity $r_c\Omega$ multiplied by $2\pi r_c$ since the constant velocity $\mathbf{V}$ is tangent to the cylinder's surface. The circulation is calculated to be

$$\Gamma = 2\pi r_c^2 \Omega = 2\pi \times 0.2^2 \times (800 \times 2\pi/60) = 21.1 \,\text{m}^2/\text{s}$$

This is slightly greater than $4\pi r_c U_\infty = 20.1 \,\text{m}^2/\text{s}$ so a single stagnation point is off the cylinder at $\theta = -90°$ [Fig. 8.10(b)].

The minimum pressure exists at the very top of the cylinder (Fig. 8.10), so let us apply Bernoulli's equation (Eq. (8.38)) between the free stream where $p = 0$ and the point on the top where $\theta = 90°$:

$$p_c = \cancel{p}_\infty + \rho\frac{U_\infty^2}{2} - \rho\frac{U_\infty^2}{2}\left(2\sin\theta + \frac{\Gamma}{2\pi r_c U_\infty}\right)^2$$

$$= 0 + 1.2 \times \frac{8^2}{2}\left[1 - \left(2\sin 90° + \frac{21.1}{2\pi \times 0.2 \times 8}\right)^2\right] = -607 \,\text{Pa}$$

using $\rho = 1.2 \,\text{kg/m}^3$ assuming atmospheric air.

8.7   Move the $U(x)$ under the integral symbol and rewrite the von Karman integral equation (8.43).

We differentiate a product: $(fg)' = fg' + gf'$. For the present equation, we let $g(x) = \int_0^{\delta(x)} \rho u\, dy$ (the y dependency integrates out) so that

$$\frac{d}{dx}U(x)\int_0^\delta \rho u\, dy = U(x)\frac{d}{dx}\int_0^\delta \rho u\, dy + \left[\int_0^\delta \rho u\, dy\right]\frac{dU(x)}{dx}$$

We can move $U(x)$ under the integral since it is a function of x and the integration is on y so that Eq. (8.43) takes the form

$$\tau_0(x) = -\delta\frac{dp}{dx} + \frac{d}{dx}U(x)\int\limits_0^\delta \rho u\,dy - \left[\int\limits_0^\delta \rho u\,dy\right]\frac{dU(x)}{dx} - \frac{d}{dx}\int\limits_0^\delta \rho u^2\,dy$$

$$= -\delta\frac{dp}{dx} + \rho\frac{d}{dx}\int\limits_0^\delta u(U-u)dy - \rho\frac{dU}{dx}\int\limits_0^\delta u\,dy$$

This is an equivalent form of the von Karman integral equation. The density ρ is assumed constant in the thin boundary layer.

8.8 Estimate the drag force on one side of the flat plate of Example 8.9: (*a*) Assuming turbulent flow from the leading edge. (*b*) Including the laminar portion of the boundary layer.

(*a*) Assuming a turbulent flow from the leading edge, the boundary-layer thickness after 4 m is given by Eq. (8.65) and is

$$\delta = 0.38x\left(\frac{v}{U_\infty x}\right)^{1/5} = 0.38\times 4\times\left(\frac{1.51\times 10^{-5}}{10\times 4}\right)^{0.2} = 0.0789\,\text{m}$$

The drag force on one side is then

$$\frac{1}{2}C_f\rho U_\infty^2 Lw = \frac{0.073}{2}\left(\frac{1.51\times 10^{-5}}{10\times 4}\right)^{1/5}\times 1.2\times 10^2\times 4\times 2 = 1.82\,\text{N}$$

Check the Reynolds number: $\text{Re} = 10\times 4/1.51\times 10^{-5} = 2.65\times 10^6$. $\therefore$ OK.

(*b*) First, the boundary layers are sketched with the appropriate distances in Fig. 8.16. The laminar boundary-layer length is found using $\text{Re}_{\text{crit}} = 5\times 10^5$:

$$x_L = \frac{\text{Re}\times v}{U_\infty} = \frac{5\times 10^5\times 1.51\times 10^{-5}}{10} = 0.755\,\text{m}$$

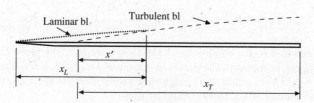

Figure 8.16

The laminar boundary-layer thickness at x_L is, Eq. (8.57),

$$\delta = 5\sqrt{\frac{xv}{U_\infty}} = 5\sqrt{\frac{0.755\times 1.51\times 10^{-5}}{10}} = 0.00534\,\text{m}$$

The location of the fictitious origin of the turbulent boundary layer is found by calculating x' in Fig. 8.16. It is found using Eq. (8.65) to be

$$x'^{4/5} = \frac{\delta}{0.38}\left(\frac{U_\infty}{v}\right)^{1/5} = \frac{0.00534}{0.38}\left(\frac{10}{1.51\times 10^{-5}}\right)^{1/5}.\qquad \therefore\ x' = 0.205\,\text{m}$$

The distance x_T is then $x_T = L - x_L + x' = 4 - 0.755 + 0.205 = 3.45\,\text{m}$. The boundary-layer thickness at the end of the plate is

$$\delta = 0.38x\left(\frac{v}{U_\infty x}\right)^{1/5} = 0.38\times 3.45\times\left(\frac{1.51\times 10^{-5}}{10\times 3.45}\right)^{0.2} = 0.070\,\text{m}$$

The drag force using Eq. (8.68) is found to be

$$\frac{1}{2}C_f\rho U_\infty^2 Lw = \frac{1}{2}\left[0.073\left(\frac{v}{U_\infty L}\right)^{1/5} - 1700\frac{v}{U_\infty L}\right]\rho U_\infty^2 Lw$$

$$= \left[\frac{0.073}{2}\left(\frac{1.51\times10^{-5}}{10\times4}\right)^{0.2} - \frac{1700}{2}\frac{1.51\times10^{-5}}{10\times4}\right]\times 1.2\times10^2\times4\times2 = 1.51\,\text{N}$$

The drag force in (a) is about 20 percent too high. The laminar portion with its smaller shear stress reduces the overall drag force for short distances.

8.9   Verify that the velocity components are in fact given by Eq. (8.82) when solving the x-component Navier–Stokes equation for a laminar boundary-layer flow.

The x-component of the velocity is given by (use Eqs. (8.77) and (8.79))

$$u = \frac{\partial\psi}{\partial y} = \frac{\partial\psi}{\partial\xi}\frac{\partial\xi}{\partial y} + \frac{\partial\psi}{\partial\eta}\frac{\partial\eta}{\partial y} = \sqrt{U_\infty v\xi}F'(\eta)\times\sqrt{\frac{U_\infty}{v\xi}} = U_\infty F'(\eta)$$

The y-component of the velocity is (use Eqs. (8.77) and (8.79))

$$v = -\frac{\partial\psi}{\partial x} = -\frac{\partial\psi}{\partial\xi}\frac{\partial\xi}{\partial x} - \frac{\partial\psi}{\partial\eta}\frac{\partial\eta}{\partial x} = -\frac{1}{2}\sqrt{\frac{vU_\infty}{x}}F(\eta) - \sqrt{U_\infty vx}F'(\eta)\left(-\frac{1}{2}x^{-3/2}y\sqrt{\frac{U_\infty}{v}}\right)$$

$$= -\frac{1}{2}\sqrt{\frac{vU_\infty}{x}}F(\eta) + \frac{1}{2}\sqrt{U_\infty vx}y^{-1}\sqrt{\frac{U_\infty}{vx}}F'(\eta) = \frac{1}{2}\sqrt{\frac{vU_\infty}{x}}(\eta F' - F)$$

where we have used x and ξ interchangeably since they are equal, as defined.

8.10   Using the x-component Navier–Stokes equation (8.72), determine an additional boundary condition for laminar flow over a flat plate with zero pressure gradient.

At the wall $u = v = 0$ so that the left-hand side of Eq. (8.72) is 0 for $y = 0$. Also, on the wall since $u = 0$, then $\partial u/\partial x = 0$ and $\partial^2 u/\partial x^2 = 0$. So, on the wall where $y = 0$, Eq. (8.72) provides

$$u\frac{\partial u}{\partial x} + v\frac{\partial u}{\partial y} = -\frac{1}{\rho}\frac{\partial p}{\partial x} + v\left(\frac{\partial^2 u}{\partial x^2} + \frac{\partial^2 u}{\partial y^2}\right) \qquad\text{or}\qquad 0 = \frac{\partial^2 u}{\partial y^2}$$

Therefore, in addition to the conditions of Eq. (8.49), we have the above condition at the wall. This condition could not be met with the parabolic profile of Sec. 8.5.3, but if a cubic profile were assumed, this condition would be required. If a straight-line profile were assumed, a rather poor assumption, only the first two conditions of Eq. (8.49) would be used.

Supplementary Problems

Flow Around Blunt Bodies

8.11   Wind is blowing parallel to the long side of a large building with a flat roof. Sketch the expected flow looking down on the building from the top and looking at the building from the side. Show the expected regions of separation and reattachment points.

8.12   One semitruck has a wind deflector on top of the tractor and another one does not. Sketch a side view of the airflow for both semitrucks showing expected regions of separation, boundary layers, points of reattachment, and wakes.

8.13   Sketch the expected flow around a sphere if (a) Re = 4, (b) Re = 4000, and (c) 40 000. Identify the separated region, the wake, any laminar or turbulent boundary layers, and the free stream.

8.14   Describe the flow to be expected around each of the following and estimate the drag coefficient.

   (*a*)   Air at 10°C flowing around a 4.1-cm-diameter golf ball traveling at 35 m/s.
   (*b*)   A 2-mm-diameter hailstone traveling through air at −10°C at a speed of 5 m/s.
   (*c*)   A 1-mm-diameter grain of sand falling in stagnant 20°C water at 1 m/s.
   (*d*)   Air at 20°C flowing over an 8-cm-diamter sphere at a speed of 1 m/s.
   (*e*)   Air at 0°C moving past a 10-cm-diameter, 4-m-high pole at 2 m/s.

8.15   A 2-cm-diameter sphere moves at Re = 10 (separation is just occurring). What is its speed if it is submerged in (*a*) water at 10°C, (*b*) air at 40°C and 400 kPa, and (*c*) water at 90°C.

8.16   Water at 10°C flows by a 2-mm-diameter wire with a speed of 2 m/s. Sketch the expected flow showing the separated region, the wake, the boundary layer, and the free stream, should any of these exist.

8.17   A fluid flows by a flat circular disk with velocity V normal to the disk at Re > 10^3. Estimate the drag coefficient if (*a*) the pressure is assumed constant over the face of the disk and (*b*) if a parabolic pressure profile exists on the front of the disk. Assume the pressure is 0 on the backside. Explain the results in light of the drag coefficient of Table 8.1.

8.18   Atmospheric air at 20°C is flowing at 10 m/s. Calculate the drag force on (*a*) a 10-cm-diameter smooth sphere, (*b*) a 10-cm-diameter, 80-cm-long smooth cylinder with free ends, (*c*) a 10-cm-diameter disk, and (*d*) a 10-cm-wide, 20-cm-long rectangular plate. The velocity vector is normal to all objects.

8.19   A 220-cm-square sign is impacted straight on by a 50 m/s 10°C wind. Estimate the force on the sign. If the sign is held by a single 3-m-high post imbedded in concrete, what moment would exist at the base of the post?

8.20   A 20-cm-diameter smooth sphere is rigged with a strain gauge calibrated to measure the force on the sphere. Estimate the wind speed in 20°C air if the gauge measures (*a*) 4 N and (*b*) 0.5 N.

8.21   An automobile travels on a horizontal section of highway at sea level where the temperature is 20°C. Estimate the horsepower needed for a speed of 100 km/h. Make any needed assumptions.

8.22   Estimate the fuel savings for one year on a semitruck that travels 300 000 mi if fuel costs $2.50/gal if the truck installs both a deflector and a gap seal. Without the deflector and seal the truck averages 4 mi/gal. If an owner desires a three-year payback, how much could an owner pay for the deflector and seal?

8.23   A bike rider expends a certain amount of energy to travel 12 m/s while in the upright position. How fast will the rider travel with the same amount of energy expenditure if the bent over position is elected? Assume the rider's projected area is reduced 25 percent in the bent-over position.

8.24   Estimate the speed of fall of a 6-ft man with arms and legs outstretched. Make reasonable assumptions. Now give the man a 6-m-diameter parachute and calculate his speed of descent, again making reasonable assumptions.

8.25   A blue spruce pine tree has the shape of a triangle which is 15 cm off the ground. The triangle has a maximum diameter of 6 m and is 10 m tall. Estimate the drag on the tree if it is exposed to a 25 m/s wind. Use $C_D = 0.4$ in your calculations.

Vortex Shedding, Cavitation, and Added Mass

8.26   Vortices are observed downstream of a 2-cm-diameter cylinder in 20°C atmospheric air. How far are the shed vortices apart downstream of the cylinder for an air speed of 5 m/s?

8.27 A sensor is positioned downstream a short distance from a 4-cm-diameter cylinder in a 20°C atmospheric airflow. It senses vortex shedding at a frequency of 0.16 Hz. Estimate the airspeed.

8.28 A 10°C wind blows over 6-mm-diameter high-tension wires. Determine the range of velocities over which vortex shedding occurs. Could the shedding frequency be heard? Humans with good hearing can hear frequencies between 20 and 20 000 Hz. (Such vortex shedding when ice coats the wires can cause "galloping" where the wires actually oscillate from the usual catenaries to inverted catenaries resulting in wire failure.)

8.29 What drag force acts on a 76-cm-diameter sphere towed 2 m below the surface of water at 20 m/s?

8.30 A 2.2-m-long hydrofoil with chord length 50 cm operates 40 cm below the water's surface with an angle of attack of 4°. For a speed of 15 m/s determine the drag and lift and decide if cavitation exists on the hydrofoil.

8.31 A 40-cm-diameter sphere is released from rest under water. If it weighs 380 N in air, what is its initial acceleration under water if the added mass is neglected? If the added mass is included?

8.32 A 20-cm-diameter 4-m-long horizontal cylinder is released from rest under water. If it weighs 1500 N in air, what is its initial acceleration under water if the added mass is neglected? If the added mass is included?

Lift and Drag on Airfoils

8.33 Sketch the flow field around an airfoil that has stalled. Show the boundary layers, the separated region, and the wake.

8.34 Estimate the takeoff speed for an aircraft with conventional airfoils if the aircraft with payload weighs 120 000 N and the effective wing area is 20 m^2 assuming a temperature of (a) 30°C, (b) 10°C, and (c) −20°C. An angle of attack at takeoff of 8° is desired.

8.35 Rework Prob. 8.34 but assume a temperature of 20°C at a pressure of (a) 100 kPa, (b) 80 kPa, and (c) 60 kPa.

8.36 A 2000-kg airplane is designed to carry a 4000-N payload when cruising near sea level. For conventional airfoils with an effective wing area of 25 m^2, estimate the takeoff speed for an angle of attack of 10°, the stall speed, and the power required (the airfoils account for approximately 40 percent of the drag) for a cruising speed of 80 m/s at an elevation of 2000 m.

8.37 If the aircraft of Prob. 8.36 were to fly at 10 km, what power would be required?

8.38 The aircraft of Prob. 8.36 is to land with the airfoils at an angle of attack near stall. Estimate the minimum landing speed for no slotted flaps, flaps with one slot, and flaps with two slots. Assume the effective wing area is the same for all three situations.

Potential Flow

8.39 Show that the difference of the stream functions between two streamlines is the flow rate q per unit depth between the two streamlines, i.e., $q = \psi_2 - \psi_1$.

8.40 Show that each of the following represents an incompressible plane flow and find the associated stream function or potential function.

(a) $\phi = 10y$
(b) $\psi = 20xy$
(c) $\phi = 10\theta$ (cylindrical coordinates)
(d) $\psi = (20/r)\sin\theta$ (cylindrical coordinates)

8.41 Does the velocity field $\mathbf{V} = (x\mathbf{i} + y\mathbf{j})/(x^2 + y^2)$ represent an incompressible flow? If so, find the velocity potential ϕ and stream function ψ.

8.42 Show that the flow represented by $\psi = 10\ln(x^2 + y^2)$ m²/s is an incompressible flow. Also,

    (a) Find the velocity potential ϕ.
    (b) Find the pressure along the negative x-axis if atmospheric air is flowing and $p = 0$ at $x = -\infty$.
    (c) Find the x-component of the acceleration at $(-4, 0)$.

8.43 Show that the flow represented by $\phi = 20\ln r$ m²/s is an incompressible flow. Also,

    (a) Find the stream function ψ.
    (b) Find the pressure along the negative x-axis if water is flowing and $p = 40$ kPa at $x = -\infty$.
    (c) Find the acceleration at rectangular coordinates $(-2, 0)$.

8.44 Show that the flow represented by $\phi = 10r\cos\theta + 40\ln r$ m²/s is an incompressible flow. Also,

    (a) Find the stream function ψ.
    (b) Find the pressure along the negative x-axis if water is flowing and $p = 100$ kPa at $x = -\infty$.
    (c) Find the acceleration at rectangular coordinates $(-2, 0)$.
    (d) Locate any stagnation points.

8.45 Superimpose a uniform flow parallel to the x-axis of 10 m/s and a source at the origin of strength $q = 10\pi$ m²/s.

    (a) Write the velocity potential ϕ and stream function ψ.
    (b) Locate any stagnation points.
    (c) Sketch the body formed by the streamline that separates the source flow from the uniform flow.
    (d) Locate the positive y-intercept of the body of (c).
    (e) Determine the thickness of the body of (c) at $x = -\infty$.

8.46 A uniform flow $\mathbf{V} = 20\mathbf{i}$ m/s is superimposed on a source of strength 20π m²/s and a sink of equal strength located at $(-2$ m, $0)$ and $(2$ m, $0)$, respectively. The resulting body formed by the appropriate streamline is a *Rankine oval*. Determine the length and thickness of the oval. Find the velocity at $(0, 0)$.

8.47 A source near a wall is created by *the method of images*: superimpose two equal strength sources of strength 4π m²/s at $(2$ m, $0)$ and $(-2$ m, $0)$, respectively. Sketch the flow showing the wall and find the velocity distribution along the wall.

8.48 Superimpose a velocity of $\mathbf{V} = 20\mathbf{i}$ m/s on the flow of Prob. 8.47. Locate any stagnation points.

8.49 A uniform flow $\mathbf{V} = 10\mathbf{i}$ m/s is superimposed on a doublet with strength 40 m³/s. Find:

    (a) The radius of the cylinder that is formed.
    (b) The velocity distribution $v_\theta(\theta)$ on the cylinder.
    (c) The locations of the stagnation points.
    (d) The minimum pressure on the cylinder if the pressure at the stagnation point is 200 kPa. Water is flowing.

8.50 Superimpose a uniform flow $\mathbf{V} = 10\mathbf{i}$ m/s, a doublet $\mu = 40$ m³/s, and a vortex. Locate any stagnation points and find the minimum pressure on the cylinder if the pressure of the standard atmospheric air is zero at a large distance from the cylinder. The strength of the vortex is (a) $\Gamma = 40\pi$ m²/s, (b) $\Gamma = 80\pi$ m²/s, and (c) $\Gamma = 120\pi$ m²/s.

8.51   Assume an actual flow can be modeled with the flow of Prob. 8.49 on the front half of the cylinder and that the pressure on the back half is equal to the minimum pressure on the cylinder (the flow is assumed to separate from the cylinder on the back half). Calculate the resulting drag coefficient.

Boundary Layers

8.52   A turbulent boundary layer is studied in a zero pressure-gradient flow on a flat plate in a laboratory. Atmospheric air at 20°C flows over the plate at 10 m/s. How far from the leading edge can turbulence be expected (*a*) if the free-stream fluctuation intensity is low? (*b*) If the free-stream fluctuation intensity is high?

8.53   Respond to Prob. 8.52 if 20°C water is the fluid.

8.54   A laminar boundary layer is to be studied in the laboratory. To obtain a sufficiently thick layer, a 2-m-long laminar portion is desired. What speed should be used if (*a*) a water channel is selected? (*b*) If a wind tunnel is selected? It is assumed that the fluctuation intensity can be controlled at a sufficiently low level.

8.55   If a differential equation for the boundary layer is to be solved on the front part of a cylinder, the velocity U at the outer edge of the boundary layer is needed as is the pressure p in the boundary layer. State U and p for the circular cylinder of Prob. 8.49.

8.56   In Fig. 8.13 laminar and turbulent velocity profiles are sketched for the same boundary-layer thickness. Calculate the percentage increase of the momentum flux for a turbulent layer assuming the power-law profile $u/U_\infty = (y/\delta)^{1/7}$ compared to a laminar layer assuming the parabolic profile $u/U_\infty = 2y/\delta - (y/\delta)^2$.

8.57   Show that Eq. (8.43) follows from Eq. (8.41).

8.58   Show that the von Karman integral equation of Solved Problem 8.7 can be written in terms of the momentum and displacement thicknesses as

$$\tau_0 = \rho \frac{d}{dx}(\theta U^2) + \rho \delta_d U \frac{dU}{dx}$$

where we have differentiated Bernoulli's equation to obtain

$$\frac{dp}{dx} = -\rho U \frac{dU}{dx} = -\frac{\rho}{\delta} \frac{dU}{dx} \int_0^\delta U\,dy$$

8.59   Assume a linear velocity profile in a laminar boundary layer on a flat plate with a zero pressure gradient. Find:

   (*a*) $\delta(x)$. Compare with the more exact solution and compute the percentage error.
   (*b*) $\tau_0(x)$.
   (*c*) v at $y = \delta$ and $x = 2$ m.
   (*d*)   The drag force if the plate is 2-m wide and 4-m long.

8.60   Assume a sinusoidal velocity profile in a laminar boundary layer on a flat plate with a zero pressure gradient using 20°C water with $U_\infty = 1$ m/s. Find:

   (*a*) $\delta(x)$. Compare with the more exact solution and compute the percentage error.
   (*b*) $\tau_0(x)$.
   (*c*)   The drag force if the plate is 2-m wide and 4-m long.

8.61   Assume a cubic velocity profile in a laminar boundary layer on a flat plate with a zero pressure gradient. Use $\partial^2 u / \partial y^2 \big|_{y=0} = 0$ as an additional boundary condition (Eq. (8.75)). Find:

      (a) $\delta(x)$. Compare with the more exact solution and compute the percentage error.
      (b) $\tau_0(x)$.
      (c) The drag force if the plate is 2-m wide and 4-m long.

8.62   Sketch to scale a relatively thick laminar boundary layer over a 10-m length for the flow of 20°C water on a flat plate with zero pressure gradient for $U_\infty = 1$ m/s. Let about 15 cm represent the 10-m length of the plate. Assume a laminar layer over the entire 10-m length.

8.63   If the walls in a wind tunnel are parallel, the flow will accelerate due to the boundary layers on each of the walls. If a wind tunnel is square, how should one of the walls be displaced outward for a zero pressure gradient to exist?

8.64   A streamline in 20°C water is 2 mm from a flat plate at the leading edge. Using the parabolic velocity profile of Sec. 8.5.3 with $U_\infty = 1$ m/s, predict how far from the plate the streamline is when $x = 1$ m.

8.65   Show that the power-law form for the velocity profile in a turbulent flow is not a good approximation at the wall or at the outer edge of the boundary layer.

8.66   Show that Eq. (8.62) follows from Eq. (8.44).

8.67   Air at 20°C flows over a 3-m long and 2-m wide flat plate at 16 m/s. Assume a turbulent flow from the leading edge (a trip wire at the leading edge can be used to cause the turbulence) and calculate:

      (a) δ at $x = 3$ m.
      (b) τ_0 at $x = 3$ m.
      (c) The drag force on one side of the plate.
      (d) The displacement and momentum thicknesses at $x = 3$ m.
      (e) The shear velocity and the viscous wall-layer thickness.

8.68   Water at 20°C flows over a 3-m long and 2-m wide flat plate at 3 m/s. Assume a turbulent flow from the leading edge (a trip wire at the leading edge can be used to cause the turbulence) and calculate:

      (a) δ at $x = 3$ m.
      (b) τ_0 at $x = 3$ m.
      (c) The drag force on one side of the plate.
      (d) The displacement and momentum thicknesses at $x = 3$ m.
      (e) The shear velocity and the viscous wall-layer thickness.

8.69   Air at 20°C flows over a 2-m-long and 3-m-wide flat plate at 16 m/s. Include the laminar portion near the leading edge (Fig. 8.16) assuming low fluctuations and a smooth plate and calculate:

      (a) δ at $x = 3$ m.
      (b) The drag force on one side of the plate.

8.70   Water at 20°C flows over a 2-m-long and 3-m-wide flat plate at 3 m/s. Include the laminar portion near the leading edge (Fig. 8.16) assuming low fluctuations and a smooth plate and calculate:

      (a) δ at $x = 3$ m.
      (b) The drag force on one side of the plate.

8.71   Atmospheric air blows in toward the shore at a beach in Florida. It is assumed that a boundary layer begins to develop about 12 km from shore. If the wind speed averages 18 m/s, estimate the thickness of the boundary layer and the shear stress on the surface of the water near the shore.

8.72   A long cigar-shaped dirigible is proposed to take rich people on cruises. It is proposed to be 1000 m long and
150 m in diameter. How much horsepower is needed to move the dirigible through sea-level air at 12 m/s if
the drag on the front and rear is neglected.

8.73   Show that Eq. (8.80) follows from Eq. (8.78).

8.74   Solve Eq. (8.80) with boundary conditions given by Eq. (8.81) using an available software program such as
MATLAB.

8.75   A laminar boundary layer of 20°C atmospheric air moving at 2 m/s exists on one side of a 2-m-wide,
3-m-long flat plate. At $x = 3$ m find:

      (a)   The boundary-layer thickness.
      (b)   The wall shear stress.
      (c)   The maximum y-component of velocity.
      (d)   The drag force.
      (e)   The displacement and momentum thicknesses.
      (f)   The flow rate through the boundary layer.

8.76   A laminar boundary layer of 20°C water moving at 0.8 m/s exists on one side of a 2-m-wide, 3-m-long flat
plate. At $x = 3$ m find:

      (a)   The boundary-layer thickness.
      (b)   The wall shear stress.
      (c)   The maximum y-component of velocity.
      (d)   The drag force.
      (e)   The displacement and momentum thicknesses.
      (f)   The flow rate through the boundary layer.

Answers to Supplementary Problems

8.11   See problem

8.12   See problem

8.13   See problem

8.14   (a)   0.25   (b)   0.55   (c)   0.46   (d)   0.4   (e)   1.14

8.15   (a)   0.000654 m/s   (b)   0.00214 m/s   (c)   0.000164 m/s

8.16   Re = 3060

8.17   (a)   1.0   (b)   0.5

8.18   (a)   0.26 N   (b)   4.0 N   (c)   0.52 N   (d)   1.32 N

8.19   8300 N, 34 000 N·m

8.20   (a)   32.6 m/s   (b)   6.9 m/s

8.21   11.3 hp

8.22 $152,700

8.23　14.1 m/s

8.24　73 m/s, 5.8 m/s

8.25　4500 N

8.26　4.25 cm

8.27　0.04 m/s

8.28　0.012 m/s to 29.3 m/s. Yes, for $V > 0.57$ m/s

8.29　43.5 kN

8.30　99 kN, 2.23 kN, No

8.31　1.32 m/s^2, 0.883 m/s^2

8.32　1.74 m/s^2, 0.870 m/s^2

8.33　See problem

8.34　(a)　102 m/s　(b)　98 m/s　(c)　93 m/s

8.35　(a)　100 m/s　(b)　112 m/s　(c)　130 m/s

8.36　36.2 m/s, 33.3 m/s, 129 hp

8.37　53 hp

8.38　33.3 m/s, 25.1 m/s, 22.2 m/s

8.39　See problem

8.40　(a)　$-10x$　(b)　$10(x^2 + y^2)$　(c)　$-10\ln r$　(d)　$20r\cos\theta$

8.41　$\ln\sqrt{x^2 + y^2}$, $\tan^{-1} x/y$

8.42　(a)　$-20\tan^{-1} y/x$　(b)$-240/x^2$　(c)　-6.25m/s^2

8.43　(a)　$\psi = \theta$　(b)　$40 - 200/x^2$ kPa　(c)　-50 m/s^2

8.44　(a)　$10r\sin\theta + 40\theta$　(b)　$100\left(1 + \dfrac{1}{x} - \dfrac{8}{x^2}\right)$ kPa　(c)　200 m/s^2　(d)　$(-4, 0)$

8.45　(a)　$10r\cos\theta + 5\ln r$ and $10r\sin\theta + 5\theta$　(b)　$(-0.5, 0)$　(d)　$\pi/4$ m　(e)　π m

8.46　4.9 m by 2.1 m

8.47　$4y/(y^2 + 4)$ m/s

8.48　1.902 m, -2.102 m

8.49　(a)　2 m　(b)　$20\sin\theta$ m/s　(c)　$(2\text{ m}, 0°), (2\text{ m}, 180°)$　(d)　$200(1 - \sin^2\theta)$ kPa

8.50 (*a*) $(2, -30°), (2, 150°), -488$ Pa   (*b*) $(2, -90°), -915$ Pa   (*c*) $(2, 270°), -1464$ Pa

8.51 2.67

8.52 (*a*) 90.5 cm   (*b*) 54.3 cm

8.53 (*a*) 5 cm   (*b*) 3 cm

8.54 (*a*) 0.25 m/s   (*b*) 4.52 m/s

8.55 $2U_\infty \sin\theta$ and $p_0 - \rho U_\infty^2 \sin^2\theta$

8.56 45.8%

8.57 See problem

8.58 See problem

8.59 (*a*) $3.46\sqrt{vx/U_\infty}, -31\%$ (*b*) $0.289\rho U_\infty^2 \mathrm{Re}_x^{-1/2}$ (*c*) $0.0256\delta^2\sqrt{U_\infty^3/v}$ (*d*) $2.31\rho U_\infty\sqrt{U_\infty v}$

8.60 (*a*) $0.00479\sqrt{x}, -4.2\%$ (*b*) $0.328x^{-1/2}$ (*c*) 2.62 N

8.61 (*a*) $4.65\sqrt{vx/U_\infty}, -7\%$ (*b*) $0.323\rho U_\infty^2\sqrt{v/U_\infty x}$ (*c*) $1.29\rho U_\infty^2\sqrt{v/U_\infty}$

8.62 See problem

8.63 $4\delta_d$

8.64 2.8 mm

8.65 See problem

8.66 See problem

8.67 (*a*) 5.92 cm   (*b*) 0.47 Pa   (*c*) 3.49 N   (*d*) 7.47 mm, 5.76 mm   (*e*) 0.626 m/s, 0.145 mm

8.68 (*a*) 4.64 cm   (*b*) 10.8 Pa   (*c*) 80.1 N   (*d*) 5.85 mm, 4.51 mm   (*e*) 0.104 m/s, 0.048 mm

8.69 (*a*) 5.28 cm   (*b*) 2.90 N

8.70 (*a*) 4.45 cm   (*b*) 75 N

8.71 53 m

8.72 800 hp

8.73 See problem

8.74 See problem

8.75 (*a*) 2.61 cm   (*b*) 0.000277 Pa   (*c*) 0.003 m/s   (*d*) 0.0333 Pa   (*e*) 3.36 mm (*f*) 0.068 m³/s

8.76 (*a*) 9.7 mm   (*b*) 0.0137 Pa   (*c*) 0.00044 m/s   (*d*) 1.65 N   (*e*) 1.25 mm (*f*) 0.0102 m³/s

Chapter 9

Compressible Flow

9.1 INTRODUCTION

Compressible flows occur when the density changes are significant between two points on a streamline. Not all gas flows are compressible flows, only those that have significant density changes. Flow around automobiles, in hurricanes, around aircraft during landing and takeoff, and around buildings and communication towers are a few examples of incompressible flows in which the density of the air does not change more than 3 percent between points of interest and are consequently treated as incompressible flows. There are, however, many examples of gas flows in which the density does change more than 3 percent; they include airflow around aircraft that fly faster than a Mach number [see Eq. (3.18)] of 0.3 (about 100 m/s), through compressors, jet engines, and tornados, to name a few. There are also compressible effects in liquid flows that give rise to water hammer and underwater compression waves from blasts; they will not be considered here.

Only compressible flow problems that can be solved using the integral equations will be considered in this chapter. The simplest of these is uniform flow in a conduit. Recall that the continuity equation, the momentum equation, and the energy equation are, respectively

$$\dot{m} = \rho_1 A_1 V_1 = \rho_2 A_2 V_2 \qquad (9.1)$$

$$\sum \mathbf{F} = \dot{m}(\mathbf{V}_2 - \mathbf{V}_1) \qquad (9.2)$$

$$\frac{\dot{Q} - \dot{W}_S}{\dot{m}} = \frac{V_2^2 - V_1^2}{2} + h_2 - h_1 \qquad (9.3)$$

where the enthalpy $h = \tilde{u} + p/\rho$ is used [see Eqs. (1.20) and (4.23)]. If the gas can be approximated as an ideal gas, then the energy equation takes either of the following two forms:

$$\frac{\dot{Q} - \dot{W}_S}{\dot{m}} = \frac{V_2^2 - V_1^2}{2} + c_p(T_2 - T_1) \qquad (9.4)$$

$$\frac{\dot{Q} - \dot{W}_S}{\dot{m}} = \frac{V_2^2 - V_1^2}{2} + \frac{k}{k-1}\left(\frac{p_2}{\rho_2} - \frac{p_1}{\rho_1}\right) \qquad (9.5)$$

where we have used the thermodynamic relations

$$\Delta h = c_p \Delta T \qquad c_p = c_v + R \qquad k = \frac{c_p}{c_v} \qquad (9.6)$$

The ideal gas law will also be used; the form most used is

$$p = \rho RT \qquad (9.7)$$

We may also determine the entropy change or assume an isentropic process ($\Delta s = 0$). Then, one of the following equations may be used:

$$\Delta s = c_p \, \ln \frac{T_2}{T_1} - R \, \ln \frac{p_2}{p_1} \qquad (9.8)$$

$$\frac{T_2}{T_1} = \left(\frac{p_2}{p_1}\right)^{(k-1)/k} \qquad \frac{p_2}{p_1} = \left(\frac{\rho_2}{\rho_1}\right)^{k} \qquad \frac{T_2}{T_1} = \left(\frac{p_2}{p_1}\right)^{(k-1)} \qquad (9.9)$$

Recall that the temperatures and pressures must always be absolute quantities when using several of the above relations, therefore, it is always safe to use absolute temperature and pressure when solving problems involving a compressible flow.

9.2   SPEED OF SOUND

A pressure wave with small amplitude is called a *sound wave* and it travels through a gas with the *speed of sound*, denoted by c. Consider the small-amplitude wave shown in Fig. 9.1 traveling through a conduit. In Fig. 9.1(a) it is moving so that a stationary observer sees an unsteady motion, in Fig. 9.1(b) the observer moves with the wave so that the wave is stationary and a steady flow is observed, and in Fig. 9.1(c) shows the control volume surrounding the wave. The wave is assumed to create a small differential change in the pressure p, temperature T, density ρ, and velocity V in the gas. The continuity equation applied to the control volume provides

$$\rho A c = (\rho + d\rho)A(c + dV) \qquad (9.10)$$

which simplifies to, neglecting the higher-order term $d\rho \, dV$,

$$\rho \, dV = -c \, d\rho \qquad (9.11)$$

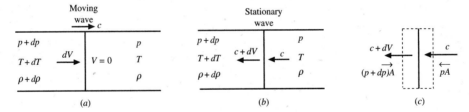

Figure 9.1   (a) A sound wave moving through a gas, (b) the gas moving through the wave, and (c) the control volume enclosing the wave of (b).

The momentum equation in the streamwise direction is written as

$$pA - (p + dp)A = \rho A c(c + dV - c) \qquad (9.12)$$

which simplifies to

$$dp = -\rho c \, dV \qquad (9.13)$$

Combining the continuity and momentum equations results in

$$c = \sqrt{\frac{dp}{d\rho}} \qquad (9.14)$$

for the small-amplitude sound waves.

The lower-frequency (less than 18 000 Hz) sound waves travel isentropically so that $p/\rho^k = $ const which, when differentiated, gives

$$\frac{dp}{d\rho} = k\frac{p}{\rho} \tag{9.15}$$

The speed of sound for such waves is then

$$c = \sqrt{\frac{kp}{\rho}} = \sqrt{kRT} \tag{9.16}$$

High-frequency waves travel isothermally resulting in a speed of sound of

$$c = \sqrt{RT} \tag{9.17}$$

For small-amplitude waves traveling through a liquid or a solid, the bulk modulus is used [see Eq. (1.13)]; it is equal to $\rho\, dp/d\rho$ and has a value of 2100 MPa for water at $20\,°C$. This gives a value of about 1450 m/s for a small-amplitude wave moving through water.

The Mach number, introduced in Chap. 3, is used for disturbances moving in a gas. It is

$$M = \frac{V}{c} \tag{9.18}$$

If $M < 1$ the flow is *subsonic* and if $M > 1$ the flow is *supersonic*. Consider the stationary source of disturbances displayed in Fig. 9.2(*a*); the sound waves are shown after three time increments. In Fig. 9.2(*b*) the source is moving at a subsonic speed, which is less than the speed of sound, so the source

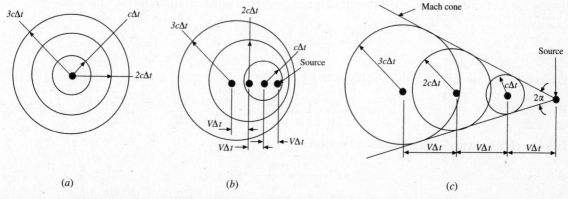

(a)   (b)   (c)

Figure 9.2   The propagation of sound waves from a source: (*a*) a stationary source, (*b*) a moving source with $M < 1$, and (*c*) a moving source with $M > 1$.

"announces" its approach to an observer to the right. In Fig. 9.2(*c*) the source moves at a supersonic speed, which is faster than the speed of the source, so an observer is unaware of the source's approach if the observer is in the *zone of silence*, which is outside the *Mach cone* shown. From the figure, the Mach cone has a *Mach angle* given by

$$\alpha = \sin^{-1}\frac{c}{V} = \sin^{-1}\frac{1}{M} \tag{9.19}$$

The small-amplitude waves discussed above are referred to as *Mach waves*. They result from sources of sound and needle-nosed projectiles and the sharp leading edge of supersonic airfoils. Large-amplitude waves, called *shock waves*, which emanate from the leading edge of blunt-nosed airfoils, also form zones of silence but the angles are larger than those created by the Mach waves. Shock waves will be studied in Secs. 9.4 and 9.5.

EXAMPLE 9.1 An electronic device is situated on the top of a hill and hears a supersonic projectile that produces Mach waves after the projectile is 500 m past the device's position. If it is known that the projectile flies at 850 m/s, estimate how high it is above the device.

Solution: The Mach number is

$$M = \frac{V}{c} = \frac{850}{\sqrt{kRT}} = \frac{850}{\sqrt{1.4 \times 287 \times 288}} = 2.5$$

where a standard temperature of 288 K has been assumed since the temperature was not given. The Mach angle relationship allows us to write

$$\sin \alpha = \frac{1}{M} = \frac{h}{\sqrt{h^2 + 500^2}} = \frac{1}{2.5}$$

where h is the height above the device [refer to Fig. 9.2(c)]. This equation can be solved for h to give $h = 218$ m.

9.3 ISENTROPIC NOZZLE FLOW

There are numerous applications where a steady, uniform, isentropic flow is a good approximation to the flow in conduits. These include the flow through a jet engine, through the nozzle of a rocket, from a broken gas line, and past the blades of a turbine. To model such situations, consider the control volume in the changing area of the conduit of Fig. 9.3. The continuity equation between two sections an infinitesimal distance dx apart is

$$\rho A V = (\rho + d\rho)(A + dA)(V + dV) \tag{9.20}$$

If only the first-order terms in a differential quantity are retained, continuity takes the form

$$\frac{dV}{V} + \frac{dA}{A} + \frac{d\rho}{\rho} = 0 \tag{9.21}$$

The energy equation (9.5) with $\dot{Q} = \dot{W}_S = 0$ is

$$\frac{V^2}{2} + \frac{k}{k-1}\frac{p}{\rho} = \frac{(V+dV)^2}{2} + \frac{k}{k-1}\frac{p+dp}{\rho+d\rho} \tag{9.22}$$

This simplifies to, neglecting higher-order terms

$$V\,dV + \frac{k}{k-1}\frac{\rho\,dp - p\,d\rho}{\rho^2} = 0 \tag{9.23}$$

Assuming an isentropic flow, Eq. (9.15) allows the energy equation to take the form

$$V\,dV + k\frac{p}{\rho^2}\,d\rho = 0 \tag{9.24}$$

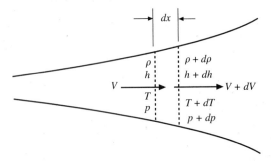

Figure 9.3 Steady, uniform, isentropic flow through a conduit.

Substitute from the continuity equation (9.21) to obtain

$$\frac{dV}{V}\left(\frac{\rho V^2}{kp}-1\right)=\frac{dA}{A}\tag{9.25}$$

or, in terms of the Mach number

$$\frac{dV}{V}(\mathrm{M}^2-1)=\frac{dA}{A}\tag{9.26}$$

This equation applies to a steady, uniform, isentropic flow.

There are several observations that can be made from an analysis of Eq. (9.26). They are as follows:

- For a subsonic flow in an expanding conduit (M < 1 and $dA>0$), the flow is decelerating ($dV<0$).
- For a subsonic flow in an converging conduit (M < 1 and $dA<0$), the flow is accelerating ($dV>0$).
- For a supersonic flow in an expanding conduit (M > 1 and $dA>0$), the flow is accelerating ($dV>0$).
- For a supersonic flow in an converging conduit (M > 1 and $dA<0$), the flow is decelerating ($dV<0$).
- At a throat where $dA=0$, either M = 1 or $dV=0$ (the flow could be accelerating through M = 1 or it may reach a velocity such that $dV=0$).

Observe that a nozzle for a supersonic flow must increase in area in the flow direction and a diffuser must decrease in area, opposite to a nozzle and diffuser for a subsonic flow. So, for a supersonic flow to develop from a reservoir where the velocity is 0, the subsonic flow must first accelerate through a converging area to a throat followed by continued acceleration through an enlarging area. The nozzles on a rocket designed to place satellites in orbit are constructed using such converging–diverging geometry, as shown in Fig. 9.4.

The energy and continuity equations can take on particularly helpful forms for the steady, uniform, isentropic flow through the nozzle of Fig. 9.4. Apply the energy equation (9.4) with $\dot{Q}=\dot{W}_S=0$ between the reservoir and some location in the nozzle to obtain

$$c_pT_0=\frac{V^2}{2}+c_pT\tag{9.27}$$

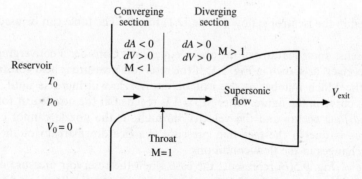

Figure 9.4   A supersonic nozzle.

Any quantity with a 0 subscript refers to a stagnation point where the velocity is 0, such as in the reservoir. Using several thermodynamic relations, Eqs. (9.6), (9.9), (9.16), and (9.18), Eq. (9.27) can be put in the forms

$$\frac{T_0}{T}=1+\frac{k-1}{2}\mathrm{M}^2\qquad\frac{p_0}{p}=\left(1+\frac{k-1}{2}\mathrm{M}^2\right)^{k/(k-1)}\qquad\frac{\rho_0}{\rho}=\left(1+\frac{k-1}{2}\mathrm{M}^2\right)^{1/(k-1)}\tag{9.28}$$

If the above equations are applied at the throat where M = 1, the *critical area* signified by an asterisk (*) superscript, the energy equation takes the forms

$$\frac{T^*}{T_0} = \frac{2}{k+1} \qquad \frac{p^*}{p_0} = \left(\frac{2}{k+1}\right)^{k/(k-1)} \qquad \frac{\rho^*}{\rho_0} = \left(\frac{2}{k+1}\right)^{1/(k-1)} \qquad (9.29)$$

The critical area is often referenced even though a throat does not exist, as in Table D.1. For air with $k = 1.4$, Eqs. (9.29) provide

$$T^* = 0.8333T_0 \qquad p^* = 0.5283p_0 \qquad \rho^* = 0.6340\rho_0 \qquad (9.30)$$

The mass flux through the nozzle is of interest and is given by

$$\dot{m} = \rho A V = \frac{p}{RT} \times AM\sqrt{kRT} = p\sqrt{\frac{k}{RT}} AM \qquad (9.31)$$

With the use of Eqs. (9.28), the mass flux, after some algebra, can be expressed as

$$\dot{m} = p_0 MA\sqrt{\frac{k}{RT_0}}\left(1 + \frac{k-1}{2}M^2\right)^{(k+1)/2(1-k)} \qquad (9.32)$$

If the critical area is selected where M = 1, this takes the form

$$\dot{m} = p_0 A^*\sqrt{\frac{k}{RT_0}}\left(1 + \frac{k-1}{2}\right)^{(k+1)/2(1-k)} \qquad (9.33)$$

which, when combined with Eq. (9.32), provides

$$\frac{A}{A^*} = \frac{1}{M}\left[\frac{2 + (k-1)M^2}{k+1}\right]^{(k+1)/2(1-k)} \qquad (9.34)$$

This ratio is included in the isentropic flow (Table D.1) for air. The table can be used in the place of the above equations.

Now we will discuss some features of the above equations. Consider a converging nozzle connecting a reservoir with a receiver, as shown in Fig. 9.5. If the reservoir pressure is held constant and the receiver pressure reduced, the Mach number at the exit of the nozzle will increase until $M_e = 1$ is reached, indicated by the left curve in the figure. After $M_e = 1$ is reached at the nozzle exit for $p_r = 0.5283p_0$, the condition of *choked flow* occurs and the velocity throughout the nozzle cannot change with further decreases in p_r. This is due to the fact that pressure changes downstream of the exit cannot travel upstream to cause changes in the flow conditions.

The right curve of Fig. 9.5(*b*) represents the case when the reservoir pressure in increased and the receiver pressure is held constant. When $M_e = 1$ the condition of choked flow also occurs but Eq. (9.33) indicates that the mass flux will continue to increase as p_0 is increased. This is the case when a gas line ruptures.

It is interesting that the exit pressure p_e is able to be greater than the receiver pressure p_r. Nature allows this by providing the streamlines of a gas the ability to make a sudden change of direction at the exit and expand to a much greater area resulting in a reduction of the pressure from p_e to p_r.

The case of a converging–diverging nozzle allows a supersonic flow to occur providing the receiver pressure is sufficiently low. This is shown in Fig. 9.6 assuming a constant reservoir pressure with a decreasing receiver pressure. If the receiver pressure is equal to the reservoir pressure, no flow occurs, represented by curve *A*. If p_r is slightly less than p_0, the flow is subsonic throughout with a minimum pressure at the throat, represented by curve B. As the pressure is reduced still further, a pressure is reached that results in $M = 1$ at the throat with subsonic flow throughout the remainder of the nozzle.

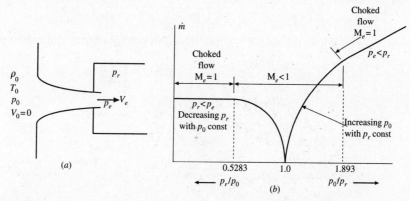

Figure 9.5 (*a*) A converging nozzle and (*b*) the pressure variation in the nozzle.

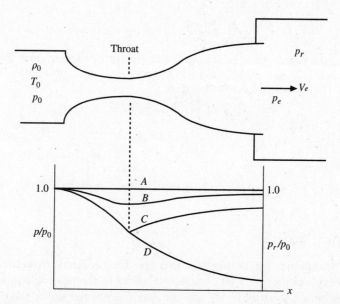

Figure 9.6 A converging–diverging nozzle with reservoir pressure fixed.

There is another receiver pressure substantially below that of curve C that also results in isentropic flow throughout the nozzle, represented by curve D; after the throat the flow is supersonic. Pressures in the receiver in between those of curves C and D result in non-isentropic flow (a shock wave occurs in the flow) and will be considered in Sec. 9.4. If p_r is below that of curve D, the exit pressure p_e is greater than p_r. Once again, for receiver pressures below that of curve C, the mass flux remains constant since the conditions at the throat remain unchanged.

It may appear that the supersonic flow will tend to separate from the nozzle, but just the opposite is true. A supersonic flow can turn very sharp angles, as will be observed in Sec. 9.6, since nature provides expansion fans that do not exist in subsonic flows. To avoid separation in subsonic nozzles, the expansion angle should not exceed $10°$. For larger angles, vanes are used so that the angle between the vanes does not exceed $10°$.

EXAMPLE 9.2   Air flows from a reservoir maintained at 300 kPa absolute and 20°C into a receiver maintained at 200 kPa absolute by passing through a converging nozzle with an exit diameter of 4 cm. Calculate the mass flux through the nozzle. Use (a) the equations and (b) the isentropic flow table.

Solution: (a) The receiver pressure that would give M = 1 at the nozzle exit is

$$p_r = 0.5283 \text{ kPa} \qquad p_0 = 0.5283 \times 300 = 158.5 \text{ kPa absolute}$$

The receiver pressure is greater than this, so $M_e < 1$. The second equation of Eqs. (9.28) can be put in the form

$$\frac{k-1}{2}M^2 = \left(\frac{p_0}{p}\right)^{(k-1)/k} - 1 \qquad \text{or} \qquad 0.2M^2 = \left(\frac{300}{200}\right)^{0.4/1.4} - 1$$

This gives M = 0.784. The mass flux is found from Eq. (9.32) to be

$$\dot{m} = p_0 M A \sqrt{\frac{k}{RT_0}} \left(1 + \frac{k-1}{2}M^2\right)^{(k+1)/2(1-k)}$$

$$= 300\,000 \times 0.784 \times \pi \times 0.02^2 \sqrt{\frac{1.4}{287 \times 293}} \left(1 + \frac{0.4}{2} \times 0.784^2\right)^{-2.4/0.8} = 0.852 \text{ kg/s}$$

For the units to be consistent, the pressure must be in Pa and R in J/(kg·K).
(b) Now use Table D.1. For a pressure ratio of $p/p_0 = 200/300 = 0.6667$, the Mach number is found by interpolation to be

$$M_e = \frac{0.6821 - 0.6667}{0.6821 - 0.6560}(0.8 - 0.76) + 0.76 = 0.784$$

To find the mass flux, the velocity must be known which requires the temperature since $V = M\sqrt{kRT}$. The temperature is interpolated (similar to the interpolation for the Mach number) from Table D.1 to be $T_e = 0.8906 \times 293 = 261$ K. The velocity and density are then

$$V = M\sqrt{kRT} = 0.784\sqrt{1.4 \times 287 \times 261} = 254 \text{ m/s}$$

$$\rho = \frac{p}{RT} = \frac{200}{0.287 \times 261} = 2.67 \text{ kg/m}^3$$

The mass flux is found to be

$$\dot{m} = \rho A V = 2.67 \times \pi \times 0.02^2 \times 254 = 0.852 \text{ kg/s}$$

9.4 NORMAL SHOCK WAVES

Shock waves are large-amplitude waves that exist in a gas. They emanate from the wings of a supersonic aircraft, from a large explosion, from a jet engine, and ahead of the projectile in a gun barrel. They can be oblique waves or normal waves. First, we will consider the normal shock wave, as shown in Fig. 9.7. In this figure, it is stationary so that a steady flow exists. If V_1 were superimposed to the left, the shock would be traveling in stagnant air with velocity V_1 and the *induced velocity* behind the shock wave would be $(V_1 - V_2)$. The shock wave is very thin, on the order of $\sim 10^{-4}$ mm, and in that short distance large pressure changes occur causing enormous energy dissipation. The continuity equation with $A_1 = A_2$ is

$$\rho_1 V_1 = \rho_2 V_2 \tag{9.35}$$

The energy equation with $\dot{Q} = \dot{W}_S = 0$ takes the form

$$\frac{V_2^2 - V_1^2}{2} + \frac{k}{k-1}\left(\frac{p_2}{\rho_2} - \frac{p_1}{\rho_1}\right) = 0 \tag{9.36}$$

The only forces in the momentum equation are pressure forces, so

$$p_1 - p_2 = \rho_1 V_1 (V_2 - V_1) \tag{9.37}$$

where the areas have divided out since $A_1 = A_2$. Assuming that the three quantities ρ_1, p_1, and V_1 before the shock waves are known, the above three equations allow us to solve for three unknowns ρ_2, p_2, and V_2 since, for a given gas, k is known.

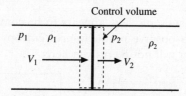

Figure 9.7 A stationary shock wave.

Rather than solve Eqs. (9.35) to (9.37) simultaneously, we write them in terms of the Mach numbers M_1 and M_2 and put them in more convenient forms. First, the momentum equation (9.37), using Eq. (9.35) and $V^2 = M^2 pk/\rho$, can be written as

$$\frac{p_2}{p_1} = \frac{1 + kM_1^2}{1 + kM_2^2} \tag{9.38}$$

In like manner, the energy equation (9.36), with $p = \rho RT$ and $V^2 = M^2 kRT$, can be written as

$$\frac{T_2}{T_1} = \frac{1 + \dfrac{k-1}{2}M_1^2}{1 + \dfrac{k-1}{2}M_2^2} \tag{9.39}$$

The continuity equation (9.35) with $\rho = p/RT$ and $V = M\sqrt{kRT}$ becomes

$$\frac{p_2}{p_1}\frac{M_2}{M_1}\sqrt{\frac{T_1}{T_2}} = 1 \tag{9.40}$$

If the pressure and temperature ratios from Eqs. (9.38) and (9.39) are substituted into Eq. (9.40), the downstream Mach number is related to the upstream Mach number by (the algebra to show this is not shown here)

$$M_2^2 = \frac{M_1^2 + \dfrac{2}{k-1}}{\dfrac{2k}{k-1}M_1^2 - 1} \tag{9.41}$$

This allows the momentum equation (9.38) to be written as

$$\frac{p_2}{p_1} = \frac{2k}{k+1}M_1^2 - \frac{k-1}{k+1} \tag{9.42}$$

and the energy equation (9.39) as

$$\frac{T_2}{T_1} = \frac{\left(1 + \dfrac{k-1}{2}M_1^2\right)\left(\dfrac{2k}{k-1}M_1^2 - 1\right)}{\dfrac{(k+1)^2}{2(k-1)}M_1^2} \tag{9.43}$$

For air, the preceding equations simplify to

$$M_2^2 = \frac{M_1^2 + 5}{7M_1^2 - 1} \qquad \frac{p_2}{p_1} = \frac{7M_1^2 - 1}{6} \qquad \frac{T_2}{T_1} = \frac{(M_1^2 + 5)(7M_1^2 - 1)}{36M_1^2} \tag{9.44}$$

Several observations can be made from these three equations:

- If $M_1 = 1$, then $M_2 = 1$ and no shock wave exists.
- If $M_1 > 1$, then $M_2 < 1$. A supersonic flow is always converted to a subsonic flow when it passes through a normal shock wave.
- If $M_1 < 1$, then $M_2 > 1$ and a subsonic flow appears to be converted to a supersonic flow. This is impossible since it results in a positive production of entropy, a violation of the second law of thermodynamics; this violation will not be proven here.

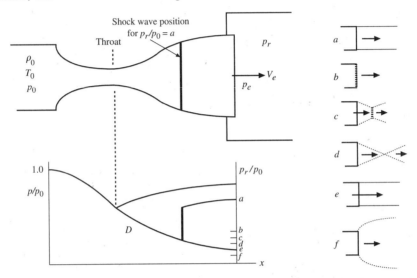

Figure 9.8   Flow with shock waves in a nozzle.

Several normal shock flow relations for air have been presented in Table D.2. The use of that table allows one to avoid using Eqs. (9.44). In addition, the ratio p_{02}/p_{01} of the stagnation point pressures in front of and behind the shock wave is listed.

Return to the converging–diverging nozzle and focus attention on the flow below curve C of Fig. 9.6. If the receiver pressure decreases to $p_r/p_0 = a$ in Fig. 9.8, a normal shock wave would be positioned somewhere inside the nozzle as shown. If the receiver pressure decreased still further, there would be some ratio $p_r/p_0 = b$ that would position the shock wave at the exit plane of the nozzle. Pressure ratios c and d would result in oblique shock wave patterns similar to those shown. Pressure ratio e is associated with isentropic flow throughout, and pressure ratio f would provide an exit pressure greater than the receiver pressure resulting in a billowing out, as shown, of the exiting flow, as seen on the rockets that propel satellites into space.

EXAMPLE 9.3   A normal shock wave travels at 600 m/s through stagnant $20°\text{C}$ air. Estimate the velocity induced behind the shock wave. (*a*) Use the equations and (*b*) use the normal shock flow (Table D.2). Refer to Fig. 9.7.

    Solution: Superimpose a velocity of 600 m/s so that the shock wave is stationary and $V_1 = 600$ m/s, as displayed in Fig. 9.7. The upstream Mach number is

$$M_1 = \frac{V_1}{\sqrt{kRT}} = \frac{600}{\sqrt{1.4 \times 287 \times 293}} = 1.75$$

(*a*) Using the equations, the downstream Mach number and temperature are, respectively

$$M_2 = \left(\frac{M_1^2 + 5}{7M_1^2 - 1}\right)^{1/2} = \left(\frac{1.75^2 + 5}{7 \times 1.75^2 - 1}\right)^{1/2} = 0.628$$

$$T_2 = \frac{T_1(M_1^2 + 5)(7M_1^2 - 1)}{36M_1^2} = \frac{293(1.75^2 + 5)(7 \times 1.75^2 - 1)}{36 \times 1.75^2} = 438\,\text{K}$$

The velocity behind the shock wave is then

$$V_2 = M_2\sqrt{kRT_2} = 0.628\sqrt{1.4 \times 287 \times 438} = 263\,\text{m/s}$$

If V_1 is superimposed to the left in Fig. 9.7, the induced velocity is

$$V_{\text{induced}} = V_1 - V_2 = 600 - 263 = 337\,\text{m/s}$$

which would act to the left, in the direction of the moving shock wave.

(b) Table D.2 is interpolated at $M_1 = 1.75$ to find

$$M_2 = \frac{1.75 - 1.72}{1.76 - 1.72}(0.6257 - 0.6355) + 0.6355 = 0.6282$$

$$\frac{T_2}{T_1} = \frac{1.75 - 1.72}{1.76 - 1.72}(1.502 - 1.473) + 1.473 = 1.495. \qquad \therefore\ T_2 = 438\,\text{K}$$

The velocity V_2 is then

$$V_2 = M_2\sqrt{kRT_2} = 0.628\sqrt{1.4 \times 287 \times 438} = 263\,\text{m/s}$$

and the induced velocity due to the shock wave is

$$V_{\text{induced}} = V_1 - V_2 = 600 - 263 = 337\,\text{m/s}$$

EXAMPLE 9.4 Air flows from a reservoir maintained at $20\,°\text{C}$ and 200 kPa absolute through a converging–diverging nozzle with a throat diameter of 6 cm and an exit diameter of 12 cm to a receiver. What receiver pressure is needed to locate a shock wave at a position where the diameter is 10 cm? Refer to Fig. 9.8.

Solution: Let us use the isentropic flow (Table D.1) and the normal shock (Table D.2) Tables. At the throat for this supersonic flow $M_t = 1$. The Mach number just before the shock wave is interpolated from Table D.1 where $A_1/A^* = 10^2/6^2 = 2.778$ to be

$$M_1 = 2.556$$

From Table D.2

$$M_2 = 0.5078 \qquad \frac{p_{02}}{p_{01}} = 0.4778$$

so that

$$p_{02} = 0.4778 \times 200 = 95.55\,\text{kPa}$$

since the stagnation pressure does not change in the isentropic flow before the shock wave so that $p_{01} = 200$ kPa. From just after the shock wave to the exit, isentropic flow again exists so that from Table D.1 at $M_2 = 0.5078$

$$\frac{A_2}{A^*} = 1.327$$

We have introduced an imaginary throat between the shock wave and the exit of the nozzle. The exit area A_e is introduced by

$$\frac{A_e}{A^*} = \frac{A_2}{A^*} \times \frac{A_e}{A_2} = 1.327 \times \frac{12^2}{10^2} = 1.911$$

Using Table D.1 at this area ratio (make sure the subsonic part of the table is used), we find

$$M_e = 0.3223 \qquad \text{and} \qquad \frac{p_e}{p_{0e}} = 0.9305$$

so that

$$p_e = 0.9305 \times 95.55 = 88.9\,\text{kPa}$$

using $p_{0e} = p_{02}$ for the isentropic flow after the shock wave. The exit pressure is equal to the receiver pressure for this isentropic subsonic flow.

9.5 OBLIQUE SHOCK WAVES

Oblique shock waves form on the leading edge of a supersonic sharp-edged airfoil or in a corner, as shown in Fig. 9.9. A steady, uniform plane flow exists before and after the shock wave. The oblique shock waves also form on axisymmetric projectiles.

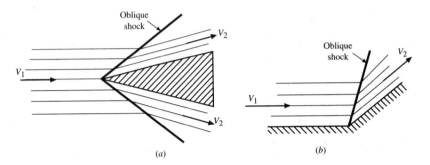

Figure 9.9   Oblique shock waves (a) flow over a wedge and (b) flow in a corner.

The oblique shock wave turns the flow so that V_2 is parallel to the plane surface. Another variable, the angle through which the flows turns, is introduced but the additional tangential momentum equation allows a solution. Consider the control volume of Fig. 9.10 surrounding the oblique shock wave. The velocity vector V_1 is assumed to be in the x-direction and the oblique shock wave turns the flow through the *wedge angle* or *deflection angle* θ so that V_2 is parallel to the wall. The oblique shock wave makes an angle of β with V_1. The components of the velocity vectors are shown normal and tangential to the oblique shock. The tangential components of the velocity vectors do not cause fluid to flow into or out of the control volume, so continuity provides

$$\rho_1 V_{1n} = \rho_2 V_{2n} \tag{9.45}$$

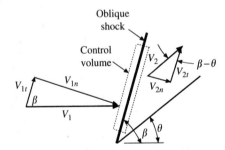

Figure 9.10   Oblique shock wave control volume.

The pressure forces act normal to the control volume and produce no net force tangential to the oblique shock. This allows the tangential momentum equation to take the form

$$\dot{m}_1 V_{1t} = \dot{m}_2 V_{2t} \tag{9.46}$$

Continuity requires $\dot{m}_1 = \dot{m}_2$ so that

$$V_{1t} = V_{2t} \tag{9.47}$$

The momentum equation normal to the oblique shock is

$$p_1 - p_2 = \rho_2 V_{2n}^2 - \rho_1 V_{1n}^2 \qquad (9.48)$$

The energy equation, using $V^2 = V_n^2 + V_t^2$, can be written in the form

$$\frac{V_{1n}^2}{2} + \frac{k-1}{k}\frac{p_1}{\rho_1} = \frac{V_{2n}^2}{2} + \frac{k-1}{k}\frac{p_2}{\rho_2} \qquad (9.49)$$

since the tangential velocity terms cancel.

Observe that the tangential velocity components do not enter Eqs. (9.45), (9.48), and (9.49). They are the same three equations used to solve the normal shock wave problem. Therefore, the components V_{1n} and V_{2n} can be replaced with V_1 and V_2, respectively, of the normal shock wave problem and a solution obtained. Table D.2 may also be used. We also replace M_{1n} and M_{2n} with M_1 and M_2 in the equations and table.

To often simplify a solution, we relate the oblique shock angle β to the deflection angle θ. This is done using Eq. (9.45) to obtain

$$\frac{\rho_2}{\rho_1} = \frac{V_{1n}}{V_{2n}} = \frac{V_{1t}\tan\beta}{V_{2t}\tan(\beta-\theta)} = \frac{\tan\beta}{\tan(\beta-\theta)} \qquad (9.50)$$

Using Eqs. (9.42) and (9.43), this density ratio can be written as

$$\frac{\rho_2}{\rho_1} = \frac{p_2 T_1}{p_1 T_2} = \frac{(k+1)M_{1n}^2}{(k-1)M_{1n}^2 + 2} \qquad (9.51)$$

Using this density ratio in Eq. (9.50) allows us to write

$$\tan(\beta-\theta) = \frac{\tan\beta}{k+1}\left(k-1+\frac{2}{M_1^2\sin^2\beta}\right) \qquad (9.52)$$

With this relationship, the oblique shock angle β can be found for a given incoming Mach number and wedge angle θ. A plot of Eq. (9.52) is useful to avoid a trial-and-error solution. It is included as Fig. 9.11. Three observations can be made by studying the figure.

- For a given Mach number M_1 and wedge angle θ there are two possible oblique shock angles β. The larger one is the "strong" oblique shock wave and the smaller one is the "weak" oblique shock wave.
- For a given wedge angle θ, there is a minimum Mach number for which there is only one oblique shock angle β.
- If the Mach number is less than the minimum for a particular θ, but greater than one, the shock wave is detached as shown in Fig. 9.12. Also, for a given M_1, there is a sufficiently large θ that will result in a detached shock wave.

The required pressure rise determines if a weak shock or a strong shock exists. The pressure rise is determined by flow conditions.

For a detached shock wave around a blunt body or a wedge, a normal shock wave exists on the stagnation streamline; the normal shock is followed by a strong oblique shock, then a weak oblique shock, and finally a Mach wave, as shown in Fig. 9.12. The shock wave is always detached on a blunt object.

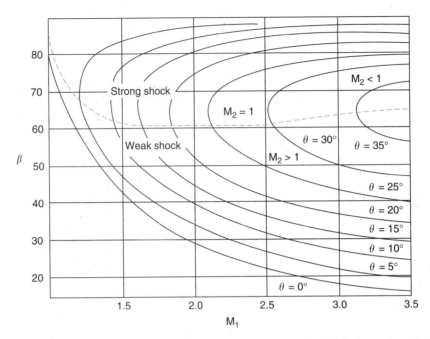

Figure 9.11 Oblique shock wave angle β related to wedge angle θ and Mach number M_1 for air.

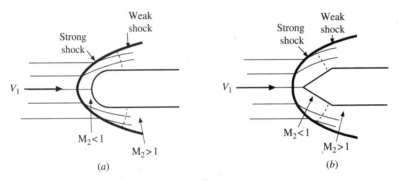

Figure 9.12 Detached shock waves around (a) a plane, blunt object and (b) a wedge.

EXAMPLE 9.5 Air at 30°C flows around a wedge with an included angle of 60° [Fig. 9.9(a)]. An oblique shock emanates from the wedge at an angle of 50°. Determine the approach velocity of the air. Also find M_2 and T_2.

Solution: From Fig. 9.11 the Mach number, at $\theta = 30°$ and $\beta = 50°$, is

$$M_1 = 3.1$$

The velocity is then

$$V_1 = M_1 \sqrt{kRT} = 3.1\sqrt{1.4 \times 87 \times 303} = 1082 \, \text{m/s}$$

If Eq. (9.52) were used for greater accuracy, we have

$$\tan(50° - 30°) = \frac{\tan 50°}{1.4 + 1}\left(1.4 - 1 + \frac{2}{M_1^2 \sin^2 50°}\right). \qquad \therefore M_1 = 3.20$$

The velocity would be $V_1 = 1117$ m/s.
To find M_2, the approaching normal velocity and Mach number are

$$V_{1n} = V_1 \sin \beta = 1117 \sin 50° = 856 \, \text{m/s}. \qquad \therefore M_{1n} = \frac{856}{\sqrt{1.4 \times 287 \times 303}} = 2.453$$

From Table D.2 interpolation provides $M_{2n} = 0.5176$ so that

$$M_2 = \frac{M_{2n}}{\sin(50° - 30°)} = \frac{0.5176}{\sin 20°} = 1.513$$

The temperature behind the oblique shock is interpolated to be

$$T_2 = T_1 \times 2.092 = 303 \times 2.092 = 634\,\text{K}$$

9.6 EXPANSION WAVES

Supersonic flow exits a nozzle, as for the pressure ratio f in Fig. 9.8, and billows out into a large exhaust plume. Also, supersonic flow does not separate from the wall of a nozzle that expands quite rapidly, as in the sketch of Fig. 9.8. How is this accomplished? Consider the possibility that a single finite wave, such as an oblique shock, is able to turn the flow around the convex corner, as shown in Fig. 9.12(a). From the tangential momentum equation, the tangential component of velocity must remain the same on both sides of the finite wave. For this to be true, $V_2 > V_1$ as is obvious from the simple sketch. As before, this increase in velocity as the fluid flows through a finite wave requires an increase in entropy, a violation of the second law of thermodynamics, making a finite wave an impossibility.

A second possibility is to allow an infinite fan of Mach waves, called an *expansion fan*, emanating from the corner, as shown in Fig. 9.13(b). This is an ideal isentropic process so the second law is not violated; such a process may be approached in a real application. Let us consider the single infinitesimal Mach wave displayed in Fig. 9.14, apply our fundamental laws, and then integrate around the corner. Since the tangential velocity components are equal, the velocity triangles yield

$$V_t = V\cos\mu = (V + dV)\cos(\mu + d\theta) \tag{9.53}$$

This can be written as,[*] neglecting higher-order terms, Eq. (9.53) becomes

$$V\,d\theta\sin\mu = \cos\mu\,dV \tag{9.54}$$

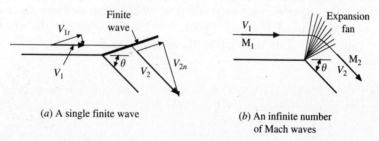

(a) A single finite wave   (b) An infinite number
                                      of Mach waves

Figure 9.13   Supersonic flow around a convex corner. (a) A single finite wave. (b) An infinite number of Mach waves.

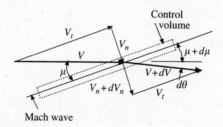

Figure 9.14   A Mach wave in an expansion fan.

Substitute $\sin\mu = 1/M$ [see Eq. (9.19)] and $\cos\mu = \sqrt{M^2 - 1}/M$, to obtain

$$d\theta = \sqrt{M^2 - 1}\,\frac{dV}{V} \tag{9.55}$$

---

[*] Recall that $\cos(\mu + d\theta) = \cos\mu\cos d\theta - \sin\mu\sin d\theta = \cos\mu - d\theta\sin\mu$, since $\cos d\theta \approx 1$ and $\sin d\theta \approx d\theta$.

Differentiate the equation $V = M\sqrt{kRT}$ and put in the form

$$\frac{dV}{V} = \frac{dM}{M} + \frac{1}{2}\frac{dT}{T} \tag{9.56}$$

The energy equation $V^2/2 + kRT/(k-1) = \text{const}$ can also be differentiated to yield

$$\frac{dV}{V} + \frac{1}{(k-1)M^2}\frac{dT}{T} = 0 \tag{9.57}$$

Combine Eqs. (9.56) and (9.57) to obtain

$$\frac{dV}{V} = \frac{2}{2+(k-1)M^2}\frac{dM}{M} \tag{9.58}$$

Substitute this into Eq. (9.55) to obtain a relationship between θ and M

$$d\theta = \frac{2\sqrt{M^2-1}}{2+(k-1)M^2}\frac{dM}{M} \tag{9.59}$$

This is integrated from $\theta = 0$ and M = 1 to a general angle θ called the *Prandtl–Meyer function*, and Mach number M [this would be M_2 in Fig. 9.12(b)] to find that

$$\theta = \left(\frac{k+1}{k-1}\right)^{1/2}\tan^{-1}\left[\frac{k-1}{k+1}(M^2-1)\right]^{1/2} - \tan^{-1}(M^2-1)^{1/2} \tag{9.60}$$

The solution to this relationship is presented for air in Table D.3 to avoid a trial-and-error solution for M given the angle θ. If the pressure or temperature is desired, the isentropic flow table can be used. The Mach waves that allow the gas to turn the corner are sometimes referred to as *expansion waves*.

Observe from Table D.3 that the expansion fan that turns the gas through the angle θ results in M = 1 before the fan to a supersonic flow after the fan. The gas speeds up as it turns the corner and it does not separate. A slower moving subsonic flow would separate from the corner and would slow down. If M = ∞ is substituted into Eq. (9.60), $\theta = 130.5°$, which is the maximum angle through which the flow could turn. This shows that turning angles greater than 90° are possible, a rather surprising result.

EXAMPLE 9.6 Air at 150 kPa and 140°C flows at M = 2 and turns a convex corner of 30°. Estimate the Mach number, pressure, temperature, and velocity after the corner.

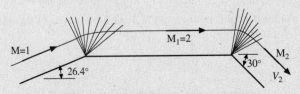

Figure 9.15

Solution: Table D.3 assumes the air is initially at M = 1. So, assume the flow originates from M = 1 and turns a corner to $M_1 = 2$ and then a second corner to M_2, as shown in Fig. 9.15. From Table D.3, an angle of 26.4° is required to accelerate the flow from M = 1 to M = 2. Add another 30 to 26.4° and at $\theta = 56.4°$ we find that

$$M_2 = 3.37$$

Using the isentropic flow table (Table D.1), the entries from the reservoir to state 1 and also to state 2 can be used to find

$$p_2 = p_1 \frac{p_0}{p_1} \frac{p_2}{p_0} = 150 \times \frac{1}{0.1278} \times 0.01580 = 18.54 \, \text{kPa}$$

$$T_2 = T_1 \frac{T_0}{T_1} \frac{T_2}{T_0} = 413 \times \frac{1}{0.5556} \times 0.3058 = 227 \, \text{K} \quad \text{or} \quad -46 \, °\text{C}$$

The velocity after the corner is then

$$V_2 = M_2 \sqrt{kRT_2} = 3.37 \sqrt{1.4 \times 287 \times 227} = 1018 \, \text{m/s}$$

Solved Problems

9.1 Two boys decide to estimate how far it is across a lake. One bangs two rocks together underwater on one side and the other estimates that it takes 0.4 s for the sound to reach the other side. What is the distance across the lake?

Using the bulk modulus as $2110 \times 10^6 \, \text{Pa}$, the speed of sound in water is

$$c = \sqrt{\frac{dp}{d\rho}} = \sqrt{\frac{1}{\rho} \left(\rho \frac{dp}{d\rho} \right)} = \sqrt{\frac{1}{1000} \times 2110 \times 10^6} = 1453 \, \text{m/s}$$

At this speed, the distance is

$$d = c \, \Delta t = 1453 \times 0.4 = 581 \, \text{m}$$

9.2 Show that Eq. (9.26) follows from Eq. (9.22).

The right-hand side of Eq. (9.22) is expanded so that

$$\frac{V^2}{2} + \frac{k}{k-1} \frac{p}{\rho} = \frac{V^2 + 2V \, dV + (dV)^2}{2} + \frac{k}{k-1} \frac{p+dp}{\rho + d\rho}$$

$$\frac{V^2}{2} = \frac{V^2}{2} + \frac{2V \, dV + (dV)^2}{2} + \frac{k}{k-1} \left(\frac{p+dp}{\rho + d\rho} - \frac{p}{\rho} \right)$$

$$0 = \frac{2V \, dV}{2} + \frac{k}{k-1} \left(\frac{p\rho + \rho \, dp - p\rho - p \, d\rho}{\rho(\rho + d\rho)} \right) = V \, dV + \frac{k}{k-1} \left(\frac{kp \, d\rho - p \, d\rho}{\rho^2} \right)$$

using $\rho \, dp = kp \, d\rho$ for an isentropic process [see Eq. (9.15)]. This is then

$$0 = V \, dV + k \left(\frac{p \, d\rho}{\rho^2} \right) = V \, dV + k \frac{p}{\rho} \left(-\frac{dV}{V} - \frac{dA}{A} \right)$$

when $d\rho/\rho$ is substituted from Eq. (9.21). This can be written as

$$\frac{dA}{A} = \left(\frac{V^2 \rho}{kp} - 1 \right) \frac{dV}{V}$$

Using $c^2 = kp/\rho$ and $M = V/c$, this is put in the form

$$\frac{dA}{A} = (M^2 - 1) \frac{dV}{V}$$

9.3 A converging nozzle with an exit diameter of 6 cm is attached to a reservoir maintained at 30°C and 150 kPa absolute. Determine the mass flux of air flowing through the nozzle if the receiver exits to the atmosphere. (a) Use equations and (b) use the appropriate table.

Is the flow choked?

$$0.5283 \times 150 = 79.2 \, \text{kPa}. \quad \therefore \; p_r > 0.5283 p_0$$

and the flow is not choked and $M_e < 1$. (The receiver pressure, the atmosphere, is assumed to be at 100 kPa.)

a) Using the equations, we have energy and the isentropic relation providing

$$\frac{\cancel{V_0^2}}{2} + 1000 \times 303 = \frac{V_e^2}{2} + \frac{1.4}{0.4}\frac{100\,000}{\rho_e} \qquad \frac{100}{150} = \left(\frac{\rho_e}{1.725}\right)^{1.4}$$

where

$$\rho_0 = \frac{150}{0.287 \times 303} = 1.725 \qquad \therefore \rho_e = 1.291\,\text{kg/m}^3 \qquad \text{and} \qquad V_e = 253\,\text{m/s}$$

The mass flux follows:

$$\therefore \dot{m} = \rho_e A_e V_e = 1.291 \times \pi \times 0.03^2 \times 253 = 0.925\,\text{kg/s}$$

b) Use Table D.1 and find

$$\frac{p_e}{p_0} = \frac{100}{150} = 0.6667$$

Interpolation gives $M_e = 0.784$ and $T_e = 0.8906 \times 303 = 270\,\text{K}$. Then

$$\rho_e = \frac{100}{0.287 \times 270} = 1.290\,\text{kg/m}^3 \qquad \text{and} \qquad V_e = 0.784\sqrt{1.4 \times 287 \times 270} = 258\,\text{m/s}$$

$$\therefore \dot{m} = \rho_e A_e V_e = 1.290 \times \pi \times 0.03^2 \times 258 = 0.941\,\text{kg/s}$$

9.4   Air flows through a converging–diverging nozzle, with a throat diameter of 10 cm and an exit diameter of 20 cm, from a reservoir maintained at 20°C and 300 kPa absolute. Estimate the two receiver pressures that provide an isentropic flow throughout (curves *C* and *D* of Fig. 9.6). Also, determine the associated exit Mach numbers.

Let us use Table D.1 rather than the equations. The area ratio is $A/A^* = 4$. There are two pressure ratios in Table D.1 corresponding to this area ratio. They are interpolated to be

$$\left(\frac{p}{p_0}\right)_C = \frac{4.0 - 3.6727}{4.8643 - 3.6727}(0.9823 - 0.990) + 0.990 = 0.988$$

$$\left(\frac{p}{p_0}\right)_D = \frac{4.0 - 3.924}{4.076 - 3.924}(0.02891 - 0.03071) + 0.03071 = 0.0298$$

The two pressures are

$$p_r = 0.988 \times 300 = 296.4\,\text{kPa} \qquad \text{and} \qquad p_r = 0.0298 \times 300 = 8.94\,\text{kPa}$$

The two Mach numbers are interpolated in Table D.1 to be

$$M_e = 0.149 \quad \text{and} \quad 2.94$$

9.5   Gas flows can be considered to be incompressible if the Mach number is less than 0.3. Estimate the error in the stagnation pressure in air if $M = 0.3$.

The stagnation pressure is found by applying the energy equation between the free stream and the stagnation point where $V_0 = 0$. Assuming incompressible flow, Eq. (9.3) with $\dot{Q} = \dot{W}_S = 0$ and $\tilde{u}_2 = \tilde{u}_1$ (no losses) provides

$$p_0 = p + \rho\frac{V^2}{2}$$

For isentropic flow with $k = 1.4$, the energy equation (9.28) can be put in the form

$$p_0 = p(1 + 0.2M^2)^{3.5}$$

This can be expanded using the binomial theorem $(1 + x)^n = 1 + nx + n(n-1)x^2/2 + \dots$ by letting $x = 0.2M^2$. We then have

$$p_0 = p(1 + 0.7M^2 + 0.175M^4 + \dots) \qquad \text{or} \qquad p_0 - p = pM^2(0.7 + 0.175M^2 + \dots)$$

Using Eqs. (9.16) and (9.18), this takes the form

$$p_0 - p = \rho \frac{V^2}{2}(1 + 0.25M^2 + \ldots)$$

Let $M = 0.3$ so that

$$p_0 - p = \rho \frac{V^2}{2}(1 + 0.0225 + \ldots)$$

Compare this to the incompressible flow equation above and we see that the error is about 2.25 percent. So, if $M < 0.3$ (about 100 m/s for air at standard conditions), the flow of a gas is considered to be incompressible.

9.6 A pitot probe (Fig. 3.11) is used to measure the stagnation pressure in a supersonic flow. The stagnation pressure is measured to be 360 kPa absolute in an airflow where the pressure is 90 kPa absolute and the temperature is 15°C. Find the free-stream velocity V_1. (A shock wave will be positioned in front of the probe. Let state 2 be located just after the shock and p_3 be the stagnation pressure at the probe opening.)

The pressure ratio across the shock is given by Eq. (9.42)

$$\frac{p_2}{p_1} = \frac{2k}{k+1}M_1^2 - \frac{k-1}{k+1}$$

The Mach numbers are related by Eq. (9.41)

$$M_2^2 = \frac{(k-1)M_1^2 + 2}{2kM_1^2 - k + 1}$$

The isentropic flow from behind the shock to the stagnation point provides the pressure ratio as [see Eq. (9.28)]

$$\frac{p_3}{p_2} = \left(1 + \frac{k-1}{2}M_2^2\right)^{k/(k-1)}$$

The above three equations can be combined to eliminate p_2 and M_2 to yield the *Raleigh pitot-tube formula* for a supersonic flow, namely

$$\frac{p_3}{p_1} = \frac{\left(\dfrac{k+1}{2}M_1^2\right)^{k/(k-1)}}{\left(\dfrac{2k}{k+1}M_1^2 - \dfrac{k-1}{k+1}\right)^{1/(k-1)}}$$

Using $k = 1.4$, $p_1 = 90$ kPa, and $p_3 = 350$ kPa, the above equation becomes

$$\frac{360}{90} = \frac{(1.2M_1^2)^{3.5}}{(1.167M_1^2 - 0.1667)^{2.5}}$$

A trial-and-error solution results in

$$M_1 = 1.65$$

The free-stream velocity is then

$$V_1 = M_1\sqrt{kRT_1} = 1.65\sqrt{1.4 \times 287 \times 288} = 561$$

9.7 Air flows with $M_1 = 2$ such that a weak oblique shock wave at $\beta_1 = 40°$ reflects from a plane wall, as shown in Fig. 9.16. Estimate M_3 and β_3. (Note V_3 must be parallel to the wall.)

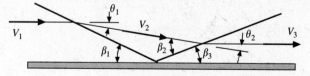

Figure 9.16

The various angles are shown in Fig. 9.16. From Fig. 9.11 with $\beta = 40°$ and $M = 2$, we see that $\theta = 11°$. The normal shock wave is of magnitude

$$M_{1n} = 2 \sin 40° = 1.28$$

From the shock (Table D.2), we find M_{2n} and then M_2 to be

$$M_{2n} = 0.7963 = M_2 \sin(40° - 11°). \qquad \therefore M_2 = 1.64$$

The reflected wave must then turn the flow through $11°$ to be parallel to the wall so that θ_2 is also $11°$. Using Fig. 9.11 again at $M_2 = 1.64$, we find $\beta_2 \cong 50°$. So, M_{2n} relative to the reflected shock is

$$M_{2n} = 1.64 \sin 50° = 1.26$$

From the shock table, we find M_{3n} and then M_3 to be

$$M_{3n} = 0.807 = M_3 \sin(50° - 11°). \qquad \therefore M_3 = 1.28$$

Since $\beta_3 + 11° = 50°$, we see that $\beta_3 = 39°$.

Supplementary Problems

9.8   Show that $c_v = R/(k-1)$.

9.9   Using English units, $c_p = 0.24$ Btu/(lbm-°R). Show that this is equivalent to 1.0 kJ/ (kg·K).

9.10   Show that Eq. (9.8) implies Eq. (9.9) providing $\Delta s = 0$.

9.11   Show that the energy equation (9.5) follows from Eq. (9.3).

9.12   Differentiate $p/\rho^k = $ const and show that Eq. (9.15) results.

Speed of a Small Disturbance

9.13   Show that the energy equation relates the temperature rise to the velocity change for a small adiabatic disturbance traveling in a gas by $c_p \Delta T = -c \Delta V$.

9.14   Show that a small disturbance travels in water at about 1450 m/s and in air at standard conditions at about 340 m/s.

9.15   Two rocks are slammed together by a friend on one side of a lake. A listening device picks up the wave generated 0.75 s later. How far is it across the lake?

9.16   An underwater animal generates a signal that travels through water until it hits an object and then echoes back to the animal 0.46 s later. How far is the animal from the object?

9.17   Estimate the Mach number for a projectile flying at:
    (a) 1000 m at 100 m/s
    (b) 10 000 m at 200 m/s
    (c) 30 000 m at 300 m/s
    (d) 10 000 m at 250 m/s

9.18   A bolt of lightning lights up the sky and 1.5 s later you hear the thunder. How far did the lightning strike from your position?

9.19   A supersonic aircraft passes 200 m overhead on a day when the temperature is 26°C.
        Estimate how long it will be before you hear its sound after it passes directly overhead and how far the aircraft is from you if its Mach number is
        (a) 1.68
        (b) 2.02
        (c) 3.49

9.20   A small-amplitude wave travels through the atmosphere creating a pressure rise of 5 Pa. Estimate the temperature rise across the wave and the induced velocity behind the wave.

Isentropic Nozzle Flow

9.21   Show that
        (a) Eq. (9.21) follows from Eq. (9.20)
        (b) Eq. (9.24) follows from Eq. (9.22)
        (c) Eq. (9.26) follows from Eq. (9.24)
        (d) Eq. (9.33) follows from Eq. (9.31)
        (e) Eq. (9.34) follows from Eq. (9.33)

9.22   A pitot probe is used to measure the speed of a ground vehicle on the Salt Lake flats. It measures 3400 Pa in 28°C air. Estimate its speed assuming:
        (a) An isentropic process
        (b) The air to be incompressible

9.23   Rework Example 9.2 but assume the receiver pressure to be 100 kPa absolute. Use
        (a) The equations
        (b) Table D.1

9.24   A converging nozzle with an exit area of 10 cm^2 is attached to a reservoir maintained at 250 kPa absolute and 20°C. Using equations only, calculate the mass flux if the receiver pressure is maintained at:
        (a) 150 kPa absolute
        (b) 100 kPa absolute
        (c) 50 kPa absolute

9.25   Solve Prob. 9.24b using Table D.1.

9.26   A converging nozzle with an exit area of 10 cm^2 is attached to a reservoir maintained at 350 kPa absolute and 20°C. Determine the receiver pressure that would just provide M$_e$ = 1 and the mass flux from the nozzle for that receiver pressure. Use (a) the equations and (b) Table D.1.

9.27   A converging nozzle with an exit area of 5 cm^2 is attached to a reservoir maintained at 20°C and exhausts directly to the atmosphere. What reservoir pressure would just result in M$_e$ = 1? Calculate the mass flux for that pressure. Use (a) the equations and (b) Table D.1.

9.28   Double the reservoir pressure of Prob. 9.27 and calculate the increased mass flux. Use the equations or Table D.1.

9.29   A large 25°C airline pressurized to 600 kPa absolute suddenly bursts. A hole with area 20 cm^2 is measured from which the air escaped. Assuming the airline pressure remained constant, estimate the cubic meters of air that was lost to the atmosphere in the first 30 s. (The same analysis can be used for a gas line that bursts.)

9.30   If hydrogen was contained in the reservoir of Prob. 9.27, calculate the mass flux for the condition of that problem. The equations must be used.

9.31   A *Venturi tube*, shown in Fig. 9.17, is used to measure the mass flux of air through the pipe by measuring the pressures before the reduced section and at the minimum area. If the temperature before the reduced section is 30°C, determine the mass flux.

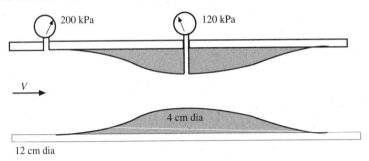

Figure 9.17

9.32   Air flows through a converging–diverging nozzle attached from a reservoir maintained at 400 kPa absolute and 20°C to a receiver. If the throat and exit diameters are 10 and 24 cm, respectively, what two receiver pressures will result in isentropic flow throughout such that M = 1 at the throat? Use (*a*) equations only and (*b*) Table D.1.

9.33   Air flows from a converging–diverging nozzle from a reservoir maintained at 400 kPa absolute and 20°C through a 12-cm-diameter throat. At what diameter in the diverging section will M = 2? Use the equations or the tables.

9.34   Calculate the exit velocity and mass flux for both pressures of Prob. 9.32.

9.35   Air enters a diffuser at 50 kPa absolute and 120°C with a M = 2.4 and a mass flux of 8.5 kg/s. Sketch the general shape of the diffuser and then determine the throat diameter and the exit pressure assuming isentropic flow throughout. Neglect the exiting kinetic energy.

Normal Shock Wave

9.36   The temperature, pressure, and velocity before a normal shock wave are 20°C, 100 kPa absolute, and 600 m/s, respectively. Determine the temperature, pressure, velocity, and Mach number after the shock wave. Assume air and use (*a*) the basic equations (9.35)–(9.37), (*b*) the specialized equations, and (*c*) the normal shock table.

9.37   Air flows through a shock wave. Given the quantities in the first parentheses before the shock and the quantities in the second parentheses after the shock, find the unknown quantities. (Pressures are absolute.)
   (*a*) (20°C, 400 kPa, 480 m/s, M_1) (T_2, p_2, M_2, V_2)
   (*b*) (20°C, 400 kPa, V_1, M_1) (T_2, p_2, 0.5, V_2)
   (*c*) (20°C, 400 kPa, V_1, M_1) (T_2, 125 kPa, M_2, V_2)

9.38   A large explosion occurs on the earth's surface producing a shock wave that travels radially outward. At a particular location, the Mach number of the wave is 2.0. Determine the induced velocity behind the shock wave. Assume standard conditions.

9.39 A pitot probe is used to measure the pressure in a supersonic pipe flow (review Solved Problem 9.6). If the pressure in the pipe is 120 kPa absolute, the temperature is 30°C, and the Mach number is 2.0, what pressure is measured by the pitot probe?

9.40 Air flows from a reservoir maintained at 400 kPa absolute and 20°C out a nozzle with a 10-cm-diameter throat and a 20-cm-diameter exit into a receiver. Estimate the receiver pressure needed to locate a shock wave at a diameter of 16 cm. Also, find the mass flux and the velocity just before the shock.

9.41 Air flows from a reservoir through a nozzle into a receiver. The reservoir is maintained at 400 kPa absolute and 20°C. The nozzle has a 10-cm-diameter throat and a 20-cm-diameter exit. Determine the receiver pressure needed to locate a shock wave at the exit. For that pressure, calculate the mass flux and the velocity just before the shock.

Oblique Shock Wave

9.42 A supersonic airflow changes direction 20° due to a sudden corner [Fig. 9.9(b)]. If $T_1 = 40°C$, $p_1 = 60$ kPa absolute, and $V_1 = 900$ m/s, calculate M_2, p_2, and V_2 assuming (a) a weak shock and (b) a strong shock.

9.43 An airflow at 25°C and 50 kPa absolute with a velocity of 900 m/s is turned by an abrupt 25° corner with a weak oblique shock. Estimate the pressure, velocity, and Mach number after the corner.

9.44 A weak oblique shock reflects from a plane wall (Fig. 9.15). If $M_1 = 3$ and $\beta_1 = 35°$, find the angle of the reflected oblique shock and M_3.

9.45 If $T_1 = 10°C$, find V_3 for the reflected oblique shock of Prob. 9.44.

9.46 A strong oblique wave is reflected from a corner. If the upstream Mach number is 2.5 and the flow turns through an angle of 25°, find the obtuse angle the wave makes with the wall and the downstream Mach number.

9.47 If $T_1 = 10°C$ in Prob. 9.46, calculate the downstream velocity.

Expansion Waves

9.48 An airflow with a Mach number of 2.4 turns a convex corner of 40°. If the temperature and pressure are 5°C and 60 kPa absolute, respectively, determine the Mach number, pressure, and velocity after the corner.

9.49 An airflow with $M = 3.6$ is desired by turning a 20°C supersonic flow with a Mach number of 1.8 around a convex corner. If the upstream pressure is 40 kPa absolute, what angle should the corner possess? What is the velocity after the corner?

9.50 A flat plate, designed to fly at an angle of 6°, is used as an airfoil in a supersonic flow. Sketch the flow pattern to be expected on the airfoil.

9.51 The airfoil of Prob. 9.50 is to fly at $M = 2.4$ at 16 000 m elevation. Find (a) the pressures on the upper and lower surfaces of the plate, (b) the Mach numbers (on the upper and lower parts) after the plate assuming the flow to be parallel to the original direction, and (c) the lift coefficient, defined by

$$C_L = \text{lift} \Big/ \left(\frac{1}{2} \rho_1 V_1^2 A \right)$$

Answers to Supplementary Problems

9.8   See problem.

9.9   See problem.

9.10   See problem.

9.11   See problem.

9.12   See problem.

9.13   See problem.

9.14   See problem.

9.15   1087 m

9.16   333 m

9.17   (*a*) 0.297   (*b*) 0.668   (*c*) 0.996   (*d*) 0.835

9.18   510 m

9.19   (*a*) 336 m, 0.463 s   (*b*) 404 m, 0.501 s   (*c*) 672 m, 0.552 s

9.20   0.00406°C, 1.19×10^{-5} m/s

9.21   See problem.

9.22   (*a*) 75.6 m/s   (*b*) 76.2 m/s

9.23   (*a*) 0.890 kg/s   (*b*) 0.890 kg/s

9.24   (*a*) 0.584 kg/s   (*b*) 0.590 kg/s   (*c*) 0.590 kg/s

9.25   0.590 kg/s

9.26   (*a*) 0.826 kg/s   (*b*) 0.826 kg/s

9.27   (*a*) 0.226 kg/s   (*b*) 0.226 kg/s

9.28   0.452 kg/s

9.29   44.5 m^3

9.30   0.00187 kg/s

9.31   0.792 kg/s

9.32   (*a*) 388 and 6.69 kPa   (*b*) 397 and 6.79 kPa

9.33   15.59 cm

9.34　34.8 m/s and 7.41 kg/s, 632 m/s and 7.42 kg/s

9.35　9.23 cm, 731 kPa absolute

9.36　(*a*)　145°C, 340 kPa, 264 m/s, 0.629

9.37　(*a*)　1.4, 95°C, 848 kPa, 0.628, 264 m/s　　　(*b*)　906 m/s, 2.64, 395°C, 319 kPa, 259 m/s
　　　　(*c*)　583 m/s, 1.7, 154°C, 0.640, 265 m/s

9.38　425 m/s

9.39　677 kPa

9.40　192 kPa, 7.41 kg/s, 569 m/s

9.41　118.2 kPa, 7.41 kg/s, 611 m/s

9.42　(*a*)　1.71, 192 kPa, 733 m/s　　　(*b*)　0.585, 431 kPa, 304 m/s

9.43　124 kPa absolute, 602 m/s, 1.51

9.44　47°, 1.39

9.45　667 m/s

9.46　104°, 0.67

9.47　324 m/s

9.48　4.98, 1.697 kPa, 998 m/s

9.49　39.4°, 835 m/s

9.51　(*a*)　6.915 kPa, 14.52 kPa　　　(*b*)　2.33, 2.46　　　(*c*)　0.182

Chapter 10

Flows in Pipes and Pumps

10.1 INTRODUCTION

Flows in pipes and ducts occur throughout the world. They are used to convey potable water, wastewater, crude oil, gasoline, chemicals, and many other liquids. They vary in size from large piping—e.g., the Alaska pipeline—to medium-sized ducts found in heating and air-conditioning systems to very small tubing found in cardiovascular and pulmonary systems. In this chapter we begin with analysis of the hydraulics associated with a single pipe, followed with a brief introduction to pumps, since they often are an integral part of pipelines. Then we focus on the analysis of steady flows in more complex systems that are best solved with an iterative technique called the Hardy Cross method. We conclude with a brief introduction to unsteady flows in pipelines.

10.2 SIMPLE PIPE SYSTEMS

10.2.1 Losses

In Secs. 7.6.3 to 7.6.5 we represent piping losses with the Darcy–Weisbach relation, Eq. (7.78) to account for friction and Eq. (7.86) to handle minor losses. They are repeated here for convenience:

$$h_L = f\frac{L}{D}\frac{V^2}{2g} \tag{7.78}$$

$$h_L = K\frac{V^2}{2g} \tag{7.86}$$

Since use will be made of those concepts, the reader should review those sections in their entirety before proceeding. Figure 7.10 may be utilized to determine the friction factor. In addition to the Darcy–Weisbach formulation, the Hazen–Williams formula has found wide use among practitioners. It is

$$h_L = \frac{K_1 L}{C^{1.85} D^{4.87}} Q^{1.85} \tag{10.1}$$

where Q = discharge
L = length of the pipe element
C = coefficient dependent on the pipe roughness
K_1 = 10.59 (for SI units) and 4.72 (for English units); note that K_1 depends on the system of units

Values of C are provided in Table 10.1. The Hazen–Williams loss formula is empirically based, and is less accurate than the Darcy–Weisbach relation; hence Eq. (7.78) is preferred.

Table 10.1   Values of the Hazen–Williams Coefficient

Type of pipe	C
Extremely smooth; fibrous cement	140
New or smooth cast iron; concrete	130
Newly welded steel	120
Average cast iron; newly riveted steel; vitrified clay	110
Cast iron or riveted steel after some years of use	95–100
Deteriorated old pipes	60–80

EXAMPLE 10.1   A cast iron pipe ($L = 400$ m, $D = 150$ mm) is carrying 0.05 m^3/s of water at 15°C. Compare the loss due to friction using the Darcy–Weisbach and the Hazen–Williams formulas.

Solution: First determine the friction factor and find the Hazen–Williams coefficient:

$$V = \frac{Q}{A} = \frac{0.05}{\pi \times 0.15^2/4} = 2.83 \text{m/s} \qquad v = 1.141 \times 10^{-6} \text{ m}^2/\text{s} \qquad \text{Re} = \frac{VD}{v} = 3.72 \times 10^5$$

$\dfrac{e}{d} = \dfrac{0.26}{150} = 0.00173$, and from Fig. 7.10: $f = 0.024$, and from Table 10.1: $C = 100$.

Using the Darcy–Weisbach relation, Eq. (7.78):

$$h_L = 0.024 \times \frac{400}{0.15} \times \frac{2.83^2}{2 \times 9.81} = 26.2 \text{ m}$$

With the Hazen–Williams formula, Eq. (10.1):

$$h_L = \frac{10.59 \times 400}{100^{1.85} \times 0.15^{4.87}} \times 0.05^{1.85} = 34.1 \text{ m}$$

The Darcy–Weisbach relation provides the more accurate result.

10.2.2 Hydraulics of Simple Pipe Systems

Flows in single-pipe reaches are examined in Secs. 7.6.3 to 7.6.5. The reader should pay particular attention to the use of the Moody diagram (Fig. 7.10) and to the three categories of pipe problems given by Eqs. (7.81) to (7.83). Examples 7.6 and 7.7 illustrate how the analysis proceeds. In this section, we study flows in three relatively simple pipe systems: series, parallel, and branching. Solution techniques are simplified due to the exclusion of pumps and lack of complexity of the piping; they are suitable for use with calculators, spreadsheet algorithms, and computational software. The fundamental principle in this so-called ad hoc approach is to identify the unknowns and write an equivalent number of independent equations to be solved. Subsequent simplification by eliminating as many variables as possible results in a series of single-pipe problems to be solved simultaneously; these may be solved by trial and error or by use of an equation system solver.

The energy and continuity equations are employed to analyze pipe systems. Normally, the predicted parameters are discharge Q and *piezometric head* $H = p/\gamma + z$. Throughout this chapter we assume that the kinetic energy term is negligible compared to the magnitude of the hydraulic grade line, that is, $V^2/2g \ll p/\gamma + z$. Referring to Fig. 10.1(a), the energy equation for a single reach of pipe is

$$H_A - H_B = \sum h_f = \left(f \frac{L}{D} \frac{1}{2gA^2} + \frac{\Sigma K}{2gA^2} \right) Q^2 \qquad (10.2)$$

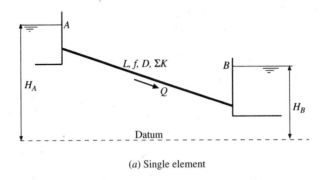

(a) Single element

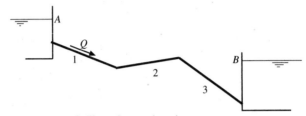

(b) Three elements in series

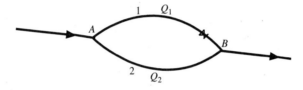

(c) Two parallel elements

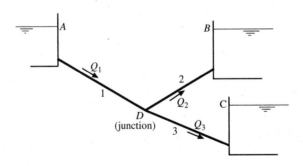

(d) Three branch elements

Figure 10.1 Simple pipe systems.

Here the Darcy–Weisbach equation is used to represent friction losses and ΣK is the sum of the minor loss coefficients in the reach. More simply, if we let the friction and minor loss terms be represented by a resistant, or loss, coefficient R, defined as

$$R = \frac{1}{2gA^2}\left(f\frac{L}{D} + \sum K\right) \qquad (10.3)$$

Then Eq. (10.2) becomes

$$H_A - H_B = RQ^2 \qquad (10.4)$$

This simplified relation contains all of the information necessary to solve a simple pipe problem that employs the Darcy–Weisbach relation for pipe friction. We make use of that relationship exclusively throughout the developments in this chapter.

EXAMPLE 10.2   Compute the discharge using the pipe data of Example 10.1 if the difference in piezometric head is 20 m. Assume the minor loss coefficients amount to $\Sigma K = 2.5$ and $f = 0.025$.

Solution: First compute the resistance coefficient with Eq. (10.3):

$$R = \frac{1}{2gA^2}\left(f\frac{L}{D} + \Sigma K\right) = \frac{1}{2 \times 9.81(\pi \times 0.15^2/4)^2}\left(0.025 \times \frac{400}{0.15} + 2.5\right) = 1.13 \times 10^4\,\text{s}^2/\text{m}^5$$

Then find the discharge using Eq. (10.4):

$$Q = \sqrt{\frac{H_A - H_B}{R}} = \sqrt{\frac{20}{1.13 \times 10^4}} = 0.042\,\text{m}^3/\text{s}$$

Figure 10.1(b) shows a series pipe system consisting of three reaches, each with a specified loss coefficient. Since the same discharge Q exists in each reach, the energy equation from location A to location B is

$$H_A - H_B = R_1Q^2 + R_2Q^2 + R_3Q^2 = (R_1 + R_2 + R_3)Q^2 \qquad (10.5)$$

Obviously, this relation can be expanded to any number of piping elements. To evaluate the loss coefficients, one can substitute Eq. (10.3), and employ Fig. 7.10 for the friction factor f along with Table 7.2 for the minor loss coefficient K. Note that a trial-and-error solution results, since f is dependent on Q. In many situations, the friction factor can be assumed constant; hence R can be evaluated with Eq. (10.3) prior to determining either the discharge or the change in piezometric head. We proceed with that assumption throughout this chapter.

EXAMPLE 10.3   For the three pipes in series shown in Fig. 10.1(b), determine the discharge if the difference in piezometric head is $H_A - H_B = 10$ m. Use $L_1 = 2000$ m, $D_1 = 450$ mm, $L_2 = 650$ m, $D_2 = 150$ mm, $K_2 = 2.0$, $L_3 = 1650$ m, $D_3 = 300$ mm, and $f_1 = f_2 = f_3 = 0.03$.

Solution: First compute the resistance coefficients using Eq. (10.3):

$$R_1 = \frac{1}{2 \times 9.81(\pi \times 0.45^2/4)^2}\left(0.03 \times \frac{2000}{0.45}\right) = 269\,\text{s}^2/\text{m}^5$$

$$R_2 = \frac{1}{2 \times 9.81(\pi \times 0.15^2/4)^2}\left(0.03 \times \frac{650}{0.15} + 2.0\right) = 21\,550\,\text{s}^2/\text{m}^5$$

$$R_3 = \frac{1}{2 \times 9.81(\pi \times 0.30^2/4)^2}\left(0.03 \times \frac{1650}{0.30}\right) = 1684\,\text{s}^2/\text{m}^5$$

Calculate the discharge with Eq. (10.4):

$$Q = \sqrt{\frac{H_A - H_B}{R_1 + R_2 + R_3}} = \sqrt{\frac{10}{269 + 21\,550 + 1684}} = 0.0206\,\text{m}^3/\text{s}$$

Parallel piping is illustrated in Fig. 10.1(c); even though only two pipes are shown, any number of pipes can be placed in parallel. The pipes are joined at locations A and B, and each pipe has its own unique geometry and minor loss term. The continuity balance at either junction A or B requires that

$$Q = Q_1 + Q_2 \qquad (10.6)$$

For the two pipe elements, the required energy equations from location A to B are

$$H_A - H_B = R_1Q_1^2$$
$$H_A - H_B = R_2Q_2^2 \qquad (10.7)$$

Assuming that Q is known, the unknowns in the above equations are Q_1, Q_2, and $\Delta H = H_A - H_B$. They are solved simultaneously in the manner shown in the following example.

EXAMPLE 10.4   Find the flow distribution and change in hydraulic grade line for the parallel piping shown in Fig. 10.1(c) using the following data: $L_1 = 50$ m, $D_1 = 100$ mm, $K_1 = 2$, $L_2 = 75$ m, $D_2 = 150$ mm, $K_2 = 10$. The total discharge in the two pipes is $Q = 0.04$ m³/s.

 Solution: Combine Eqs. (10.6) and (10.7) in the manner

$$Q = Q_1 + Q_2 = \sqrt{\frac{H_A - H_B}{R_1}} + \sqrt{\frac{H_A - H_B}{R_2}} = \sqrt{H_A - H_B}\left(\frac{1}{\sqrt{R_1}} + \frac{1}{\sqrt{R_2}}\right)$$

Note that $H_A - H_B$ is also an unknown, and we first solve for that value:

$$R_1 = \frac{1}{2 \times 9.81(\pi \times 0.10^2/4)^2}\left(0.025 \times \frac{50}{0.10} + 2\right) = 11\,980 \text{ s}^2/\text{m}^5$$

$$R_2 = \frac{1}{2 \times 9.81(\pi \times 0.15^2/4)^2}\left(0.030 \times \frac{75}{0.15} + 10\right) = 4082 \text{ s}^2/\text{m}^5$$

$$H_A - H_B = \frac{Q^2}{\left(\frac{1}{\sqrt{R_1}} + \frac{1}{\sqrt{R_2}}\right)^2} = \frac{0.04^2}{\left(\frac{1}{\sqrt{11\,980}} + \frac{1}{\sqrt{4082}}\right)^2} = 2.604 \text{ m}$$

Lastly, the flows in the two parallel pipes are computed using Eq. (10.7):

$$Q_1 = \sqrt{\frac{H_A - H_B}{R_1}} = \sqrt{\frac{2.604}{11\,980}} = 0.0147 \text{ m}^3/\text{s}$$

$$Q_2 = \sqrt{\frac{H_A - H_B}{R_2}} = \sqrt{\frac{2.604}{4082}} = 0.0253 \text{ m}^3/\text{s}$$

An example of branch piping is illustrated in Fig. 10.1(d); it is made up of three elements connected to a single junction. Typically, the piezometric heads at locations A to C are considered to be known, and the unknowns are the discharges Q_1, Q_2, and Q_3 in each line and the piezometric head at location D. Analysis proceeds by assuming the direction of flow and writing the energy balance in each element:

$$H_A - H_D = R_1 Q_1^2 \qquad\qquad (10.8)$$

$$H_D - H_B = R_2 Q_2^2 \qquad\qquad (10.9)$$

$$H_D - H_C = R_3 Q_3^2 \qquad\qquad (10.10)$$

The continuity balance at location D is

$$Q_1 - Q_2 - Q_3 = 0 \qquad\qquad (10.11)$$

Note that a flow direction was assumed in each pipe. One method of solution is as follows:

1. Assume H_D at the junction.
2. Compute Q_1, Q_2, and Q_3 in the three branches using Eqs. (10.8) to (10.10).
3. Substitute Q_1, Q_2, and Q_3 into Eq. (10.11) to check for continuity balance. Generally, the flow imbalance $\Delta Q = Q_1 - Q_2 - Q_3$ will be nonzero at the junction.
4. Adjust the head H_D and repeat steps 2 and 3 until ΔQ is within desired limits. It may be necessary to correct the sign in one or more of the equations if during the iterations H_D moves from above or below one of the reservoirs or vice versa.

An alternative method of solution is to combine the equations and eliminate all of the variables except one (commonly H_D) and employ a numerical solving technique. Example 10.5 represents a level of complexity that represents a limit using a calculator-based solution. For more complex systems that might include pumps, additional reservoirs, or pipe elements, the network analysis described in Sec. 10.4 is recommended.

EXAMPLE 10.5   Determine the flow rates and the piezometric head at the junction for the branch system of Fig. 10.1(d). Assume constant friction factors. $H_A = 12$ m, $H_B = 15$ m, $H_C = 5$ m, $f_1 = f_2 = f_3 = 0.02$, $L_1 = 200$ m, $D_1 = 100$ mm, $L_2 = 150$ m, $D_2 = 100$ mm, $L_3 = 750$ m, $D_3 = 150$ mm.

   Solution: Use the four-step procedure outlined above. First compute the resistance coefficients using Eq. (10.3); the result is $R_1 = 33\,890$ s^2/m^5, $R_2 = 27\,280$ s^2/m^5, and $R_3 = 16\,570$ s^2/m^5. We assume that H_D is lower than H_A and H_B, but higher than H_C. Consequently, the resulting flow directions are Q_1 from A to D, Q_2 from B to D, and Q_3 from D to C. An iterative solution is shown in the accompanying table. Iteration ceases when $|\Delta Q| < 0.001$ units.

Iteration	H_D (assumed)	Q_1 (Eq. (10.8))	Q_2 (Eq. (10.10))	Q_3 (Eq. (10.9))	$Q_1 + Q_2 - Q_3$ (Eq. (10.11))
1	12	0	0.01049	0.02055	-0.01006
2	11	0.00543	0.01211	0.01903	-0.00149
3	10	0.00768	0.01354	0.01737	$+0.00385$
4	10.74	0.00610	0.01250	0.01861	-0.00002

Hence the resulting solution is $H_D = 10.7$ m, $Q_1 = 0.0061$ m^3/s, $Q_2 = 0.0125$ m^3/s, and $Q_3 = 0.0186$ m^3/s.

10.3   PUMPS IN PIPE SYSTEMS

Until now, we have considered systems that have not involved a pump. If a pump is included in the pipe system and the flow rate is specified, the solution is straightforward using the methods we have already developed. On the other hand, if the discharge is not known, which is commonly the case, a trial-and-error solution is required. The reason for this is that the pump head H_P depends upon the discharge, as shown by the *pump characteristic curve*, the solid curve in Fig. 10.2. Pump manufacturers provide the characteristic curves. Figure 10.3 shows a complete set of curves for a manufactured centrifugal pump; included are sets of head versus discharge curves for various impeller sizes, as well as efficiency and power curves. The power requirement for a pump is given by the expression (see Eq. (4.25))

$$\dot{W}_P = \frac{\gamma Q H_P}{\eta} \qquad (10.12)$$

Determining the discharge in a pumped line requires an additional relation, namely the *demand curve*, which is generated by writing the energy balance across the system for varying discharges. Referring to the pump–pipe system in Fig. 10.2, the energy equation (see Eq. (10.4)) for the pipe including the pump is a quadratic in Q:

$$H_P = (H_B - H_A) + RQ^2 \qquad (10.13)$$

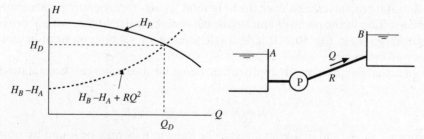

Figure 10.2   Pump and system demand curves.

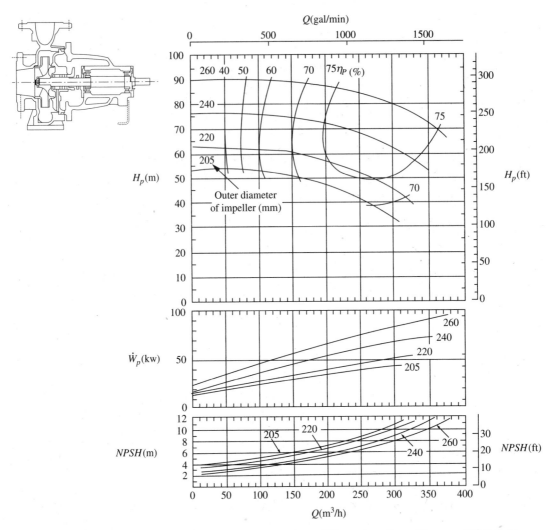

Figure 10.3   Centrifugal pump and performance curves for four different impellers. Water at 20°C is the pumped liquid. (Courtesy Sulzer Pumps Ltd).

The demand curve is illustrated in Fig. 10.2 by the dashed line (see Eq. (10.13)). The first term on the right-hand side of Eq. (10.13) is the static head and the second term accounts for the system losses. The steepness of the demand curve depends on the losses in the piping; as the losses increase, the required pumping head increases and vice versa. Piping may experience short-term alterations in the demand curve such as throttling of valves, and over the long term, aging of pipes may permanently increase the demand. The intersection of the pump characteristic curve and the demand curve will provide the design head H_D and discharge Q_D in Fig. 10.2. It is desirable to have the solution occur at or close to the point of best efficiency of the pump.

Instead of an actual pump curve, an approximate pump head–discharge representation is sometimes used:

$$H_P(Q) = a_0 + a_1 Q + a_2 Q^2 \qquad (10.14)$$

where the coefficients a_0, a_1, and a_2 are assumed to be known; they may be found by substituting three known data points from a specified pump curve into Eq. (10.14) and solving the three resulting equations simultaneously.

EXAMPLE 10.6   Estimate the discharge in the pipe system shown in Fig. 10.2, and in addition find the required pump power. For the pipe, $L = 700$ m, $D = 300$ mm, $f = 0.02$, and $H_B - H_A = 30$ m. Use the 240-mm curve in Fig. 10.3 for the pump head–discharge relation.

   Solution: First determine R from Eq. (10.3):

$$R = \frac{1}{2 \times 9.81(\pi \times 0.30^2/4)^2}\left(0.02 \times \frac{700}{0.30}\right) = 476\,\text{s}^2/\text{m}^5$$

A trial solution is utilized to determine the pump head and discharge. The procedure is as follows: (1) guess a discharge; (2) compute H_P with Eq. (10.13); and (3) compare that value with the one from the 240-mm pump curve of Fig. 10.3. Continue estimating values of Q until the two pump heads agree. The solution is shown in the table.

Q, m³/h	Q, m³/s	H_P, m (Eq. (10.13))	H_P, m (Fig. 10.3)
150	0.042	70.8	74
250	0.069	72.3	67
200	0.056	71.5	72

Hence, the approximate solution is $Q = 200$ m³/h and $H_P = 72$ m. From Fig. 10.3, the efficiency is approximately 75%, so that the required power is

$$\dot{W}_P = \frac{\gamma Q H_P}{\eta} = \frac{9800 \times 0.056 \times 72}{0.75} = 52\,700\,\text{W or 706 hp}$$

In some instances, pumping installations may require a wide range of heads or discharges, so that one pump cannot meet the required demand range. In such situations, the pumps may be staged in series or in parallel to provide operation at greater efficiency. When a large variation in flow demand occurs, two or more pumps may be placed in parallel, as in Fig. 10.4(a). The combined characteristic curve is determined by recognizing that the head H_P across each pump is identical, and the total discharge through the system ΣQ is the sum of the discharges through each pump for a given head. For demands that require high heads, placing the pumps in series will provide a head greater than the pumps individually (Fig. 10.4(b)). Since the discharge through each pump in series is the same, the combined characteristic curve is found by summing the heads ΣH_P across the pumps for a given discharge.

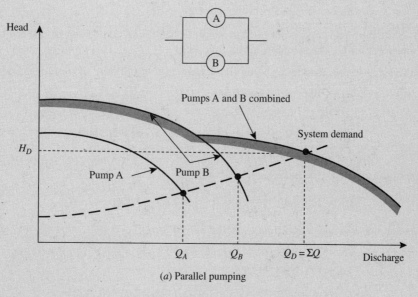

(a) Parallel pumping

Figure 10.4   Continued

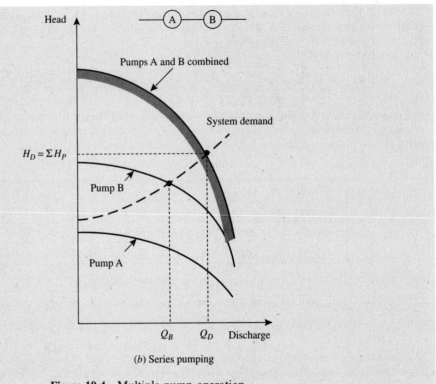

(b) Series pumping

Figure 10.4   Multiple pump operation.

EXAMPLE 10.7   Water is pumped between two reservoirs in a single pipe with the value of $R = 85$ s^2/m^5. For the pump characteristic curve, use $H_P = 22.9 + 10.7Q - 111Q^2$. Compute the discharge Q and pump head H_P for:

(a) $H_B - H_A = 15$ m with one pump placed in operation
(b) $H_B - H_A = 15$ m with two identical pumps operating in parallel
(c) $H_B - H_A = 25$ m with two pumps operating in series

Solution: Since the pump curve is provided in quadratic form, Eqs. (10.13) and (10.14) can be combined to eliminate H_P and solve for Q. The solutions are as follows:

(a) Equate the system demand curve to the pump characteristic curve and solve the resulting quadratic equation:

$$15 + 85Q^2 = 22.9 + 10.7Q - 111Q^2$$
$$195Q^2 - 10.7Q - 7.9 = 0$$
$$Q = \frac{1}{2 \times 195}\left(10.7 + \sqrt{10.7^2 + 4 \times 195 \times 7.9}\right) = 0.23 \text{ m}^3/\text{s}$$
$$H_P = 15 + 85 \times 0.23^2 = 19.5 \text{ m}$$

(b) For two pumps in parallel, the characteristic curve is

$$H_P = 22.9 + 10.7\left(\frac{Q}{2}\right) - 111\left(\frac{Q}{2}\right)^2 = 22.9 + 5.35Q - 27.75Q^2$$

The system demand curve is equated to this result and solved for Q:

$$15 + 85Q^2 = 22.9 + 5.35Q - 27.75Q^2$$
$$112.8Q^2 - 5.35Q - 7.9 = 0$$
$$Q = \frac{1}{2 \times 112.8}\left(5.35 + \sqrt{5.35^2 + 4 \times 112.8 \times 7.9}\right) = 0.29 \text{ m}^3/\text{s}$$
$$H_P = 15 + 85 \times 0.29^2 = 22.2 \text{ m}$$

(c) With two pumps in series, the characteristic curve becomes

$$H_P = 2(22.9 + 5.35Q - 111Q^2) = 45.8 + 21.4Q - 222Q^2$$

Equate this to the system demand curve and solve for Q:

$$25 + 85Q^2 = 45.8 + 21.4Q - 222Q^2$$
$$307Q^2 - 21.4Q - 20.8 = 0$$
$$Q = \frac{1}{2 \times 307}\left(21.4 + \sqrt{21.4^2 + 4 \times 307 \times 20.8}\right) = 0.30\,\text{m}^3/\text{s}$$
$$H_P = 25 + 85 \times 0.30^2 = 32.5\,\text{m}$$

10.4 PIPE NETWORKS

10.4.1 Network Equations

Simple pipe system solution techniques, as outlined above, are limited to the size and complexity of a piping system that can be analyzed. Indeed, it is advantageous to seek a more generalized method that can handle a system, or so-called network, consisting of a number of pipe elements, one or more pumps and perhaps several reservoirs. Quite often the objective of the analysis would be to predict the discharges in the network. There are a number of pipe network solutions available, and nearly all of them make use of a trial-and-error method. The one technique that we utilize herein is called the Hardy Cross method; it can easily be adapted to a computer-based algorithm; however, we make use of spreadsheet software as an alternate.

Consider piping as shown in Fig. 10.5(a); it is more complex than those analyzed in Secs. 10.2 and 10.3, hence it would be difficult to solve in an ad hoc method. Any of the piping systems we have previously studied in this chapter can be solved using the *Hardy Cross technique*, but first we need to state the problem in a consistent and systematic manner.

Piping networks such as those shown in Fig. 10.5(a) can be viewed to be made up of *interior nodes, interior loops*, and paths that connect two fixed-grade nodes (sometimes these paths are called *pseudoloops*). An interior node is a location where two or more pipes connect and the head is unknown, and fixed-grade nodes are reservoirs and locations of constant pressure. Figure 10.5(b) shows the nodes and loops for the piping system of Fig. 10.5(a). Nodes A and E are fixed-grade nodes, and nodes B, C, and D are interior nodes. Loop I is an interior loop and loop II is a pseudoloop. For this or any other pipe system, the generalized network equations are as follows.

- Energy balance in a clockwise positive manner around an interior loop or along a unique path or pseudoloop which connects fixed-grade nodes:

$$\sum (\pm)_i[R_iQ_i^2 - (H_P)_i] + \Delta H = 0 \qquad (10.15)$$

where i = pipe elements that make up the loop or path
$(H_P)_i$ = head across a pump that may exist in pipe i
ΔH = difference in magnitude of the two fixed-grade nodes in the path ordered in a clockwise fashion across an imaginary pipe (the dashed line in Fig. 10.5(b))

For an interior loop, $\Delta H = 0$, and if no pump is located in the loop or path, $(H_P)_i = 0$. The plus or minus sign pertains to the assumed flow direction in each pipe relative to clockwise positive.

- Continuity at an interior node:

$$\sum (\pm)_jQ_j - Q_e = 0 \qquad (10.16)$$

where the subscript j refers to all pipes connected to node j and Q_e is the external demand. The plus or minus sign pertains to the assumed flow direction (positive for flow into the node and negative for flow out).

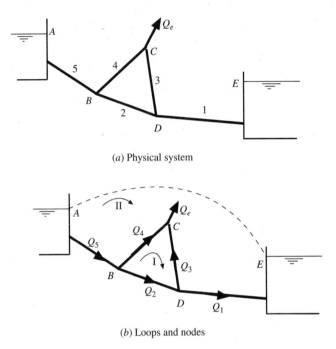

(a) Physical system

(b) Loops and nodes

Figure 10.5 Representative pipe network.

To determine if the network is properly defined, one can use the following rule. Let F be the number of fixed-grade nodes, P the number of pipe elements, J the number of interior nodes, and L the number of interior loops. Then, if the network is properly defined the following relation will hold:

$$P = J + L + F - 1 \qquad (10.17)$$

In Fig. 10.4, $J = 3$, $F = 2$, and $P = 5$, so that $L = 1$.

10.4.2 Hardy Cross Method

The Hardy Cross solution is a trial-and-error technique and it requires that the network equations be linear. Equation (10.15) is a general relation that can be applied to any path or closed loop in a network; as stated earlier, if there is no pump $(H_P)_i = 0$, and for an interior loop $\Delta H = 0$. Let the variable discharge $\tilde{Q}$ be determined from a previous estimate, and Q be the new estimate. Then the nonlinear terms in Eq. (10.15) are linearized in the following manner:

$$RQ^2 = R\tilde{Q}^2 + \frac{d(RQ^2)}{dQ}(Q - \tilde{Q}) + \cdots$$
$$\approx R\tilde{Q}^2 + 2R\tilde{Q}(Q - \tilde{Q}) \qquad (10.18)$$

$$H_P(Q) = H_P(\tilde{Q}) + \frac{dH_P(Q)}{dQ}(Q - \tilde{Q}) + \cdots$$
$$\approx a_0 + a_1\tilde{Q} + a_2\tilde{Q}^2 + (a_1 + 2a_2\tilde{Q})(Q - \tilde{Q}) \qquad (10.19)$$

In expanding Eq. (10.19), we have made use of Eq. (10.14). The loop or path energy relation (Eq. (10.15)) now becomes

$$\sum (\pm)_i [R_i \tilde{Q}_i^{\,2} - (a_0 + a_1 \tilde{Q}_i + a_2 \tilde{Q}_i^{\,2})]$$
$$+ \sum [2R_i \tilde{Q}_i - (a_1 + 2a_2 \tilde{Q}_i)](Q_i - \tilde{Q}_i) + \Delta H = 0 \qquad (10.20)$$

Note that the second term does not contain the plus or minus sign. Defining a flow adjustment for a given loop or path to be $\Delta Q = Q - \tilde{Q}$, substituting it into Eq. (10.20) and solving for ΔQ, we obtain

$$\Delta Q = \frac{-\sum (\pm)_i [R_i \tilde{Q}_i^{\,2} - (a_0 + a_1 \tilde{Q}_i + a_2 \tilde{Q}_i^{\,2})] - \Delta H}{\sum [2R_i \tilde{Q}_i - (a_1 + 2a_2 \tilde{Q}_i)]} \qquad (10.21)$$

In the Hardy Cross method, it is assumed that the flow adjustment ΔQ is applied independently to all pipes in a given loop. It is required that the algebraic sign of Q be positive in the direction of normal pump operation; otherwise, the pump curve will not be represented properly and Eq. (10.21) will not be valid. In addition, it is important that the discharge through the pump remains within the limits of the data employed to generate the pump curve. For a closed loop in which no pumps or fixed-grade nodes exist, Eq. (10.21) reduces to the simpler form

$$\Delta Q = \frac{-\sum (\pm)_i R_i \tilde{Q}_i^{\,2}}{\sum 2R_i \tilde{Q}_i} \qquad (10.22)$$

In the Hardy Cross iterative solution, continuity (Eq. (10.16)) is satisfied initially with assumed flows that are assigned and remains satisfied throughout the solution process. The method is summarized in the following steps:

1. Assume an initial flow distribution in the network that satisfies Eq. (10.16). The closer the initial estimates are to the true values, fewer iterations will be necessary for convergence. One rule to follow is to recognize that as R increases for a pipe element, Q will decrease.
2. Determine ΔQ in each path or loop using either Eq. (10.21) or (10.22) as appropriate. The numerators will approach zero as the paths or loops become balanced.
3. Adjust the flows in each pipe element in all loops and paths using the relation

$$Q_i = \tilde{Q}_i + \sum \Delta Q \qquad (10.23)$$

Here the term $\sum \Delta Q$ is employed as a correction because a given pipe may be part of more than one loop or path. As a result, the correction will be the sum of corrections from all loops for which the pipe element is common.
4. Repeat steps 2 and 3 until a desired accuracy is reached. One convergence criterion is

$$\frac{\sum |Q_i - \tilde{Q}_i|}{\sum |Q_i|} \le \varepsilon \qquad (10.24)$$

where ε is an arbitrarily small number, say $0.001 < \varepsilon < 0.005$. Another criterion is to continue iteration until each of the loop ΔQ's reach an arbitrarily small value.

EXAMPLE 10.8 Determine the flow distribution and piezometric heads at the junctions using the Hardy Cross method for the network shown in Fig. 10.5(a). $H_A = 45$ m, $H_E = 0$, $Q_e = 0.025$ m^3/s.

Pipe	L, m	D, mm	f	$\sum K$
1	100	100	0.02	1
2	75	100	0.02	0
3	120	150	0.02	0
4	80	150	0.02	0
5	20	300	0.02	1

Solution: There are three junctions ($J = 3$), five pipes ($P = 5$), and two fixed-grade nodes ($F = 2$), hence $L = 5 - 3 - 2 + 1 = 1$ interior loop. In addition, there is one pseudoloop. The two loops and assumed flow directions (clockwise positive) are shown in Fig. 10.5(b). Equation (10.22) is applied to loop I and Eq. (10.21) to loop II:

$$\Delta Q_I = \frac{-(R_2 \tilde{Q}_2 |\tilde{Q}_2| + R_3 \tilde{Q}_3 |\tilde{Q}_3| \pm R_4 \tilde{Q}_4 |\tilde{Q}_4|)}{2(R_1 |\tilde{Q}_1| + R_2 |\tilde{Q}_2| + R_3 |\tilde{Q}_3|)}$$

$$\Delta Q_{II} = \frac{-(R_1 \tilde{Q}_1 |\tilde{Q}_1| + R_3 \tilde{Q}_3 |\tilde{Q}_3| + R_4 \tilde{Q}_4 |\tilde{Q}_4| + R_5 \tilde{Q}_5 |\tilde{Q}_5|) - (H_A - H_E)}{2(R_1 |\tilde{Q}_1| + R_3 |\tilde{Q}_3| + R_4 |\tilde{Q}_4| + R_5 |\tilde{Q}_5|)}$$

Note that the terms $\pm R\tilde{Q}^2$ and $R\tilde{Q}$ have been replaced by $R\tilde{Q}|\tilde{Q}|$ and $R|\tilde{Q}|$ in the equations which allow for the correct sign to be automatically attributed. The values of Q take on a positive or negative sign depending on the assumed flow direction relative to the assigned positive clockwise direction for each loop. A spreadsheet solution is illustrated in the two tables, which show respectively the spreadsheet formulas and the numerical solution. Values of R are calculated with the given data using Eq. (10.3) and are entered in column B. Initial estimates of Q are provided in column C, and updated values are shown for four iterations in columns F, I, L, and O. The convergence criterion used here is to cease iterations when the absolute value of ΔQ falls below 0.001. Note that Q_3 changes direction by the final iteration. The discharges after the fourth iteration are shown in Fig. 10.6. The piezometric heads at junctions B, C, and D are

$$H_B = H_A - R_5 Q_5^2 = 45 - 24 \times 0.0725^2 = 44.9 \text{ m}$$

$$H_C = H_B - R_4 Q_4^2 = 44.9 - 1741 \times 0.050^2 = 40.5 \text{ m}$$

$$H_D = H_E + R_1 Q_1^2 = 0 + 17\,350 \times 0.0475^2 = 39.2 \text{ m}$$

Spreadsheet Formulas

	A	B	C	D	E	F
1		R	Q	$RQ|Q|$	$2R|Q|$	Q
2						
3					*Iteration 1*	
4						
5	*Loop 1*					
6	*Pipe 2*	12390	−0.035	=B6*C6*ABS(C6)	=2*B6*ABS(C6)	=C6+E12
7	*Pipe 3*	2611	−0.015	=B7*C7*ABS(C7)	=2*B7*ABS(C7)	=C7+E12−E22
8	*Pipe 4*	1741	0.01	=B8*C8*ABS(C8)	=2*B8*ABS(C8)	=C8+E12−E22
9						
10				=SUM(D6:D8)	=SUM(E6:E8)	
11						
12					$\Delta Q_I=$ =−D10/E10	
13	*Loop 2*					
14	$H_A - H_E$			45		
15	*Pipe 1*	17350	−0.02	=B15*C15*ABS(C15)	=2*B15*ABS(C15)	=C15+E22
16	*Pipe 3*	2611	0.015	=B15*C15*ABS(C16)	=2*B16*ABS(C16)	=C16+E22−E12
17	*Pipe 4*	1741	−0.01	=B17*C17*ABS(C17)	=2*B17 ABS(C17)	=C17+E22−E12
18	*Pipe 5*	24	−0.045	=B18*C18*ABS(C18)	=2*B18 ABS(C18)	=C18+E22
19						
20				=SUM(D14:D18)	=SUM(E15:E18)	
21				=B16*C15*ABS(C16)		
22				$\Delta Q_{II}=$ =−D20/E20		

Spreadsheet Solution

	A	B	C	D	E	F	G	H	I	J	K	L	M	N	O
1		R	Q	RQ|Q|	2R|Q|	Q	2R|Q|	2R|Q|	Q	RQ|Q|	2R|Q|	Q	RQ|Q|	2R|Q|	Q
2															
3					Iteration 1			Iteration 2			Iteration 3			Iteration 4	
4															
5	Loop 1														
6	Pipe 2	12390	-0.035	-15.178	867.300	-0.0191	-4.519	473.248	-0.0303	-11.355	750.165	-0.0235		582.306	-0.0225
7	Pipe 3	2611	-0.015	-0.587	78.330	0.0484	6.111	252.643	0.0198	1.023	103.368	0.0250	1.631	130.533	-0.0250
8	Pipe 4	1741	0.01	-0.174	34.820	0.0734	9.375	255.511	0.0448	3.493	155.975	0.0500	4.352	174.089	-0.0500
9															
10				-15.591	980.450		10.967	981.402		-6.838	1009.508		-0.858	886.928	
11										-6.838					
12				$\Delta Q_I=$	1.59E-02		$\Delta Q_I=$	1.12E-02		$\Delta Q_I=$	6.77E-03		$\Delta Q_I=$	9.68E-04	
13	Loop 2														
14	H_A-H_E			45.000			45.000			45.000			45.000		
15	Pipe 1	17350	-0.02	-6.940	694.000	-0.0675	-79.000	2341.502	-0.0501	-43.493	1737.349	-0.0485	-40.804	1682.804	-0.0475
16	Pipe 3	2611	-0.015	0.587	78.330	-0.0484	-6.111	252.643	-0.0198	-1.023	103.368	-0.0250	-1.631	130.533	-0.0250
17	Pipe 4	1741	-0.01	-0.174	34.820	-0.0734	-9.375	255.511	-0.0448	-3.493	155.975	-0.0500	-4.352	174.089	-0.0500
18	Pipe 5	24	-0.045	-0.049	2.160	-0.0925	-0.205	4.439	-0.0751	-0.135	3.603	-0.0735	-0.130	3.528	-0.0725
19															
20				38.425	809.310		-49.692	2854.094		-3.144	2000.295		-1.917	1990.954	
21															
22				$\Delta Q_{II}=$	-4.75E-02		$\Delta Q_{II}=$	1.74E-02		$\Delta Q_{II}=$	1.57E-03		$\Delta Q_{II}=$	9.63E-04	
23															

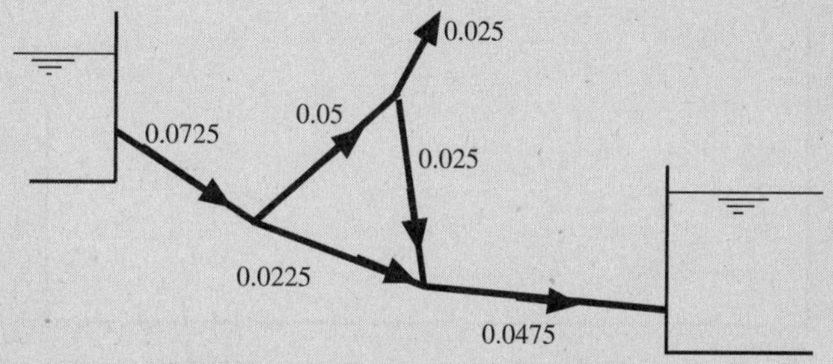

Figure 10.6   Flow after four iterations.

10.4.3   Computer Analysis of Network Systems

The Hardy Cross analysis is a modified version of the method of successive approximations used to solve a set of linear equations. Since it does not require the inversion of a matrix, the Hardy Cross method can be used to solve relatively small networks using a calculator or spreadsheet software. However, for larger networks that contain multiple loops and branches, spreadsheet programming becomes time-consuming as well as cumbersome. Generalized pipe network solution software is now available that has robust solutions and provides convenient input and output schemes. For example, the source code and users manual for the cost-free network analysis program EPANET can be obtained from the website of the United States Environmental Protection Agency (www.epa.gov). EPANET is a comprehensive program that simulates both flow and water quality in pressurized pipe networks. For the hydraulic analysis, it incorporates a hybrid node–loop algorithm termed the gradient method. In addition to piping, systems can include pumps, valves, reservoirs, and storage water tanks. Pipe friction losses are represented with the Darcy–Weisbach, Hazen–Williams, or Chezy–Manning formulations.

10.5   UNSTEADY FLOW

While many transient, or unsteady, flows have in the past focused on problems dealing with hydropower systems, and water and oil pipelines, more recently applications have expanded to include control system operation, and nuclear and thermal power plant piping. The excitation that generates the transient can be one of a number of causes, but typically it is rapid valve closing or opening, pipe rupture or break, pump or turbine operation, or cavitation phenomenon. In this section we focus on a single horizontal pipe and examine two fundamental types of flow: (1) incompressible and inelastic unsteady flow, called *surging*, and (2) slightly compressible and elastic unsteady flow, termed *water hammer*.

10.5.1 Incompressible Flow

We consider a horizontal length L of pipe with constant diameter D, shown in Fig. 10.6. The upstream end of the pipe is connected to a reservoir with head H_1, and at the downstream end there is a valve exiting to a reservoir with head H_3. Both H_1 and H_3 are time invariant. Initially, in the pipe, there is a steady velocity V_0 and the valve is partially open. Then the valve is opened to a new position, resulting ultimately in a new and increased steady-state velocity. For situations where the valve is closed, either partially or completely, one must consider the possible occurrence of water hammer (see Sec. 10.5.2).

In Fig. 10.7, location 2 is upstream of the valve and location 1 is just inside the pipe. Applying the linear momentum equation to the water between the two locations, we have

$$A(p_1 - p_2) - \tau_0 \pi DL = \rho AL \frac{dV}{dt} \qquad (10.25)$$

where A = pipe cross section
$\quad\quad V$ = time-variable velocity
$\quad\quad \tau_0$ = wall shear stress

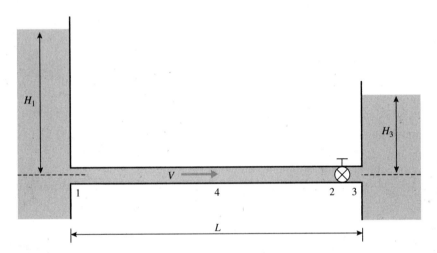

Figure 10.7   Unsteady flow in a horizontal pipe.

It is reasonable to assume quasi-steady flow conditions across the valve so that

$$p_2 = p_3 + K \frac{V^2}{2g} \qquad (10.26)$$

where K is the valve loss coefficient. We additionally assume that the Darcy–Weisbach friction factor f based on steady-state flow can be employed without serious error, and that it is furthermore constant. The shear stress is (see Eq. (7.74)):

$$\tau_0 = \frac{\rho f V^2}{8} \qquad (10.27)$$

Substituting Eqs. (10.26) and (10.27) into Eq. (10.25), dividing by the liquid column mass ρAL, and recognizing that $p_1 - p_3 = \rho g(H_1 - H_3)$, we have, after rearranging,

$$\frac{dV}{dt} + \left(\frac{f}{D} + \frac{K}{L}\right)\frac{V^2}{2} - g\frac{H_1 - H_3}{L} = 0 \qquad (10.28)$$

This relation for unsteady incompressible flow has the initial condition $V = V_0$ at time $t = 0$. After the valve is opened further by altering the coefficient K, the velocity accelerates to a final steady-state velocity V_{ss}. Since at steady-state conditions, $dV/dt = 0$, we can determine V_{ss} by setting the derivative in Eq. (10.28) to zero and solving for $V = V_{ss}$:

$$V_{ss} = \sqrt{\frac{2g(H_1 - H_3)}{fL/D + K}} \qquad (10.29)$$

Substituting Eq. (10.29) into Eq. (10.28), we can separate variables and express the result in integral form:

$$\int_0^t dt = \frac{V_{ss}^2 L}{g(H_1 - H_3)} \int_{V_0}^V \frac{dV}{V_{ss}^2 - V^2} \qquad (10.30)$$

After integrating, the result provides a relation between the velocity and the time subsequent to the valve excitation:

$$t = \frac{V_{ss}L}{2g(H_1 - H_3)} \ln \frac{(V_{ss} + V)(V_{ss} - V_0)}{(V_{ss} - V)(V_{ss} + V_0)} \qquad (10.31)$$

According to the result, infinite time is required for steady-state velocity V_{ss} to be reached. In reality, it will be attained somewhat sooner since losses have not been completely accounted for. However, we can state for engineering purposes that the time when a percentage of V_{ss} has been reached is adequate using the equation. Note that V_0 may be equal to zero, i.e., initially the fluid is at rest. Remember that Eq. (10.31) is based on assuming that the liquid is incompressible and the pipe is inelastic; Sec. 10.5.2 addresses the situation in which those assumptions are not valid.

EXAMPLE 10.9   A horizontal pipe ($L = 500$ m, $D = 250$ mm, $V_0 = 0.35$ m/s) is suddenly subjected to a new head differential $H_1 - H_3 = 15$ m when a valve at the downstream end is suddenly opened wider and its coefficient changes to $K = 0.2$. If the friction factor is $f = 0.02$, determine the final steady-state velocity and the time when the actual velocity is 99% of that final value.

  Solution: The final steady-state velocity is determined by substituting the given data into Eq. (10.29):

$$V_{ss} = \sqrt{\frac{2 \times 9.81 \times 15}{0.02 \times 500/0.250 + 0.2}} = 2.705 \text{ m/s}$$

Ninety-nine percent of V_{ss} is $V_{99} = 0.99 V_{ss} = 0.99 \times 2.705 = 2.68$ m/s. The time corresponding to that velocity is found using Eq. (10.31):

$$t_{99} = \frac{2.705 \times 500}{2 \times 9.81 \times 15} \ln \frac{(2.705 + 2.68)(2.705 - 0.35)}{(2.705 - 2.68)(2.705 + 0.35)} = 23.5 \text{ s}$$

The final steady-state velocity is 2.70 m/s and the time when the velocity is 99% of that value is approximately 23.5 s.

10.5.2   Compressible Flow of Liquids

There are a number of situations in which a liquid-filled pipe, when subjected to an excitation, may not react in an incompressible, rigid manner. Indeed, both liquid compressibility and pipe elasticity play a significant role in the nature of the response. This action is traditionally caller water hammer, but it can occur in piping that contains any type of liquid. Herein, we focus our attention to a simple situation in which a valve at the downstream end of a pipe closes rapidly to initiate water hammer.

  There are two fundamental equations to develop that will enable us to understand the nature in which pressure waves travel in the pipe due to water hammer. Again, consider the horizontal pipe shown in Fig. 10.7. In contrast to the development in Sec. 10.5.1, we now assume that the valve closes rapidly so that the fluid compressibility and pipe elasticity occur. The valve movement causes an acoustic, or pressure, wave to propagate upstream with speed a. Figure 10.8(a) is a control volume of liquid in the pipe upstream of the valve, showing the pressure wavefront located at a given instant. Since unsteady flow is taking place inside the control volume, the velocity upstream of the front is V, while at the exit to the control volume it is $V + \Delta V$. The wavefront can be made to appear stationary by superposing to the right the acoustic velocity a; the result is shown in Fig. 10.8(b), where the entrance velocity is now $V + a$ and the exit velocity is $V + \Delta V + a$. Due to the passage of the wave, the pressure p, pipe area A, and liquid density ρ are altered to $p + \Delta p$, $A + \Delta A$, and $\rho + \Delta \rho$, respectively.

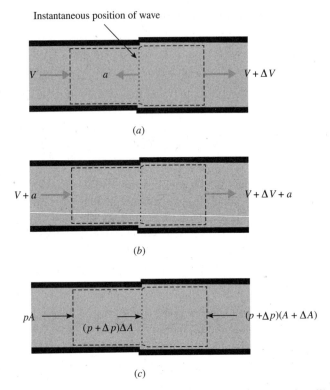

Figure 10.8   Pressure wave traveling in a pipe: (*a*) wave moving to the left at speed *a*; (*b*) wave appears stationary using superposition principle; (*c*) pressure forces acting on control volume.

Apply the conversation of mass to the control volume of Fig. 10.8(*b*):

$$0 = (\rho + \Delta\rho)(V + \Delta V + a)(A + \Delta A) - \rho(V + a)A \tag{10.32}$$

With reference to Fig. 10.8(*c*), we neglect frictional and gravitational forces so that only pressure forces act on the control volume in the flow direction, and apply the conversation of momentum with the result

$$pA + (p + \Delta p)A - (p + \Delta p)(A + \Delta A) = \rho A(V + a)[V + \Delta V + a - (V + a)] \tag{10.33}$$

Expanding Eqs. (10.32) and (10.33) and deleting terms containing Δ^2 and Δ^3 since they are smaller than the remaining ones, we have the results

$$\rho A\Delta V + (V + a)(A\Delta\rho + \rho\Delta A) = 0 \tag{10.34}$$

$$-A\Delta p = \rho A(V + a)\Delta V \tag{10.35}$$

In almost all industrial flow situations, $V \ll a$, so that Eq. (10.35) reduces to

$$\Delta p = -\rho A\Delta V \tag{10.36}$$

Equation (10.36) is labeled the *Joukowsky equation*; it relates the pressure change to the identity, the acoustic wave speed and the change in velocity. It is clear that a velocity reduction (negative ΔV) results in a pressure rise (positive Δp) and a velocity increase results in a pressure drop. The wave, passing through the control volume, yields the altered conditions $p + \Delta p$, $V + \Delta V$, $A + \Delta A$, and $\rho + \Delta\rho$. These conditions will persist in the pipe until the moment when the wave reflects from the upstream boundary and returns to the given position; this wave motion will be discussed later.

In order to determine magnitude of the acoustic wave speed, we combine Eqs. (10.34) and (10.35) and eliminate ΔV, recognizing again that $V \ll a$:

$$\frac{\Delta p}{\rho a^2} = \frac{\Delta\rho}{\rho} + \frac{\Delta A}{A} \tag{10.37}$$

The relative change in density is related to the change in pressure by the relation $\Delta\rho/\rho = \Delta p/B$, where B is the bulk modulus of elasticity for the liquid. Also, the relative change in pipe area is related to the change in pressure using $\Delta A/A = \Delta p(D/eE)$. In this latter relation we have assumed instantaneous elastic response of a thin-wall circular pipe to pressure changes, with E being the elastic modulus of the pipe wall material and e the pipe thickness. Substituting these two relations into Eq. (10.37) and rearranging, we have the expression for the pressure pulse wave speed:

$$a = \sqrt{\frac{B/\rho}{1 + (D/e)(B/E)}} \qquad (10.38)$$

It can be seen that a depends upon the properties of the liquid contained in the pipe (ρ and B) and those of the pipe wall (D, e, and E). If the pipe is very rigid, then the denominator approaches unity and Eq. (10.38) becomes $a = \sqrt{B/\rho}$, which is the speed of sound in an unbounded liquid. Note that the effect of pipe elasticity is to reduce the magnitude of the pressure wave.

In addition to using Eqs. (10.36) and (10.38) to predict the pressure rise and pressure pulse wave speed, it is necessary to understand the periodic nature of water hammer in a pipe of length L. Consider the case in which a downstream valve is suddenly closed in a horizontal, frictionless pipe with an open reservoir at the upstream end. One cycle of motion is illustrated in Fig. 10.9 and described as follows:

- A steady-state velocity V_0 exists throughout, the hydraulic grade line is horizontal, and the valve is suddenly closed at time zero.
- The wave travels upstream at speed a, subsequent to valve closure, and behind the wave the velocity is zero, the pressure rises by amount Δp, the liquid is compressed, and the pipe slightly expanded.
- The wave reaches the reservoir at time L/a, and an unbalanced force acts at the pipe inlet. At that location, the pipe pressure reduces to the reservoir pressure and the velocity reverses direction.
- The wave propagates downstream to the valve.
- The wave reaches the valve at time $2L/a$ and the velocity has magnitude $-V_0$ throughout the pipe.
- The velocity reduces to zero and the pressure reduces by magnitude Δp, adjacent to the closed valve. Behind the wave, the liquid is expanded and the pipe wall is contracted. (If the pressure behind the wave reduces to vapor pressure, cavitation will occur causing the liquid to vaporize, a condition termed column separation.)
- The pressure wave reaches the reservoir at time $3L/a$, where an unbalanced condition again occurs, and opposite in magnitude to that at time L/a.
- Forces equilibrate and a wave travels downstream with elevated pressure Δp and liquid velocity $+V_0$ behind the front.
- The wave reaches the valve at time $4L/a$ with initial steady-state conditions once again prevailing throughout the pipe.

The process repeats itself every $4L/a$ seconds and for the ideal frictionless sequence described herein the motion will be cyclic. The pressure waveform at the valve and midpoint of the pipe, and the velocity at the pipe entrance are shown in Fig. 10.9. In a real pipe, the action of liquid friction, pipe motion, and inelastic behavior of the pipe material will eventually cause the water hammer oscillation to dissipate.

The pressure rise Δp predicted by Eq. (10.36) is based on the assumption that the valve closes instantaneously, but it can also be used to predict the maximum pressure rise for valve closure in any time less than $2L/a$, the travel time for the pressure wave to travel from the valve to the reservoir and back again. For valve closure times greater than $2L/a$, and for more complicated piping systems that may contain multiple piping elements, addition of friction, and more complicated boundary conditions such as pumps and surge suppressors, computer-based analyses are required.

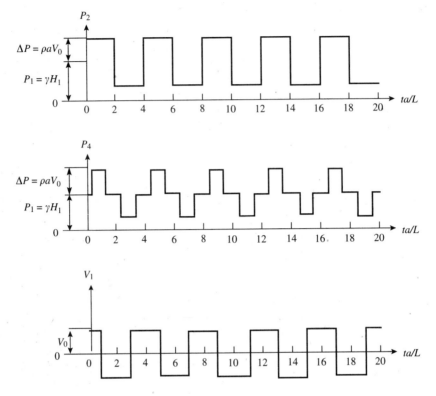

Figure 10.9   Pressure waves at the valve (p_2) and pipe midpoint (p_4), and velocity at the pipe entrance (V_1).

EXAMPLE 10.10   A steel pipe ($E = 220 \times 10^6$ kPa, $L = 2300$ m, $D = 500$ mm, $e = 10$ mm) conveys water with an initial velocity of $V_0 = 0.75$ m/s. A valve at the downstream end of the horizontal pipe is closed suddenly so that the excitation is considered instantaneous, reducing the velocity to zero. Determine (a) the pressure pulse wave speed in the pipe, (b) the speed of sound in an unbounded water medium, (c) the pressure rise at the closed valve, (d) the time it takes for the wave to travel to the reservoir and return to the valve, and (e) the period of water hammer oscillation.

   Solution: Since the water temperature is not specified, assume that $B = 210 \times 10^7$ Pa and $\rho = 1000$ kg/m^3.

(a) The wave speed is computed using Eq. (10.38):

$$a = \sqrt{\frac{210 \times 10^7 / 1000}{1 + \dfrac{500}{10} \times \dfrac{210 \times 10^7}{220 \times 10^9}}} = 1190 \, \text{m/s}$$

(b) The speed of sound in an unbounded medium is

$$a = \sqrt{\frac{B}{\rho}} = \sqrt{\frac{210 \times 10^7}{1000}} = 1450 \, \text{m/s}$$

Note that the speed of sound in the water medium is about 23% higher than the wave speed in the pipe.

(c) Equation (10.36) is employed to find the pressure rise, noting that the reduction in velocity is $\Delta V = -V_0$:

$$\Delta p = -1000 \times 1190(-0.75) = 8.92 \times 10^5 \, \text{Pa or 892 kPa}$$

(d) The time of wave to travel two pipe lengths is $2L/a = 2 \times 2300/1190 = 3.87$ s

(e) The period of oscillation is $4L/a = 4 \times 2300/1190 = 7.73$ s.

Solved Problems

10.1 Find the flow distribution in the three-parallel pipe system shown in Fig. 10.10. $Q_{in} =$ 2500 L/min.

Element	L, m	D, mm	f	$\sum K$
1	50	75	0.02	2
2	80	85	0.03	4
3	120	100	0.025	2

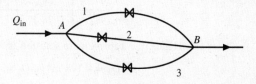

Figure 10.10

The resistance coefficients are computed using Eq. (10.3); the results are $R_1 = 40\,060$, $R_2 = 34\,280$, $R_3 = 22\,320$. Then write Eqs. (10.6) and (10.7) for three elements, combine, and solve for $H_A - H_B$:

$$H_A - H_B = \frac{Q_{in}^2}{(R_1^{-1/2} + R_2^{-1/2} + R_3^{-1/2})^2} = \frac{(2.50/60)^2}{(40\,060^{-1/2} + 34\,280^{-1/2} + 22\,320^{-1/2})^2} = 5.94\,\text{m}$$

The individual discharges are

$$Q_1 = \sqrt{\frac{H_A - H_B}{R_1}} = \sqrt{\frac{5.94}{40\,060}} = 0.0122\,\text{m}^3/\text{s or } 731\,\text{L/min}$$

$$Q_2 = \sqrt{\frac{5.94}{34\,280}} = 0.0132\,\text{m}^3/\text{s or } 790\,\text{L/min}$$

$$Q_3 = \sqrt{\frac{5.94}{22\,320}} = 0.0163\,\text{m}^3/\text{s or } 979\,\text{L/min}$$

10.2 Find the water flow distribution and the piezometric head at the junction of Fig. 10.11. The pump head–discharge curve is $H_P = 20 - 30Q^2$, where the head is in m and the discharge is in m³/s. $H_A = 10$ m, $H_B = 20$ m, $H_C = 18$ m, $R_1 = 112.1$ s²/m⁵, $R_2 = 232.4$ s²/m⁵, $R_3 = 1773$ s²/m⁵.

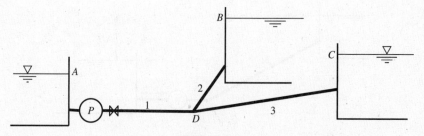

Figure 10.11

Let H_D be the hydraulic grade line at the junction. Write the energy equations for each branch:

Pipe 1 : $10 + H_P = R_1 Q_1^2 + H_D$ or $H_D = 10 + (20 - 30Q_1^2) - 112.1Q_1^2 = 30 - 142.1Q_1^2$

Pipe 2 : $H_D = R_2 Q_2^2 + 20$ or $Q_2 = \sqrt{\frac{(H_D - 20)}{232.4}}$

Pipe 3 : $H_D = R_3 Q_3^2 + 18$ or $Q_3 = \sqrt{\frac{(H_D - 18)}{1773}}$

Continuity at junction: $\sum Q = Q_1 - Q_2 - Q_3 = 0$

Assume Q_1, then solve for H_D, Q_2, and Q_3 until $|\sum Q|$ is acceptably low:

Q_1, m³/s	H_D, m	Q_2, m³/s	Q_3, m³/s	$\sum Q$, m³/s
0.15	26.80	0.1711	0.0705	−0.0916
0.20	24.32	0.1363	0.0597	+0.004
0.1982	24.42	0.1379	0.0602	−0.0001

Hence, the approximate solution is

$$Q_1 = 0.198 \text{ m}^3/\text{s}, \ Q_2 = 0.138 \text{ m}^3/\text{s}, \ Q_3 = 0.060 \text{ m}^3/\text{s}, \text{ and } H_D = 24.4 \text{ m}$$

10.3 Water is pumped through three pipes in series. Compute the discharge if the power delivered to the pump is $\dot{W}_P = 1920$ kW and the pump efficiency is $\eta = 82\%$. The pipe resistance coefficients are $R_1 = 13.2$ s²/m⁵, $R_2 = 204.1$ s²/m⁵, $R_3 = 25.8$ s²/m⁵, and the head difference between the downstream and upstream reservoirs is 50 m.

Employ Eq. (10.12) for the pump head and write the energy equation from the lower to the upper reservoir as a function of the unknown discharge Q:

$$F(Q) = \frac{\dot{W}_P \eta}{\gamma Q} - [H_B - H_A + (R_1 + R_2 + R_3)Q^2]$$

Substituting the given data into the relations results in

$$F(Q) = \frac{160.5}{Q} - 243.1Q^2 - 50$$

A root-solving algorithm, trial method, or exact solution can be used to find the result $Q = 0.792$ m³/s.

10.4 Determine the flow of water in the system shown in Fig. 10.12 using the Hardy Cross method. The pump curve is $H_P = 30 - 8.33Q^2$, with H_P in m and Q in m³/s. Use $R_1 = 30$ s²/m⁵, $R_2 = 20$ s²/m⁵, $\Delta H = 20$ m, and $Q_e = 0.25$ m³/s.

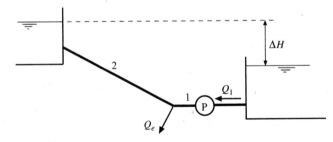

Figure 10.12

Consider the flow correction in a clockwise sense through the system:

$$\Delta Q = \frac{-[R_1 Q_1 |Q_1| - (a_0 - a_2 Q_1^2) + R_2 Q_2 |Q_2|] - \Delta H}{2(R_1 |Q_1| - a_2 Q_1 + R_2 |Q_2|)}$$

$$= \frac{-[30 Q_1 |Q_1| - (30 - 8.33 Q_1^2) - 20 Q_2 |Q_2|] - 20}{2(30 |Q_1| + 8.33 Q_1 + 40 |Q_2|)}$$

Assume $Q_1 = 0.650$ m³/s, $Q_2 = 0.400$ m³/s, and iterate until $|\Delta Q| \leq 0.001$ m³/s.

Q_1, m³/s	Q_2, m³/s	ΔQ, m³/s
0.640	0.390	−0.030
0.610	0.360	−0.011
0.599	0.349	−0.004
0.595	0.345	−0.001

Hence $Q_1 = 0.595$ m³/s and $Q_2 = 0.345$ m³/s.

10.5 Water ($B = 2.20$ GPa) is flowing in a pipe ($D = 200$ mm, $L = 800$ m). The pipe is cast iron ($E = 150$ GPa, $e = 12$ mm). A reservoir is situated at the upstream end of the pipe, and downstream there is a valve. At steady-state conditions, the discharge is $Q = 50$ L/s, and then a valve is rapidly actuated.

(a) Determine the time it takes for the pressure wave to travel from the valve to the reservoir and back to the valve.

(b) What is the change in pressure at the valve if the valve is partially opened and the original discharge is doubled?

(c) What is the change in pressure at the valve if the valve is partially closed and the original discharge is halved?

Compute the acoustic wave speed:

$$a = \sqrt{\frac{B}{\rho\left(1 + \dfrac{DB}{eE}\right)}} = \sqrt{\frac{220 \times 10^7}{1000\left(1 + \dfrac{200 \times 2.20 \times 10^9}{12 \times 150 \times 10^9}\right)}} = 1330 \,\text{m/s}$$

(a) $t = 2 \times \dfrac{L}{a} = 2 \times \dfrac{800}{1330} = 1.203 \,\text{s}$

(b) First compute the change in velocity and then use Eq. (10.36) to compute the pressure change (assume water hammer occurs):

$$\Delta V = \frac{Q}{\pi D^2/4} = \frac{0.05}{0.7854 \times 0.2^2} = 1.592 \,\text{m/s}$$

$$\Delta p = -\rho a \Delta V = -1000 \times 1330 \times 1.592 = -2.12 \times 10^6 \,\text{Pa or } -2120 \,\text{kPa}$$

Note the large pressure reduction due to the water hammer effect. The original pressure at the valve must be sufficiently large so that cavitation will not occur. Cavitation at the valve could be avoided by opening the valve slowly.

(c) The change in velocity and the pressure rise are

$$\Delta V = -\frac{0.05/2}{0.7854 \times 0.2^2} = -0.796 \,\text{m/s}$$

$$\Delta p = -1000 \times 1330(-0.796) = 1.06 \times 10^6 \,\text{Pa}$$

Hence the pressure rise will be 1060 kPa.

Supplementary Problems

10.6   A pump is located between two reaches of horizontal piping. The conditions upstream of the pump are $D_1 = 75$ mm and $p_1 = 450$ kPa, and downstream of the pump $D_2 = 100$ mm and $p_2 = 900$ kPa. For a discharge of $Q = 100$ L/min and a loss across the pump of $h_L = 7$ m, what is the required input power of the pump if its efficiency is 78%?

10.7   Two pipes in series have the following properties: $L_1 = 200$ m, $D_1 = 400$ mm, $K_1 = 2$, $L_2 = 650$ m, $D_2 = 350$ mm, $K_2 = 3$. The upstream piezometric head is $H_A = 200$ m and downstream $H_B = 57$ m. For both pipes, the friction factor is $f = 0.025$. Estimate the discharge flowing in the two pipes.

10.8   Water flows in the system shown. The pump curve is approximated by $H_P = 150 - 5Q_1^2$, with H_P in m and Q in m^3/s. Find (a) the flow distribution. (b) If the pump efficiency is 75%, what is the required pump power? Use $R_1 = 400$ s^2/m^5, $R_2 = 1000$ s^2/m^5, $R_3 = 1500$ s^2/m^5, $H_A = 10$ m, and $H_B = 40$ m.

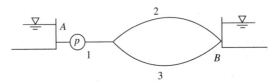

10.9   An oil pipeline ($S = 0.86$) has three segments as shown, with a booster pump for each segment employed to overcome pipe friction. Reservoirs A and B are at the same elevation. Find the discharge for the following conditions:

Pipe	R, s^2/m^5	$\dot{W}_P$, kW	η, %
1	40 000	200	80
2	30 000	200	80
3	200 000	250	70

10.10   Oil ($S = 0.92$) is pumped from a storage tank and discharges into a reservoir through a pipe whose length is $L = 550$ m and diameter is $D = 350$ mm. The pump efficiency is $\eta = 80\%$ and its power output is $\dot{W}_P = 10$ kW. Determine the discharge Q if the elevation in the tank is $z_A = 24$ m, the tank pressure is $p_A = 110$ kPa, and the lower reservoir elevation is $z_B = 18$ m. The sum of the minor losses in the pipe is $\Sigma K = 4.5$ and the friction factor is $f = 0.015$.

10.11   Determine the total discharge and the individual flows in the four parallel pipes shown. The hydraulic grade line difference between A and B is $H_A - H_B = 60$ m. The following data apply:

Pipe	L, m	D, mm	f	ΣK
1	650	850	0.02	1
2	1000	1000	0.025	3
3	500	750	0.015	0
4	750	1000	0.03	2

10.12 What is the required head and discharge of water to be handled by the pump for the branching system? The flow in pipe 3 is $Q_3 = 40$ L/s in the direction shown. Use $H_A = 3$ m, $H_B = 11.5$ m, $H_C = 12$ m, $H_D = 10$ m, $R_1 = 1400$ s²/m⁵, $R_2 = 2000$ s²/m⁵, $R_3 = 1500$ s²/m⁵, and $R_4 = 1000$ s²/m⁵.

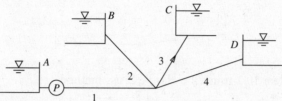

10.13 An irrigation system lies in a horizontal plane, with a large diameter pipe delivering water through a single line to three branches. The delivery pipe has an internal pressure $p_0 = 200$ kPa and is sufficiently large that internal kinetic energy terms can be neglected. Find the flow distribution in the four irrigation pipes if $R_1 = 1.6 \times 10^4$, $R_2 = 5.3 \times 10^5$, $R_3 = 6.0 \times 10^5$, $R_4 = 8.1 \times 10^5$ (all in units of s²/m⁵).

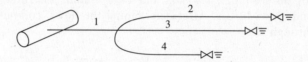

10.14 It is common in pump design and manufacturing to make use of dimensionless coefficients that relate to pump power $\dot{W}_P$, pressure rise Δp, and discharge Q. Additional variables include the density ρ, diameter of the pump impeller D, and rotational speed of the impeller ω. Using ρ, D, and ω as repeating variables, compute the three dimensionless coefficients related to power, pressure rise, and discharge.

10.15 Find the flow distribution of water in the branching system using a trial method. Assume for all pipes $f = 0.02$. The pump curve is represented by the relation $H_P = 120 - 0.5Q^2$ (head in m, discharge in m³/s), $H_A = 20$ m, $H_B = 50$ m, $H_C = 100$ m, and $H_D = 40$ m. Neglect minor losses.

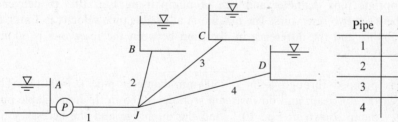

Pipe	L, m	D, mm
1	250	500
2	700	300
3	2000	300
4	1500	350

10.16 Find the flow distribution using the Hardy Cross method. Given data are $Q_B = 30$ L/s, $Q_C = 30$ L/s, $R_1 = 30$ s²/m⁵, $R_2 = 50$ s²/m⁵, and $R_3 = 20$ s²/m⁵.

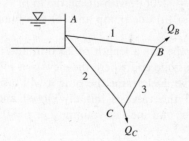

10.17 Find the discharge in the pumping system using the Hardy Cross method. The pump curve is $H_P = 100 - 50Q - 850Q^2$, with head in m and discharge in m^3/s. Use $R_1 = 5000$ s^2/m^5, $R_2 = 300$ s^2/m^5, $H_A = 35$ m, and $H_B = 10$ m.

10.18 Work Solved Problem 10.2 using the Hardy Cross method.

10.19 Add a pump to pipe 1 in Fig. 10.5(a) and solve using the Hardy Cross method. The head across the pump is represented by $H_P = 150 - 30Q^2$, with head in m and discharge in m^3/s. Use the pipe and reservoir data provided in Example 10.8.

10.20 Determine the flow distribution in the water supply system shown using the Hardy Cross method. The piezometric head at location A is $H_A = 30$ m, and the head at the reservoir is $H_F = 0$ m. For all six pipes, $f = 0.03$ and $D = 75$ mm. The flow demands at C and D are $Q_C = 5$ L/s and $Q_D = 12$ L/s. After determining the flows, compute the piezometric heads at locations B through E.

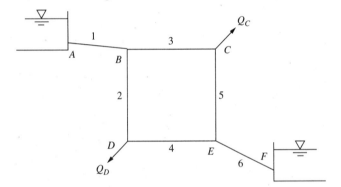

Pipe	L, m	K
1	10	1
2	30	0
3	75	0
4	60	0
5	35	0
6	80	2

10.21 Select an appropriate pipe diameter and size of pump using Fig. 10.3 to deliver water at $Q = 250$ m^3/h between two reservoirs. The maximum allowable pipe velocity is 3 m/s, the length of the line is 1500 m, and the difference in elevation between the reservoirs is 30 m. Assume $f = 0.02$ and $K = 0.5$.

10.22 Oil ($S = 0.86$) is pumped through 5 km of 500-mm-diameter pipe ($f = 0.017$). The rise in elevation between the upstream and downstream sections is 165 m. If an available pump is the 260-mm-diameter pump shown in Fig. 10.3, find the discharge and the necessary number of pumps to be placed in series. What is the power requirement? Neglect minor losses.

10.23 The 220-mm-diameter pump curve shown in Fig. 10.3 delivers water in a piping system whose system demand is $H_P = 50 + 270Q^2$, with discharge in m^3/s and head in m. Find the discharge and required pump power for (a) one pump and (b) two pumps in parallel.

10.24 A horizontal water supply pipe has a length of 2000 m and a diameter of 150 mm. The pipe is connected to an open tank at one end where the elevation of water is 4 m, and at the other end there is a quick-opening valve. Determine the time it will take for the flow to reach 99% of the final steady-state velocity if the valve is initially closed and then suddenly opened at $t = 0$. Assume incompressible water and inelastic piping. Let $f = 0.030$ and $K = 0.2$ once the valve is opened.

10.25 Oil is flowing at a discharge of 0.50 m³/s in a 4-km, 50-mm-diameter steel pipe. The elastic modulus of the pipe is 200×10^6 kPa and its thickness is 5 mm. The oil has a specific gravity of 0.86 and a bulk modulus of 1.50×10^6 kPa. A valve at the end of the pipe is partially closed in a rapid fashion so that water hammer occurs and a pressure wave propagates upstream in the pipe. The magnitude of the pressure wave is not to be greater than 600 kPa. Determine the percent decrease of flow rate tolerable during the valve closure and the period of the water hammer oscillation.

Answers to Supplementary Problems

10.6   1109 W

10.7   0.669 m³/s

10.8   (a)   0.413 m³/s, 0.227 m³/s, 0.185 m³/s   (b)   1070 hp

10.9   0.0601 m³/s

10.10   0.365 m³/s

10.11   4.82 m³/s, 5.10 m³/s, 4.79 m³/s, 5.44 m³/s

10.12   0.144 m³/s, 40.4 m

10.13   0.0154 m³/s, 0.0056 m³/s, 0.00526 m³/s, 0.00453 m³/s

10.14 $\dot{W}_P/\rho\omega^3 D^5$, $\Delta p/\rho\omega^2 D^2$, $Q/\omega D^3$

10.15   1.063 m³/s, 0.433 m³/s, 0.170 m³/s, 0.459 m³/s

10.16   33.75 L/s into B, 26.25 L/s into C, 3.75 L/s into C

10.17   0.106 m³/s

10.18   0.198 m³/s out of A, 0.138 m³/s into B, 0.060 m³/s into C

10.19   0.0749 m³/s into D, 0.0249 m³/s into B, 0.050 m³/s into C, 0.0250 m³/s into B, 0.0499 m³/s into A

10.20   27.7 L/s, 17.0 L/s, 10.7 L/s, 5.0 L/s, 10.7 L/s, 20 m, 11 m, 10.9 m, 9.35 m

10.21   200 mm, 240 mm

10.22   270 m³/h with two pumps, 186 hp

10.23   (a)   250 m³/h, 63 hp   (b)   450 m³/h, 120 hp

10.24   58.4 s

10.25   73%, 16 s

Appendix A

Units and Conversions

Table A.1 English Units, SI Units, and Their Conversion Factors

Quantity	English units	International system[a] SI	Conversion factor
Length	inch	millimeter	1 in = 25.4 mm
	foot	meter	1 ft = 0.3048 m
	mile	kilometer	1 mi = 1.609 km
Area	square inch	square centimeter	$1\ in^2 = 6.452\ cm^2$
	square foot	square meter	$1\ ft^2 = 0.09290\ m^2$
Volume	cubic inch	cubic centimeter	$1\ in^3 = 16.39\ cm^3$
	cubic foot	cubic meter	$1\ ft^3 = 0.02832\ m^3$
	gallon		$1\ gal = 0.003789\ m^3$
Mass	pound mass	kilogram	1 lb = 0.4536 kg
	slug		1 slug = 14.59 kg
Density	slug per cubic foot	kilogram per cubic meter	$1\ slug/ft^3 = 515.4\ kg/m^3$
Force	pound force	newton	1 lb = 4.448 N
Work/torque	foot pound	newton meter	1 ft-lb = 1.356 N·m
Pressure	pound per square inch	newton per square meter (pascal)	$1\ lb/in^2 = 6895\ Pa$
	pound per square foot		$1\ lb/ft^2 = 47.88\ Pa$
Temperature	degree Fahrenheit	degree Celsius	$°F = 9/5\,°C + 32$
	degree Rankine	kelvin	$°R = 9/5\ K$
Energy	British thermal unit	joule	1 Btu = 1055 J
	calorie		1 cal = 4.186 J
	foot pound		1 ft-lb = 1.356 J
Power	horsepower	watt	1 hp = 745.7 W
	foot pound per second		1 ft-lb/s = 1.356 W
Velocity	foot per second	meter per second	1 ft/s = 0.3048 m/s
Acceleration	foot per second squared	meter per second squared	$1\ ft/s^2 = 0.3048\ m/s^2$
Frequency	cycle per second	hertz	1 c/s = 1.000 Hz
Viscosity	pound second per square foot	newton second per square meter	$1\ lb\text{-}s/ft^2 = 47.88\ N·s/m^2$

[a] The reversed initials in this abbreviation come from the French form of the name: Système International.

Table A.2 Conversions of Units

Length
1 cm = 0.3937 in
1 m = 3.281 ft
1 km = 0.6214 mi
1 in = 2.54 cm
1 ft = 0.3048 m
1 mi = 1.609 km
1 mi = 5280 ft
1 mi = 1760 yd

Force
1 lb = 0.4536 kg
1 lb = 0.4448×10^6 dyn
1 lb = 32.17 pdl
1 kg = 2.205 lb
1 N = 0.2248 lb
1 dyn = 2.248×10^{-6} lb
1 lb = 4.448 N

Mass
1 oz = 28.35 g
1 lb = 0.4536 kg
1 slug = 32.17 lb
1 slug = 14.59 kg
1 kg = 2.205 lb
1 kg = 0.06852 slug

Velocity
1 mph = 1.467 ft/s
1 mph = 0.8684 kn
1 ft/s = 0.3048 m/s
1 m/s = 3.281 ft/s
1 km/h = 0.278 m/s

Work, energy, and power
1 Btu = 778.2 ft-lb
1 J = 10^7 ergs
1 J = 0.7376 ft-lb
1 cal = 3.088 ft-lb
1 cal = 0.003968 Btu
1 kWh = 3413 Btu
1 Btu = 1.055 kJ
1 ft-lb = 1.356 J
1 hp = 550 ft-lb/sec
1 hp = 0.7067 Btu/s
1 hp = 0.7455 kW
1 W = 1 J/s
1 W = 1.0×10^7 (dyn·cm)/s
1 erg = 10^{-7} J
1 quad = 10^{15} Btu
1 therm = 10^5 Btu

Pressure
1 lb/in^2 = 2.036 in Hg
1 lb/in^2 = 27.7 in H$_2$O
14.7 lb/in^2 = 22.92 in Hg
14.7 lb/in^2 = 33.93 ft H$_2$O
14.7 lb/in^2 = 1.0332 kg/cm^2
14.7 lb/in^2 = 1.0133 bar
1 kg/cm^2 = 14.22 lb/in^2
1 in Hg = 0.4912 lb/in^2
1 ft H$_2$O = 0.4331 lb/in^2
1 lb/in^2 = 6895 Pa
1 lb/ft^2 = 47.88 Pa
10^5 Pa = 1 bar
1 kPa = 0.145 lb/in^2

Volume
1 ft^3 = 28.32 L
1 ft^3 = 7.481 gal (U.S.)
1 gal (U.S.) = 231 in^3
1 gal (Brit.) = 1.2 gal (U.S.)
1 m^3 = 1000 L
1 ft^3 = 0.02832 m^3
1 m^3 = 35.31 ft^3

Flow rate
1 ft^3/min = 4.719×gal 10^{-4} m^3/s
1 ft^3/s = 0.02832 m^3/s
1 m^3/s = 35.31 ft^3/s
1 gal/min = 0.002228 ft^3/s
1 ft^3/s = 448.9 gal/min

Viscosity
1 stoke = 10^{-4} m^2/s
1 P = 0.1 (N·s)/m^2
1 (lb·s)/ft^2 = 47.88 (N·s)/m^2
1 ft^2/s = 0.0929 m^2/s

Appendix B

Vector Relationships

$$\mathbf{A} \cdot \mathbf{B} = A_x B_x + A_y B_y + A_z B_z$$

$$\mathbf{A} \times \mathbf{B} = (A_y B_z - A_z B_y)\mathbf{i} + (A_z B_x - A_x B_z)\mathbf{j} + (A_x B_y - A_y B_x)\mathbf{k}$$

$$\text{gradient operator} : \nabla = \frac{\partial}{\partial x}\mathbf{i} + \frac{\partial}{\partial y}\mathbf{j} + \frac{\partial}{\partial z}\mathbf{k}$$

$$\text{divergence of } \mathbf{V} = \nabla \cdot \mathbf{V} = \frac{\partial u}{\partial x} + \frac{\partial v}{\partial y} + \frac{\partial w}{\partial z}$$

$$\text{curl of } \mathbf{V} = \nabla \times \mathbf{V} = \left(\frac{\partial w}{\partial y} - \frac{\partial v}{\partial z}\right)\mathbf{i} + \left(\frac{\partial u}{\partial z} - \frac{\partial w}{\partial x}\right)\mathbf{j} + \left(\frac{\partial v}{\partial x} - \frac{\partial u}{\partial y}\right)\mathbf{k}$$

$$\text{Laplace's equation: } \nabla^2 \phi = 0$$

$$\text{Irrotational vector field: } \nabla \times \mathbf{V} = 0$$

Appendix C

Fluid Properties

Table C.1 Properties of Water

Temperature, T (°C)	Density, ρ (kg/m³)	Viscosity, μ [(N·s)/m²]	Kinematic viscosity, ν (m²/s)	Surface tension, σ (N/m)	Vapor pressure, p_v (kPa)	Bulk modulus, B (Pa)
0	999.9	1.792×10^{-3}	1.792×10^{-6}	0.0762	0.610	204×10^{7}
5	1000.0	1.519	1.519	0.0754	0.872	206
10	999.7	1.308	1.308	0.0748	1.13	211
15	999.1	1.140	1.141	0.0741	1.60	214
20	998.2	1.005	1.007	0.0736	2.34	220
30	995.7	0.801	0.804	0.0718	4.24	223
40	992.2	0.656	0.661	0.0701	3.38	227
50	988.1	0.549	0.556	0.0682	12.3	230
60	983.2	0.469	0.477	0.0668	19.9	228
70	977.8	0.406	0.415	0.0650	31.2	225
80	971.8	0.357	0.367	0.0630	47.3	221
90	965.3	0.317	0.328	0.0612	70.1	216
100	958.4	0.284×10^{-3}	0.296×10^{-6}	0.0594	101.3	207×10^{7}

Table C.1E English Properties of Water

Temperature (°F)	Density (slug/ft³)	Viscosity (lb·sec/ft²)	Kinematic viscosity (ft²/sec)	Surface tension (lb/ft)	Vapor pressure (lb/in²)	Bulk modulus (lb/in²)
32	1.94	3.75×10^{-5}	1.93×10^{-5}	0.518×10^{-2}	0.089	293 000
40	1.94	3.23	1.66	0.514	0.122	294 000
50	1.94	2.74	1.41	0.509	0.178	305 000
60	1.94	2.36	1.22	0.504	0.256	311 000
70	1.94	2.05	1.06	0.500	0.340	320 000
80	1.93	1.80	0.93	0.492	0.507	322 000
90	1.93	1.60	0.83	0.486	0.698	323 000
100	1.93	1.42	0.74	0.480	0.949	327 000
120	1.92	1.17	0.61	0.465	1.69	333 000
140	1.91	0.98	0.51	0.454	2.89	330 000
160	1.90	0.84	0.44	0.441	4.74	326 000
180	1.88	0.73	0.39	0.426	7.51	318 000
200	1.87	0.64	0.34	0.412	11.53	308 000
212	1.86	0.59×10^{-5}	0.32×10^{-5}	0.404×10^{-2}	14.7	300 000

Table C.2   Properties of Air at Atmospheric Pressure

Temperature, $T\,(°C)$	Density, $\rho\,(kg/m^3)$	Viscosity, $\mu\,(N\cdot s/m^2)$	Kinematic viscosity, $\nu\,(m^2/s)$	Velocity of sound, $c\,(m/s)$
-50	1.582	1.46×10^{-5}	0.921×10^{-5}	299
-30	1.452	1.56	1.08	312
-20	1.394	1.61	1.16	319
-10	1.342	1.67	1.24	325
0	1.292	1.72	1.33	331
10	1.247	1.76	1.42	337
20	1.204	1.81	1.51	343
30	1.164	1.86	1.60	349
40	1.127	1.91	1.69	355
50	1.092	1.95	1.79	360
60	1.060	2.00	1.89	366
70	1.030	2.05	1.99	371
80	1.000	2.09	2.09	377
90	0.973	2.13	2.19	382
100	0.946	2.17	2.30	387
200	0.746	2.57	3.45	436
300	0.616	2.93×10^{-5}	4.75×10^{-5}	480

Table C.2E   English Properties of Air at Atmospheric Pressure

Temperature $(°F)$	Density $(slug/ft^3)$	Viscosity $[(lb\cdot sec)/ft^2]$	Kinematic viscosity (ft^2/sec)	Velocity of sound (ft/sec)
-20	0.00280	3.34×10^{-7}	11.9×10^{-5}	1028
0	0.00268	3.38	12.6	1051
20	0.00257	3.50	13.6	1074
40	0.00247	3.62	14.6	1096
60	0.00237	3.74	15.8	1117
68	0.00233	3.81	16.0	1125
80	0.00228	3.85	16.9	1138
100	0.00220	3.96	18.0	1159
120	0.00213	4.07	18.9	1180
160	0.00199	4.23	21.3	1220
200	0.00187	4.50	24.1	1258
300	0.00162	4.98	30.7	1348
400	0.00144	5.26	36.7	1431
1000	0.000844	7.87×10^{-7}	93.2×10^{-5}	1839

Table C.3　Properties of the Standard Atmosphere

Altitude (m)	Temperature (K)	Pressure (kPa)	Density (kg/m³)	Velocity of sound (m/s)
0	288.2	101.3	1.225	340
500	284.9	95.43	1.167	338
1000	281.7	89.85	1.112	336
2000	275.2	79.48	1.007	333
4000	262.2	61.64	0.8194	325
6000	249.2	47.21	0.6602	316
8000	236.2	35.65	0.5258	308
10 000	223.3	26.49	0.4136	300
12 000	216.7	19.40	0.3119	295
14 000	216.7	14.17	0.2278	295
16 000	216.7	10.35	0.1665	295
18 000	216.7	7.563	0.1216	295
20 000	216.7	5.528	0.0889	295
30 000	226.5	1.196	0.0184	302
40 000	250.4	0.287	4.00×10^{-3}	317
50 000	270.7	0.0798	1.03×10^{-3}	330
60 000	255.8	0.0225	3.06×10^{-4}	321
70 000	219.7	0.00551	8.75×10^{-5}	297
80 000	180.7	0.00103	2.00×10^{-5}	269

Table C.3E　English Properties of the Atmosphere

Altitude (ft)	Temperature (°F)	Pressure (lb/ft²)	Density (slug/ft³)	Velocity of sound (ft/sec)
0	59.0	2116	0.00237	1117
1000	55.4	2014	0.00231	1113
2000	51.9	1968	0.00224	1109
5000	41.2	1760	0.00205	1098
10 000	23.4	1455	0.00176	1078
15 000	5.54	1194	0.00150	1058
20 000	− 12.3	973	0.00127	1037
25 000	− 30.1	785	0.00107	1016
30 000	− 48.0	628	0.000890	995
35 000	− 65.8	498	0.000737	973
36 000	− 67.6	475	0.000709	971
40 000	− 67.6	392	0.000586	971
50 000	− 67.6	242	0.000362	971
100 000	− 51.4	23.2	3.31×10^{-5}	971

Table C.4 Properties of Ideal Gases at 300 K ($c_v = c_p - k$ $k = c_p/c_v$)

Gas	Chemical formula	Molar mass	R		c_p		k
			(ft-lb)/ slug-°R	kJ/ (kg·K)	(ft-lb)/ slug-°R	kJ/ (kg·K)	
Air		28.97	1716	0.287	6012	1.004	1.40
Argon	Ar	39.94	1244	0.2081	3139	0.5203	1.667
Carbon dioxide	CO_2	44.01	1129	0.1889	5085	0.8418	1.287
Carbon monoxide	CO	28.01	1775	0.2968	6238	1.041	1.40
Ethane	C_2H_6	30.07	1653	0.2765	10 700	1.766	1.184
Helium	He	4.003	12 420	2.077	31 310	5.193	1.667
Hydrogen	H_2	2.016	24 660	4.124	85 930	14.21	1.40
Methane	CH_4	16.04	3100	0.5184	13 330	2.254	1.30
Nitrogen	N_2	28.02	1774	0.2968	6213	1.042	1.40
Oxygen	O_2	32.00	1553	0.2598	5486	0.9216	1.394
Propane	C_3H_8	44.10	1127	0.1886	10 200	1.679	1.12
Steam	H_2O	18.02	2759	0.4615	11 150	1.872	1.33

Table C.5　Properties of Common Liquids at Atmospheric Pressure and Approximately 16 to 21°C (60 to 70°F)

Liquid	Specific weight		Density		Surface tension		Vapor pressure	
	lb/ft³	N/m³	slug/ft³	kg/m³	lb/ft	N/m	lb/in² abs	kPa abs
Ethyl alcohol	49.3	7744	1.53	789	0.0015	0.022	—	—
Benzene	56.2	8828	1.75	902	0.0020	0.029	1.50	10.3
Carbon tetrachloride	99.5	15 629	3.09	1593	0.0018	0.026	12.50	86.2
Glycerin	78.6	12 346	2.44	1258	0.0043	0.063	2×10^{-6}	1.4×10^{-5}
Kerosene	50.5	7933	1.57	809	0.0017	0.025		
Mercury[a]	845.5	132 800	26.29	13 550	0.032	0.467	2.31×10^{-5}	1.59×10^{-4}
SAE 10 oil	57.4	9016	1.78	917	0.0025	0.036	—	—
SAE 30 oil	57.4	9016	1.78	917	0.0024	0.035	—	—
Water	62.4	9810	1.94	1000	0.0050	0.073	0.34	2.34

[a] In contact with air.

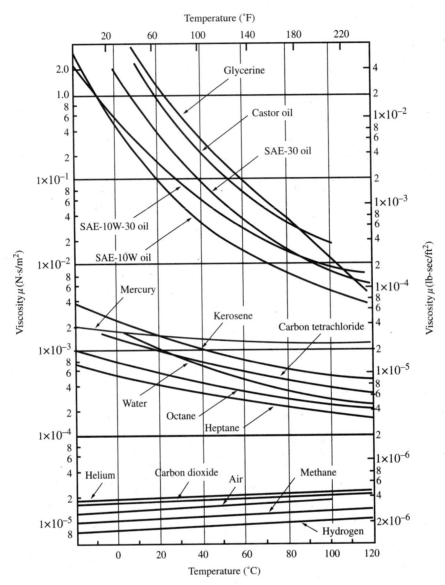

Figure C.1   Viscosity as a function of temperature. (From R.W. Fox and T.A. McDonald, *Introduction to Fluid Mechanics*, 2nd ed., John Wiley & Sons, Inc., New York, 1978.)

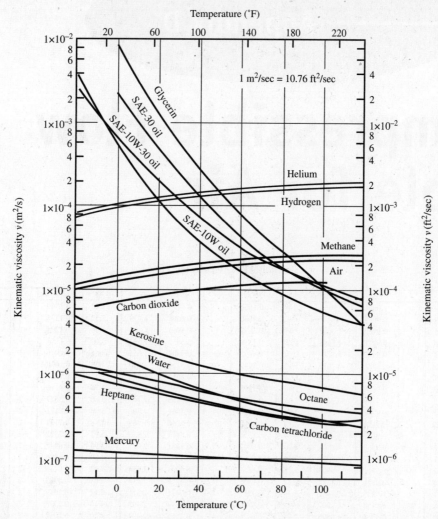

Figure C.2   Kinematic viscosity as a function of temperature at atmospheric pressure. (From R.W. Fox and T.A. McDonald, *Introduction to Fluid Mechanics*, 2nd ed., John Wiley & Sons, Inc., New York, 1978.)

Appendix D

Compressible Flow Table for Air

Table D.1   Isentropic Flow

M	p/p_0		T/T_0	A/A^*	M	p/p_0		T/T_0	A/A^*	M	p/p_0		T/T_0	A/A^*
0	1.0000		1.0000	0	1.76	0.1850		0.6175	1.397	3.48	0.1349	−1	0.2922	6.664
0.04	0.9989		0.9997	14.4815	1.80	0.1740		0.6068	1.439	3.52	0.1274	−1	0.2875	6.917
0.08	0.9955		0.9987	7.2616	1.84	0.1637		0.5963	1.484	3.56	0.1204	−1	0.2829	7.179
0.12	0.9900		0.9971	4.8643	1.88	0.1539		0.5859	1.531	3.60	0.1138	−1	0.2784	7.450
0.16	0.9823		0.9949	3.6727	1.90	0.1492		0.5807	1.555	3.64	0.1076	−1	0.2740	7.730
0.20	0.9725		0.9921	2.9635	1.92	0.1447		0.5756	1.580	3.68	0.1018	−1	0.2697	8.020
0.24	0.9607		0.9886	2.4956	1.96	0.1360		0.5655	1.633	3.72	0.9633	−2	0.2654	8.320
0.28	0.9470		0.9846	2.1656	2.00	0.1278		0.5556	1.688	3.76	0.9116	−2	0.2613	8.630
0.32	0.9315		0.9799	1.9219	2.04	0.1201		0.5458	1.745	3.80	0.8629	−2	0.2572	8.951
0.36	0.9143		0.9747	1.7358	2.08	0.1128		0.5361	1.806	3.84	0.8171	−2	0.2532	9.282
0.40	0.8956		0.9690	1.5901	2.12	0.1060		0.5266	1.869	3.88	0.7739	−2	0.2493	9.624
0.44	0.8755		0.9627	1.4740	2.16	0.9956	−1	0.5173	1.935	3.92	0.7332	−2	0.2455	9.977
0.48	0.8541		0.9560	1.3801	2.20	0.9352	−1	0.5081	2.005	3.96	0.6948	−2	0.2418	10.34
0.52	0.8317		0.9487	1.3034	2.24	0.8785	−1	0.4991	2.078	4.00	0.6586	−2	0.2381	10.72
0.56	0.8082		0.9410	1.2403	2.28	0.8251	−1	0.4903	2.154	4.04	0.6245	−2	0.2345	11.11
0.60	0.7840		0.9328	1.1882	2.32	0.7751	−1	0.4816	2.233	4.08	0.5923	−2	0.2310	11.51
0.64	0.7591		0.9243	1.1452	2.36	0.7281	−1	0.4731	2.316	4.12	0.5619	−2	0.2275	11.92
0.68	0.7338		0.9153	1.1097	2.40	0.6840	−1	0.4647	2.403	4.16	0.5333	−2	0.2242	12.35
0.72	0.7080		0.9061	1.0806	2.44	0.6426	−1	0.4565	2.494	4.20	0.5062	−2	0.2208	12.79
0.76	0.6821		0.8964	1.0570	2.48	0.6038	−1	0.4484	2.588	4.24	0.4806	−2	0.2176	13.25
0.80	0.6560		0.8865	1.0382	2.52	0.5674	−1	0.4405	2.686	4.28	0.4565	−2	0.2144	13.72
0.84	0.6300		0.8763	1.0237	2.56	0.5332	−1	0.4328	2.789	4.32	0.4337	−2	0.2113	14.20
0.88	0.6041		0.8659	1.0129	2.60	0.5012	−1	0.4252	2.896	4.36	0.4121	−2	0.2083	14.70
0.92	0.5785		0.8552	1.0056	2.64	0.4711	−1	0.4177	3.007	4.40	0.3918	−2	0.2053	15.21
0.96	0.5532		0.8444	1.0014	2.68	0.4429	−1	0.4104	3.123	4.44	0.3725	−2	0.2023	15.74
1.00	0.5283		0.8333	1.000	2.72	0.4165	−1	0.4033	3.244	4.48	0.3543	−2	0.1994	16.28
1.04	0.5039		0.8222	1.001	2.76	0.3917	−1	0.3963	3.370	4.52	0.3370	−2	0.1966	16.84
1.08	0.4800		0.8108	1.005	2.80	0.3685	−1	0.3894	3.500	4.54	0.3288	−2	0.1952	17.13
1.12	0.4568		0.7994	1.011	2.84	0.3467	−1	0.3827	3.636	4.58	0.3129	−2	0.1925	17.72
1.16	0.4343		0.7879	1.020	2.88	0.3263	−1	0.3761	3.777	4.62	0.2978	−2	0.1898	18.32
1.20	0.4124		0.7764	1.030	2.92	0.3071	−1	0.3696	3.924	4.66	0.2836	−2	0.1872	18.94
1.24	0.3912		0.7648	1.043	2.96	0.2891	−1	0.3633	4.076	4.70	0.2701	−2	0.1846	19.58
1.28	0.3708		0.7532	1.058	3.00	0.2722	−1	0.3571	4.235	4.74	0.2573	−2	0.1820	20.24
1.32	0.3512		0.7416	1.075	3.04	0.2564	−1	0.3511	4.399	4.78	0.2452	−2	0.1795	20.92
1.36	0.3323		0.7300	1.094	3.08	0.2416	−1	0.3452	4.570	4.82	0.2338	−2	0.1771	21.61
1.40	0.3142		0.7184	1.115	3.12	0.2276	−1	0.3393	4.747	4.86	0.2229	−2	0.1747	22.33
1.44	0.2969		0.7069	1.138	3.16	0.2146	−1	0.3337	4.930	4.90	0.2126	−2	0.1724	23.07
1.48	0.2804		0.6954	1.163	3.20	0.2023	−1	0.3281	5.121	4.94	0.2028	−2	0.1700	23.82
1.52	0.2646		0.6840	1.190	3.24	0.1908	−1	0.3226	5.319	4.98	0.1935	−2	0.1678	24.60
1.56	0.2496		0.6726	1.219	3.28	0.1799	−1	0.3173	5.523	6.00	0.0633	−2	0.1219	53.19
1.60	0.2353		0.6614	1.250	3.32	0.1698	−1	0.3121	5.736	8.00	0.0102	−2	0.0725	109.11
1.64	0.2217		0.6502	1.284	3.36	0.1602	−1	0.3069	5.956	10.00	0.0236	−3	0.0476	535.94
1.68	0.2088		0.6392	1.319	3.40	0.1512	−1	0.3019	6.184	∞	0		0	∞
1.72	0.1966		0.6283	1.357	3.44	0.1428	−1	0.2970	6.420					

Table D.2   Normal Shock Flow

M_1	M_2	p_2/p_1	T_2/T_1	p_{02}/p_{01}	M_1	M_2	p_2/p_1	T_2/T_1	p_{02}/p_{01}
1.00	1.000	1.000	1.000	1.000	3.12	0.4685	11.19	2.823	0.2960
1.04	0.9620	1.095	1.026	0.9999	3.16	0.4664	11.48	2.872	0.2860
1.08	0.9277	1.194	1.052	0.9994	3.20	0.4643	11.78	2.922	0.2762
1.12	0.8966	1.297	1.078	0.9982	3.24	0.4624	12.08	2.972	0.2668
1.16	0.8682	1.403	1.103	0.9961	3.28	0.4605	12.38	3.023	0.2577
1.20	0.8422	1.513	1.128	0.9928	3.30	0.4596	12.54	3.049	0.2533
1.24	0.8183	1.627	1.153	0.9884	3.32	0.4587	12.69	3.075	0.2489
1.28	0.7963	1.745	1.178	0.9827	3.36	0.4569	13.00	3.127	0.2404
1.30	0.7860	1.805	1.191	0.9794	3.40	0.4552	13.32	3.180	0.2322
1.32	0.7760	1.866	1.204	0.9758	3.44	0.4535	13.64	3.234	0.2243
1.36	0.7572	1.991	1.229	0.9676	3.48	0.4519	13.96	3.288	0.2167
1.40	0.7397	2.120	1.255	0.9582	3.52	0.4504	14.29	3.343	0.2093
1.44	0.7235	2.253	1.281	0.9476	3.56	0.4489	14.62	3.398	0.2022
1.48	0.7083	2.389	1.307	0.9360	3.60	0.4474	14.95	3.454	0.1953
1.52	0.6941	2.529	1.334	0.9233	3.64	0.4460	15.29	3.510	0.1887
1.56	0.6809	2.673	1.361	0.9097	3.68	0.4446	15.63	3.568	0.1823
1.60	0.6684	2.820	1.388	0.8952	3.72	0.4433	15.98	3.625	0.1761
1.64	0.6568	2.971	1.416	0.8799	3.76	0.4420	16.33	3.684	0.1702
1.68	0.6458	3.126	1.444	0.8640	3.80	0.4407	16.68	3.743	0.1645
1.72	0.6355	3.285	1.473	0.8474	3.84	0.4395	17.04	3.802	0.1589
1.76	0.6257	3.447	1.502	0.8302	3.88	0.4383	17.40	3.863	0.1536
1.80	0.6165	3.613	1.532	0.8127	3.92	0.4372	17.76	3.923	0.1485
1.84	0.6078	3.783	1.562	0.7948	3.96	0.4360	18.13	3.985	0.1435
1.88	0.5996	3.957	1.592	0.7765	4.00	0.4350	18.50	4.047	0.1388
1.92	0.5918	4.134	1.624	0.7581	4.04	0.4339	18.88	4.110	0.1342
1.96	0.5844	4.315	1.655	0.7395	4.08	0.4329	19.25	4.173	0.1297
2.00	0.5774	4.500	1.688	0.7209	4.12	0.4319	19.64	4.237	0.1254
2.04	0.5707	4.689	1.720	0.7022	4.16	0.4309	20.02	4.301	0.1213
2.08	0.5643	4.881	1.754	0.6835	4.20	0.4299	20.41	4.367	0.1173
2.12	0.5583	5.077	1.787	0.6649	4.24	0.4290	20.81	4.432	0.1135
2.16	0.5525	5.277	1.822	0.6464	4.28	0.4281	21.20	4.499	0.1098
2.20	0.5471	5.480	1.857	0.6281	4.32	0.4272	21.61	4.566	0.1062
2.24	0.5418	5.687	1.892	0.6100	4.36	0.4264	22.01	4.633	0.1028
2.28	0.5368	5.898	1.929	0.5921	4.40	0.4255	22.42	4.702	0.9948^{-1}
2.30	0.5344	6.005	1.947	0.5833	4.44	0.4247	22.83	4.771	0.9628^{-1}
2.32	0.5321	6.113	1.965	0.5745	4.48	0.4239	23.25	4.840	0.9320^{-1}
2.36	0.5275	6.331	2.002	0.5572	4.52	0.4232	23.67	4.910	0.9022^{-1}
2.40	0.5231	6.553	2.040	0.5401	4.56	0.4224	24.09	4.981	0.8735^{-1}
2.44	0.5189	6.779	2.079	0.5234	4.60	0.4217	24.52	5.052	0.8459^{-1}
2.48	0.5149	7.009	2.118	0.5071	4.64	0.4210	24.95	5.124	0.8192^{-1}
2.52	0.5111	7.242	2.157	0.4991	4.68	0.4203	25.39	5.197	0.7934^{-1}
2.56	0.5074	7.479	2.198	0.4754	4.72	0.4196	25.82	5.270	0.7685^{-1}
2.60	0.5039	7.720	2.238	0.4601	4.76	0.4189	26.27	5.344	0.7445^{-1}
2.64	0.5005	7.965	2.280	0.4452	4.80	0.4183	26.71	5.418	0.7214^{-1}
2.68	0.4972	8.213	2.322	0.4307	4.84	0.4176	27.16	5.494	0.6991^{-1}
2.72	0.4941	8.465	2.364	0.4166	4.88	0.4170	27.62	5.569	0.6775^{-1}
2.76	0.4911	8.721	2.407	0.4028	4.92	0.4164	28.07	5.646	0.6567^{-1}
2.80	0.4882	8.980	2.451	0.3895	4.96	0.4158	28.54	5.723	0.6366^{-1}
2.84	0.4854	9.243	2.496	0.3765	5.00	0.4152	29.00	5.800	0.6172^{-1}
2.88	0.4827	9.510	2.540	0.3639	6.00	0.4042	41.83	7.941	0.2965^{-1}
2.92	0.4801	9.781	2.586	0.3517	7.00	0.3974	57.00	10.469	0.1535^{-1}
2.96	0.4776	10.06	2.632	0.3398	8.00	0.3929	74.50	13.387	0.0849^{-1}
3.00	0.4752	10.33	2.679	0.3283	9.00	0.3898	94.33	16.693	0.0496^{-1}
3.04	0.4729	10.62	2.726	0.3172	10.00	0.3875	116.50	20.388	0.0304^{-1}
3.08	0.4706	10.90	2.774	0.3065	∞	0.3780	∞	∞	0

Table D.3   Prandtl–Meyer Function

M	θ	μ	M	θ	μ
1.00	0	90.00	3.04	50.523	19.20
1.04	0.3510	74.06	3.08	51.277	18.95
1.08	0.9680	67.81	3.12	52.020	18.69
1.12	1.735	63.23	3.16	52.751	18.45
1.16	2.607	59.55	3.20	53.470	18.21
1.20	3.558	56.44	3.24	54.179	17.98
1.24	4.569	53.75	3.28	54.877	17.75
1.28	5.627	51.38	3.32	55.564	17.53
1.32	6.721	49.25	3.36	56.241	17.31
1.36	7.844	47.33	3.40	56.907	17.10
1.40	8.987	45.58	3.44	57.564	16.90
1.44	10.146	43.98	3.48	58.210	16.70
1.48	11.317	42.51	3.52	58.847	16.51
1.52	12.495	41.14	3.56	59.474	16.31
1.56	13.677	39.87	3.60	60.091	16.13
1.60	14.861	38.68	3.64	60.700	15.95
1.64	16.043	37.57	3.68	61.299	15.77
1.68	17.222	36.53	3.72	61.899	15.59
1.72	18.397	35.55	3.76	62.471	15.42
1.76	19.565	34.62	3.80	63.044	15.26
1.80	20.725	33.75	3.84	63.608	15.10
1.84	21.877	32.92	3.88	64.164	14.94
1.88	23.019	32.13	3.92	64.713	14.78
1.92	24.151	31.39	3.96	65.253	14.63
1.96	25.271	30.68	4.00	65.785	14.48
2.00	26.380	30.00	4.04	66.309	14.33
2.04	27.476	29.35	4.08	66.826	14.19
2.08	28.560	28.74	4.12	67.336	14.05
2.12	29.631	28.14	4.16	67.838	13.91
2.16	30.689	27.58	4.20	68.333	13.77
2.20	31.732	27.04	4.24	68.821	13.64
2.24	32.763	26.51	4.28	69.302	13.51
2.28	33.780	26.01	4.32	69.777	13.38
2.32	34.783	25.53	4.36	70.245	13.26
2.36	35.771	25.07	4.40	70.706	13.14
2.40	36.746	24.62	4.44	71.161	13.02
2.44	37.708	24.19	4.48	71.610	12.90
2.48	38.655	23.78	4.52	72.052	12.78
2.52	39.589	23.38	4.56	72.489	12.67
2.56	40.509	22.99	4.60	72.919	12.56
2.60	41.415	22.62	4.64	73.344	12.45
2.64	42.307	22.26	4.68	73.763	12.34
2.68	43.187	21.91	4.72	74.176	12.23
2.72	44.053	21.57	4.76	74.584	12.13
2.76	44.906	21.24	4.80	74.986	12.03
2.80	45.746	20.92	4.84	75.383	11.92
2.84	46.573	20.62	4.88	75.775	11.83
2.88	47.388	20.32	4.92	76.162	11.73
2.92	48.190	20.03	4.96	76.544	11.63
2.96	48.980	19.75	5.00	76.920	11.54
3.00	49.757	19.47			

INDEX